ACS SYMPOSIUM SERIES **560**

Synthetic Oligosaccharides
Indispensable Probes
for the Life Sciences

Pavol Kováč, EDITOR

National Institutes of Health

Developed from a symposium sponsored
by the Division of Carbohydrate Chemistry
at the Southeast Regional Meeting
of the American Chemical Society,
Johnson City, Tennessee,
October 17–20, 1993

American Chemical Society, Washington, DC 1994

SEP/AE
CHEM

Library of Congress Cataloging-in-Publication Data

Synthetic oligosaccharides: indispensable probes for the life sciences / Pavol Kováč, editor.

  p.   cm.—(ACS symposium series, ISSN 0097–6156; 560)

"Developed from a symposium sponsored by the Division of Carbohydrate Chemistry at the Southeast Regional Meeting of the American Chemical Society, Johnson City, Tennessee, October 17–20, 1993."

Includes bibliographical references and indexes.

ISBN 0–8412–2930–9

  1. Oligosaccharides—Physiological effect—Congresses.   2. Oligosac-charides—Diagnostic   use—Congresses.   3. Oligosaccharides—Thera-peutic use—Congresses.

  I. Kováč, Pavol, 1938–   . II. American Chemical Society. Division of   Carbohydrate   Chemistry.   III. American   Chemical   Society. Southeastern Regional Meeting (1993: Johnson City, Tenn.) IV. Series.

QP702.044S96   1994
574.19′24815—dc20   94–16835
                                                               CIP

The paper used in this publication meets the minimum requirements of American National Standard for Information Sciences—Permanence of Paper for Printed Library Materials, ANSI Z39.48–1984. ∞

Foreword

THE ACS SYMPOSIUM SERIES was first published in 1974 to provide a mechanism for publishing symposia quickly in book form. The purpose of this series is to publish comprehensive books developed from symposia, which are usually "snapshots in time" of the current research being done on a topic, plus some review material on the topic. For this reason, it is necessary that the papers be published as quickly as possible.

Before a symposium-based book is put under contract, the proposed table of contents is reviewed for appropriateness to the topic and for comprehensiveness of the collection. Some papers are excluded at this point, and others are added to round out the scope of the volume. In addition, a draft of each paper is peer-reviewed prior to final acceptance or rejection. This anonymous review process is supervised by the organizer(s) of the symposium, who become the editor(s) of the book. The authors then revise their papers according to the recommendations of both the reviewers and the editors, prepare camera-ready copy, and submit the final papers to the editors, who check that all necessary revisions have been made.

As a rule, only original research papers and original review papers are included in the volumes. Verbatim reproductions of previously published papers are not accepted.

M. Joan Comstock
Series Editor

Contents

Preface

OLIGOSACCHARIDES ARE COMPONENTS of many important natural products, and their importance in biological processes has been recognized for a long time. In the past, their role could often only be assumed. Now, the involvement of carbohydrates, and particularly oligosaccharides, in a variety of vital cellular functions has been proven by solid scientific evidence. Consequently, activity in the carbohydrate field is constantly escalating. This situation is documented by the large number of manuscripts submitted to an increasing number of scientific journals devoted solely to carbohydrates, their conjugates, and their functions.

Prompted by the progress carbohydrate chemistry has made, particularly in the way we can now make synthetic oligosaccharides, together with the ever-increasing importance scientists in the life sciences ascribe to saccharides, the American Chemical Society decided to organize a symposium on synthetic oligosaccharides at its Southeastern Regional Meeting in the fall of 1993 (SERMACS '93). I feel very honored to have been asked to organize that meeting.

When I selected the title, *Synthetic Oligosaccharides: Indispensable Probes in the Life Sciences,* the decision to associate the theme of the symposium with the life sciences was not accidental. What started as an intellectual and a laboratory exercise more than a century ago has developed into a branch of carbohydrate chemistry that has given us tools contributing enormously to our better understanding of many fundamental processes in the life sciences. The source of most complex oligosaccharides used in the life sciences is synthesis. Thus, synthetic oligosaccharides have truly become indispensable tools in this area of research.

I was very lucky in the process of organizing the symposium. Some of the very best carbohydrate chemists active in the preparation and use of synthetic oligosaccharides accepted my invitation to share some of their most recent results with the attendees of SERMACS '93. This volume constitutes a compendium of fine works relevant to the theme of the symposium and reflects realistically the state of the art in the discussed topics as we knew it by the end of 1993.

The time frame of SERMACS '93 did not allow me to invite all the outstanding scientists active in oligosaccharide chemistry and application. Consequently, this symposium on synthetic oligosaccharides could not completely cover the subject. To make the permanent record of the symposium provided by this volume more comprehensive, I have included a few invited papers. Of course, no book on such a timely subject can hope to cover all facets of the topic. Nevertheless, I hope that this volume of the ACS Symposium Series will be found informative and useful in the community of researchers and scientists in both academic and industrial laboratories, and to teachers, students, and all those interested in staying abreast with the progress in the carbohydrate field.

Acknowledgments

The symposium from which this publication originated could not have been such a success without the financial assistance obtained from the Division of Carbohydrate Chemistry of the American Chemical Society, as well as from the private industry, namely Accurate Chemical & Scientific Corporation, General Mills, Inc., Glaxo Research Institute, Lederle-Praxis Biologicals, and McNeil Specialty Products Company. The interest of these organizations in the meeting and their outstanding support are hereby gratefully acknowledged.

The production of a permanent record of the symposium was undertaken by the ACS Books Department. The assistance of one of their members, Anne Wilson, stands out in particular, and I wish to hereby express my thanks for her efficacious contribution to the project.

I am also extremely indebted to Angela Karash for proofreading many of the manuscripts contained in this volume. Her timely and competent help, as well as thoughtful comments on individual chapters from the point of view of not only a linguist but also a very able carbohydrate chemist, has been truly invaluable.

PAVOL KOVÁČ
National Institutes of Health
Building 8, Room B1A24
Bethesda, MD 20892

Received January 4, 1994

REVIEWS

Chapter 1

Synthetic Oligosaccharides In Glycobiology

An Overview

Y. C. Lee

Biology Department, Johns Hopkins University,
3400 North Charles Street, Baltimore, MD   21218

Glycoconjugates play important roles in biological systems via recognition of oligosaccharides. Natural glycoconjugates are much too complicated for the currently available analytical and separation methodologies to allow dissection of their individual roles in biological systems to be clearly defined. Therefore chemical and enzymatic syntheses to produce useful quantities of oligosaccharides of definitive structures are very important in understanding functions of glycoconjugates. Refinement of oligosaccharide synthesis for mass production and cost reduction is urgently needed for development of glycobiotechnology.

The methodology for the synthesis of oligosaccharides has been evolving continually since the era of Emil Fischer. The recent upsurge of the interests in this branch of carbohydrate chemistry owes very much to the emerging awareness that carbohydrates, especially in the forms of glycoconjugates, serve many important roles in biology (1-3). The roles of carbohydrates in biology, however, are not as simple or clean-cut as those of nucleic acids or peptides/proteins. This can be attributed to many factors. Firstly, the chemistry of carbohydrates is inherently far more complex than that of other major groups of biological compounds, even at the basic monomeric level (monosaccharides). Secondly, assembly of the monomeric units into oligosaccharides raises the level of complexity. Naturally occurring oligosaccharides can take many closely related but subtly different forms (glyco-forms, ref. 1). The powerful techniques of molecular biology have only begun to be effective in elucidating complexity of carbohydrate structures. Its influence has been mainly in the site-directed

0097–6156/94/0560–0002$08.00/0

mutation to abolish or introduce oligosaccharide chains in glycoproteins or by modifying the properties of glycosyl-transferases.

Glycoconjugates are those carbohydrates which are covalently linked to peptides or lipids. The recognition of glycoconjugates in biological systems is mostly through recognition of the oligosaccharides therein. Often it is only a small segment of a large polysaccharide that is involved in recognition, as in the case of heparin. Therefore, to have any clear understanding of the roles of carbohydrates in glycoconjugates or polysaccharides, one must deal with the oligosaccharides.

For this purpose, one can isolate and purify oligosaccharides from the naturally available glycoconjugates involved in biological reactions. This analytical approach is a conventional but is a very difficult approach, requiring intensive labor and, frequently, expensive instrumentation. Another approach is to synthetically construct oligosaccharides of interest by chemical and/or enzymatic means. The oligosaccharides to be constructed can be only a partial representation of the natural carbohydrate structures, or can be artificially modified oligosaccharides substantially different from the natural progenitors. The synthetic approach does not replace the analytical approach, but rather complements and enhances it. If glycotechnology is to be elevated to a similar levels as those of peptides and nucleic acids, this is the tool it must employ.

Complexity of naturally occurring glycoconjugates

Unlike amino acids or nucleotides, monosaccharides have many ways of linking to each other. Although each monosaccharide has a single "reducing group", when it cyclizes, it can assume either a pyranose form (6-membered ring) or furanose form (5-membered ring). Both forms exist in nature. On top of the ring forms, there is the stereochemistry of glycosidic linkage to contend with, which can be either α- or ß-, depending on the configuration of the "anomeric carbon". Thus, to connect two monomeric units of A and B, in addition to having the "sequence" of A-B and B-A, there are four more variables for A and B, thus each can be in α- or ß-furanose or α- or ß-pyranose forms. Furthermore, because of polyhydroxy nature of monosaccharide, several positional isomers of glycosyl linkage are possible. When the number of sugar residues increases to more than three, the situation becomes more complicated because the presence of many hydroxyl groups on a single sugar allows formation of branched structures. This is the unique structural feature of carbohydrate as compared with peptides or nucleotides. When the branching is taking into consideration, the possible structures that can be generated by a finite number of mono-saccharides increases exponentially.

Fortunately, natural oligosaccharides do not fully utilize all the possible structural forms. The number of monosaccharides used in construction of most of the natural oligosaccharides are quite limited. The sugars used in glycopro-teins and glycolipids are mostly N-acetyl-D-glucosamine (GlcNAc), N-acetyl-D-galactosamine (GalNAc), D-mannose (Man), D-galactose (Gal), sialic acids (SA), L-fucose (Fuc), D-glucose (Glc), and D-xylose (Xyl). For glycosaminoglycans, D-glucuronic acid (GlcUA), D-iduronic acid (IdoUA), D-glucosamine (GlcN) are also found. The propensity for certain anomeric forms to be associated with certain sugars is also quite apparent. GlcNAc is usually found in ß-pyranosyl form including when it is linked to asparagine (Asn) in glycoproteins, and sialic acids are usually α-linked. In contrast, D-mannose and D-galactose in natural glycoconjugates can be linked as either α- or ß-form. The ß-D-mannopyranosyl linkage occupies an important position in the so-called N-glycosides of glycoproteins. D-Galactose can be found in nature not only in both α- and ß-anomeric configurations, but also in pyranose and furanose forms (4). In humans, The α-D-galactopyranosyl residue is a rarity, but it is very common in rodents.

Neither the overall construction pattern is randoml. There seems to be only a few "blueprints" that are used for construction of branched oligosaccharides of N-glycoside in glycoproteins. For example, from the standpoint of RP-HPLC analyses of pyridylaminated oligosaccharides, the N-glycoside structures can be classified into the following four generic patterns (5): M-series (equivalent to the "mannose-rich" type); X-series (those containing xylose); F-series (those containing fucose); and Z-series (none of the above). This classification was originally designed for the purpose of analyzing HPLC data (parameterization of elution volume) of N-linked oligosaccharide structures, but the clasification serves to demonstrate limited possibilities of the natural oligosaccharides in glycoconjugates. Most, if not all of the known "mannose-rich" type oligosaccharides can be expressed by choosing appropriate contiguous component sugars (starting from the core structure) from the generic parent structure shown in Figure 1. Likewise, the structures of the Z-series oligosaccharides can be represented by choosing the proper, contiguous components from the parent pattern shown in Figure 2.

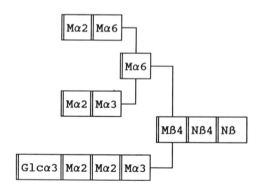

N : GlcNAc   M: Man   Glc: Glc

Figure 1. Generic Structure of M-series Oligosaccharides

O-Glycosides often found in the so-called "mucin-type" glycoproteins can be classified into several groups based on their core structures (6, 7). Unlike *N*-glycosides, however, there appears to be no consensus sequence for *O*-glycosylation sites (8), and the glycosylation sites are often very close, sometimes even contiguous.

Although the majority of the *N*- and *O*-glycosides seems to be derived from a limited number of "master plans", many "exceptions" are found quite frequently, as the analytical instrumentation becomes more sophisticated. For example, GalNAc rather than GlcNAc is sometimes found to be linked to Asn in some non-mammammalian glycoproteins (9). Fucose was found to be linked to Ser/Thr in human Factor IX (10). Glucose-Tyr is found to be the linkage between glycogen and protein (11). Moreover, there are modifications on the sugar moieties, such as sulfation, phosphorylation, and methylation. The synthetic methodologies utilized must be prepared to cope with these new structural features.

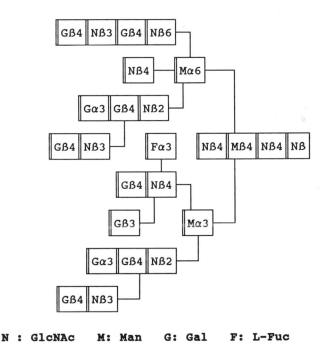

N : GlcNAc   M: Man   G: Gal   F: L-Fuc

Figure 2. Generic Structure of Z-series Oligosaccharides

Glyco-forms

Because glycoconjugates are the indirect products of gene action, their carbohydrate chains are not as tightly regulated as those of peptides and nucleic acids. The same peptides or lipids possessing different oligosaccharides are said to be in different glyco-forms (*1*). Coping with the problem of glyco-forms is a reality and a challenge for the "glyco scientists". A few simple examples of glyco-forms will be discussed here.

Ovalbumin, a favorite standard for protein chemists, has a single site of oligosaccharide attachment, but at that single site, the variation of the oligosaccharide structures are enormous. Although the different oligosaccharides (more than 20 different kinds of the "Man-rich" and the "hybrid" types) can be separated as Asn-oligosaccharides (*12*) or as peptide-free oligosaccharides (e.g., ref. *13*), separation of the individual glycoproteins containing only one oligosaccharide structure has not been accomplished. Another often cited example is ribonuclease B, which also has a single Asn site of glycosylation to which "Man-rich" oligosaccharides are attached (See Figure 1). Again, effective separation of oligosaccharides or Asn-oligosaccharides can be attained (e.g. ref. *14*), but the separation of individual glycoproteins having an oligosaccharide chain of unique structure has been very difficult.

A more complicated case is the monoclonal IgG, in which many different oligosaccharides (all in a biantennary type structure) are attached to two symmetrical glycosylation sites. The oligosaccharide mixture, at least after desialylation, can be separated into component oligosaccharides quite efficiently (e.g., ref. *15*), but not so with the "glyco-forms".

At present, how different individual glyco-forms function in a given biological reaction has not yet been clearly answered. In the examples shown above, there is only one glycosylation site per peptide chain (two per molecule of IgG, because it consists of two monomeric subunits). Obviously, when there is more than one glycosylation site, the degree of complexity increases dramatically . For example, erythropoietin contains three *N*-glycoside chains, and there is oligosaccharide heterogeneity at each glycosylation site (*16*). If there are 4 different oligosaccharide structures at each of the glycosylation sites, the total number of possible glyco-forms would be 4 x 4 x 4 = 64. There is no separation technique available today that can cope with this degree of complexity. One way to dissect the role of each glyco-form is to begin by understanding the role of each unique oligosaccharide. Only after that can combinations of different oligosaccharides on the same molecule can be investigated. For that purpose, substantial quantities of pure oligosaccharides of definitive structure must be made available, and the synthetic approach offers the most promising solution.

Recognition of carbohydrates in biological systems

Recognition of carbohydrates in biological systems is through recognition of their component oligosaccharides. Recognition of oligosaccharides can be limited to the non-reducing terminal residue. Although the actual recognition may be due to the sequence of sugars, as in the case of hepatic lectins, the branching of oligosaccharide structures is often necessary for optimal binding. The branched structures can offer multiple target sugars in a certain spatial arrangement which some lectins can bind simultaneously for enhanced binding affinity. This is termed as "glycoside clustering effect" (*17*), and it is an important consideration in developing synthetic ligands for animal lectins. Some examples will follow.

Selectins (cell adhesion molecules).   Selectins form a group of cell adhesion molecules, through which leukocytes can be localized to regions of inflammation (for a recent review, see ref. *2*). There are L-, E- and P-selectins in this group (localized in leukocytes, endothelium, and platelets, respectively), and they all require calcium for binding activity. Until recently, it was not suspected that carbohydrates were involved in the roles of selectins. However, by a diverse array of approaches, it has become clear that selectins recognize carbohydrate ligands.

Because of their role in cell adhesion indicates possible medical applications, there has been a vigorous pursuit of the structural elucidation and synthesis of ligands for these selectins. It appears that sialyl Lewis[x] or sialyl Lewis[a] (See Figure 3) can fulfill the basic structural requirement for E- and P-selectins. The endogenous carbohydrate ligand for L-selectin has not been firmly established, but it appears to require sialic acids and/or sulfates. Many synthetic variations can be introduced to simplify the structure without impairing the binding affinity. For example, replacement of sialic acid with sulfate simplifies the synthesis and reduces the production cost. Increasing the potency by increasing the valency (*18*) is another effective means of enhancing the performance/cost ratio.

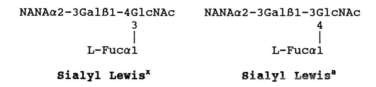

                NANAα2-3Galβ1-4GlcNAc   NANAα2-3Galβ1-3GlcNAc
                          3   4
                          | |
                      L-Fucα1   L-Fucα1

              Sialyl Lewis[x]   **Sialyl Lewis**[a]

Figure 3. Structure of sialylated Lewis[x] and Lewis[a] antigens

Influenza virus.   The earliest realization that the recognition of carbohydrates plays a role in biological reactions was the discovery that cell-surface sialic acids are necessary for influenza virus invasion (*19*). Recently, it was found that the 4-amino-2,3-anhydro-derivative of *N*-acetyl-neuraminic acid (NANA) shows a much stronger binding affinity towards the viral neuraminidase than NANA itself (*20*). This offers a possibility of using this derivative to combat influenza. However, for influenza hemagglutinin, NANA glycosides clustered at a specific distance proved to be much more potent than monomeric NANA (*21–23*). This again shows the importance of multiple valency of glycoside ligands in eliciting strong affinity.

Ceredase. In the Gaucher's disease patients, the enzyme glucocerebrosidase is genetically deficient, and accumulation of glucocerebrosides in these patients leads to serious clinical symptoms. For the enzyme replacement therapy, glucocerebrosidase isolated from human placenta is administered to the patients to allow the digestion of the glucocerebrosides which accumulate mostly in the Kupffer cells of liver. Unfortunately, the glucocerebroside isolated from placenta contains exposed galactose terminals, which serve as a marker for binding by the powerful galactose-binding receptor that exists in the liver parenchymal cells. Consequently, most of the administered enzyme does not reach the target cells (Kupffer cells) and thus is wasted.

```
Galβ(1,4)GlcNAcβ(1,2)Manα(1,6)
                                \
                                  Manβ(1,4)GlcNAcβ(1,4)GlcNAc-R
                                /
Galβ(1,4)GlcNAcβ(1,2)Manα(1,3)
```

Gal-terminated biantennary structure

```
Manα(1,6)
         \
           Manβ(1,4)GlcNAcβ(1,4)GlcNAc-R
         /
Manα(1,3)
```

Man-terminated biantennary structure

```
        /
R = Asn
        \
```

Figure 4. Carbohydrate structures in glucocerebrosidases

When the oligosaccharide chain is trimmed with exoglycosidases, not only to remove galactosyl residues but also to expose mannose residues (Figure 4), the enzyme is much more efficiently delivered to the Kupffer cells, because of the mannose-recognizing receptors of the Kupffer cells. The artificially modified enzyme is commercially known as "Ceredase." Treatment of a patient with Ceredase rapidly reversed haematopoietic failure and incapacitating skeletal disease (*24*). Although the uptake of Ceredase is mannose-dependent, there is evidence that it is not by the classic mannose-receptor (*25*). This is a good example of how manipulation of the oligosaccharide chains in glycoprotein can reap beneficial results.

Synthetic oligosaccharides for elucidating binding mechanisms

Synthetic oligosaccharides have already shown to be indispensable for glycobiology and glycotechnology. They are helpful in our understanding of the binding mode of carbohydrate-binding proteins, and some of them have the potential to become a new class of drugs. The following are some examples of how synthetic oligosaccharides have proven invaluable.

Heparin pentasaccharide. Heparin is a glycosaminoglycan with many diverse functions. The best known function of heparin is its anticoagulation activity. The realization that a pentasaccharide (**Figure 5**) is the structure responsible for the anti-AT-III activity prompted vigorous synthetic activity of this fragment and related analogues (for a review, see ref. *26*). A pentasaccharide which contains a 2-*N*-sulfate at the residue "D" (*27*) turned out to be a more effective anti-Xa agent than the naturally more abundant structure where the "D" residue contains an acetamido group (*26*). The importance of the *O*-sulfate in the residue "F" was also demonstrated by the synthetic oligosaccharides.

Hepatic carbohydrate-receptors (hepatic lectins). Liver parenchymal cells of mammalian, avian, and amphibian animals contain powerful carbohydrate-binding receptors. Interestingly, the binding specificity shows differences among the animal classes. For example, rat hepatocytes contain a galactose-binding receptor (for a recent review, see ref. *28*) but in chicken, the corresponding receptor binds *N*-acetyl-glucosamine (*29*). One unique feature of this type of lectins is that although the binding is limited to the terminal sugar residue, the higher affinity binding requires branched structure to be in the proper spatial arrangement (*28, 30, 31*). Consequently, relatively simple glycoside clusters containing only one type of sugar can be made to manifest dramatically stronger binding affinity over that of monovalent ligand, provided the target sugars are arranged in appropriate distances.

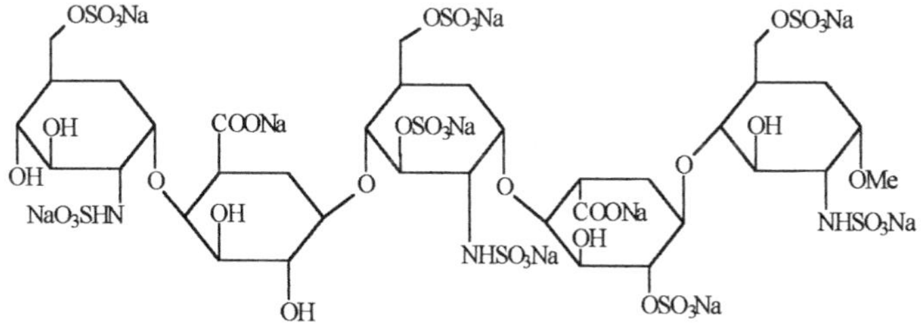

Figure 5. A pentasaccharide from heparin

A series of oligosaccharides of bi-, tri- and tetra-antennary structures synthesized by Lönngren and coworkers proved to be invaluable in dissection of binding mode of mammalian hepatic lectins (17, 28, 30, 32). Some of the examples are shown in **Figure 6**. With these synthetic oligosaccharides, it was clearly established that the biantennary structure can show a nearly 1000-fold stronger binding affinity than the monoantennary structure, and the triantennary structure an additional 1000-fold increase in affinity than the biantennary structure (17). PENTA-2,4, a biantennary oligosaccharide, gave a Kd in the range of μM range, while NONA I showed the corresponding value in the nM range. Within the groups of bi- and tri-antennary structure, some are better ligands than others. For example, NONA I is an order of magnitute better ligand than the closely similar structure of NONA II. The conformational analysis of some of these oligosaccharides revealed the optimal inter-galactose distances for maximal binding (32, 33). This, in turn, led to design of simpler cluster glycosides using peptide backbone for branching to provide the same or similar spacial arrangement of galactosyl residues (34). An example of such a cluster ligand using amino acids as a branching device is shown in Figure 7.

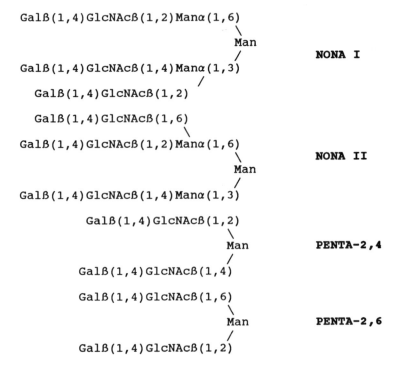

Figure 6. Some synthetic oligosaccharides of N-glycosides

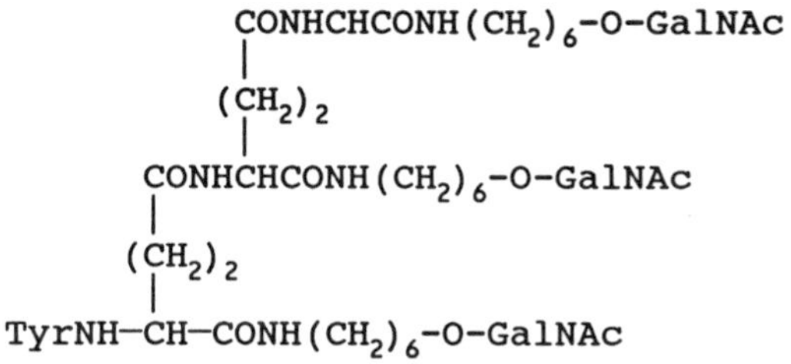

```
           CONHCHCONH(CH₂)₆-O-GalNAc
                |
              (CH₂)₂
                |
           CONHCHCONH(CH₂)₆-O-GalNAc
                |
              (CH₂)₂
                |
TyrNH—CH—CONH(CH₂)₆-O-GalNAc
```

Figure 7. A synthetic cluster ligand for mammalian hepatic lectin

Other animal lectins.   Similarly, synthetic mannose-oligosaccharides analogous to natural "mannose-rich" type N-glycosides proved to be useful for determination of mannose-ligand binding by rabbit alveolar macrophages (*35*). Some synthetic oligosaccharides having different functional groups than those found in nature can be useful in elucidating the carbohydrate-binding mode of lectins. For example, galaptin (a galactose binding protein from human spleen) binds Galß(1-4)GlcNAc with an I_{50} (the concentration that causes 50% inhibition) value of 0.13 mM. However, when the 3-OH of the GlcNAc residue is deoxygenated, its binding becomes considerably weaker, showing an I_{50} value of 6.3 mM (*36*). These results suggests the involvement of the 3-OH group in the interaction with galaptin.

Plant lectins.   As there have been voluminous studies on plant lectins in the past, examples of the usage of synthetic oligosaccharides in elucidation of binding mechanism also are quite numerous. Here only a few examples will be brought up. The oligosaccharides used in the investigation of mammalian hepatic lectins (Figure 6) have also been used to investigate plant lectins (*37*). Interestingly, the Datura lectin shows a totally different preference for the branching pattern, binding **NONA II** better than **NONA I**, and **PENTA-2,6**, a biantennary, better than any of the triantennary oligosaccharides. Ligand binding characteristics of the major mistletoe lectin were elucidated with a number of unusual synthetic galactose disaccharides (*38*).

Epitope mapping of antibodies.   Epitope mapping of immunoglobulins directed against carbohydrates has been pursued vigorously for some time. An example is the elucidation of the binding mode of ß-(1-6)-D-galactopyranan to a group of monoclonal antibodies by synthesis of an extensive array of galacto-oligosaccharides of various lengths and their deoxyfluoro analogs used as inhibitors (*39*).

Carbohydrate-specific antibodies.   Undoubtedly, this is the area where synthetic oligosaccharides have already proven their merit. The classic examples are the "synthetic" Lewis antigens and ABO-antigens by Lemieux and coworkers (e.g., ref. *40, 41*). These antigenic oligosaccharides were conjugated to serum albumin, and were used as antigens for interaction with antibodies as well as affinity ligaands for isolation of mono-specific antibodies of appropriate blood types by affinity binding.

Glycosyltransferase. Synthetic oligosaccharides have been indispensible for examination of substrate requirement by glycosyltransferases. Some of the more recent examples are: 1) Demonstration of conformational preference of *N*-acetylglucosaminyltransferase V by using conformationally restricted substrates (*42*), 2) Examination of the specificity of rat liver ß-*N*-acetyl-glucosaminyl-transferase I (*43*), 3) Characterization of ß-*N*-acetyl-glucosaminyl-transferase VI (*44*).

Problems and aspects in oligosaccharide synthesis

Although we are witnessing a tremendous progress in the art of oligosaccharide synthesis in recent years (*45—48*), the problems are far from being resolved completely. The optimal conditions for synthesis of oligosaccahrides tend to become "case-by-case" studies due to the different patterns of reactivity with different sugars. Many of the steps in the oligosaccharide synthesis are not totally specific and require some form of chromatographic separation of isomers or intermediates and final products. The heavy dependency on chromatographic separation would make scale-up to mass production more difficult.

Anomeric specificity. This is still one of the tougher problems in oligosaccharide synthesis. Newer methods using different anomeric "activators" such as "armed" and "disarmed" pentenyl glycosides (*49*), imidate (*47*), thioglycosides (*50, 51*), fluoride (*52*) and phosphite (*53*) have been devised to improve on the stereospecificity of glycoside formation, but the near quantitative specificity which is required for successful solid-phase synthesis has not been achieved consistently and universally. The use of glycosyltransferases (*54*) and glycohydro-lases which are capable of transglycosylation offers a promise in this area. Some of the more difficult problems of anomeric activation and anomeric specificity are the glycosylation of sialic acid and formation of the ß-D-mannopyranosyl linkages. These are discussed in several reviews (*55—57*).

Positional isomerism. Another area of difficulty in oligosaccharide synthesis is the necessity for selective protection of the hydroxyl or other groups. Monosaccharides used in the construction of bio-active oligosaccharides contain three or more hydroxyl groups of approximately equal reactivity. Intricate masking and unmasking techniques are necessary to isolate a single, specific hydroxyl group in an unprotected form. This requires frequent purification and identification of the intermediates. The isolated hydroxyl group surrounded by protective groups is sometimes not as reactive as that in the original sugar due to steric hindrance by the protective groups, and more drastic conditions have to be used. This usually means employment of a greater quantity of glycosylating agent, higher temperature, or longer reaction time. All of these add an extra burden on the subsequent purification. One of the more effective method for protection-activation is the use of stannylene derivatives (*58*). An interesting newer approach to this problem is seen in the use of "lightly protected" sugars (e.g., ref. *59, 60*).

Branched structures and polymeric structures. As mentioned above, carbohydrates can form branched structures, and recognition of carbohydrates often involves branching pattern. The most dramatic example is the superior binding affinity for the tri- and bi-antennary structures over the mono-valent structures by the mammalian hepatic lectin (for reviews, see ref. *61, 30*). The sialyl Lewis^x and sialyl Lewis^a structures mentioned above are also branched oligosaccharides.

From the synthetic standpoint, it is quite easy to assemble mono- or oligosaccharides as "clusters." One of the earliest and simplest examples was to attach sugars on the hydroxyl groups of amino-(tris-hydroxymethyl)methane (*62*). More effective cluster ligands were later designed with peptide backbones (*34*). The multi-valent ligands mentioned previsously which containing sialic acids are other examples. A special form of multi-layered branched structure, called dendrimers (See R. Roy's Chapter in this volume), are also used to provide multivalency. All these methodologies are for the purpose of increasing the valency of oligosaccharide ligands. When the exact structure of natural ligand for a carbohydrate-binding protein is known, the design of branched structures will become more effective.

Compatibility of protective groups. As in the peptide or nucleotide syntheses, there are a large number of protective groups that can be used to protect certain groups on the sugar moieties. Some modifiers are even used to enhance reactivity or provider of steric hindrance. As the synthetic scheme become more diverse, compatibility problem of protective groups becomes more complex. For example, the conditions required for glycosylation must not affect the integrity of the protective groups of the acceptor as well as the donor. In constructing a branched structure, it would be necessary to perform different deprotection reactions under varied conditions, if the branches to be added are not identical.

Enzymatic approach (54). The enzymatic approach excells in the control of stereospecificity of glycosylation. When a glycosyltransferase or a trans-glycosidase is used for attachment of glycosyl groups, specificity of the anomeric configuration generally is far superior to the chemical method. Most, if not all, glycosyl transferases are also specific for the position of glycosylation. For example, the galactosyltransferase from bovine milk attaches galactosyl residue only at the 4-OH of Glc/GlcNAc. The trans-glycosidases, i.e., the hydrolases with reasonable trans-glycosylation activities under certain conditions, may not be totally specific for the glycosylation positions. For example, while the sialidase/trans-sialylase from *Trypanosoma cruzi* transfers NANA residues onto the 3-position of galactose, even when lactose is used as an acceptor (*63*), sialidase from *Arthrobacter ureafaciens* performs trans-sialylation to both galactose or glucose but only at the 6-position (*64*).

Glycosyltransferases are quite specific if only the naturally occurring carbohydrates are considered. In the recent years, it has become known that these enzymes can tolerate certain variations in the substrate structure. For example, the commonly used galactosyltransferase can utilize UDP-GalNAc in addition to UDP-Gal (54). This allows placement of GalNAc residues in the oligosaccharides at the position usually occupied by Gal. There is some laxness in the acceptor specificity as well. The same galactosyltransferase can transfer Gal residue onto xylose or glucal. In an intriguing example of relaxed donor specificity of glycosyltransferase action, it was shown (65) that a preassembled trisaccharide, αGal(1-3)[αFuc(1-2)]βGal, attached to a GDP-Fuc derivative was successfully transferred to cell surface oligosaccharides by fucosyltransferase. Thus, the usage of unnatural substrates for enzymes can widen the scope of the enzymatic approach considerably.

C- and S-Analogs of oligosaccharides. There are many reasons to construct oligosaccharides in which the linking atoms are not the usual oxygen but carbon or sulfur. The major reason is that such species becomes resistant to most glycosidases. The case of isopropyl ß-thio-D-galactopyranoside which is resistant to ß-galactosidases is well known. In our construction of neoglycoproteins (66, 67), this was also a main consideration. The additional advantage of the *C*- and *S*-glycosides as probes is that they allow examination of possible involvement of the inter-residue oxygen in a given reaction or a binding process. In some cases, the more hydrophobic sulfur atom can actually enhance binding to the protein. This can be utilized for construction of ligands of higher potency. Cello-oligosaccharides containing sulfur as the inter-sugar atom has been synthesized up to trimer and have been used as affinity ligands for isolation of cellulose degrading enzymes by affinity binding (68). Disaccharides of *C*-glycoside linkages have been made (e.g., ref. 69). Interestingly, *C*-glycosides do not seem to be less active than the corresponding *O*-glycosides in at least two systems (70, 71).

Mini and macro clusters. The distinction of the "mini" and "macro" clusters has been proposed recently (72). The mini-clusters are the ligands which place target sugars at the distances of 2–3 nm. The optimal ligands for all hepatic lectins fall in this category. The enhancement of binding by this class of lectins when the sugars are clustered has already been mentioned. On the other hand, lectins such as serum mannose-binding protein (MBP-A) requires greater spacing between the target sugars. None of the "mannose-rich" type oligosaccharides manifest greatly cluster effect towards MBP-A but the lysozyme modified with as few as 4 mannosyl residues showed much enhanced activity. Such clusters are termed "macro clusters." Actually, the hepatic lectins do show large affinity enhancement by macro clusters as well as mini clusters. The spacing of the target sugars for optimal binding is dictated by the subunit arrangement of the binding proteins, and it should be no surprise that the different binding proteins require different spacings of sugars. Since clustering of oligosaccharides is often effective in increasing the binding affinity drastically, the methodology of making varied clusters with the same oligosaccharide units must be carefully developed.

Conclusion

The synthesis of oligosaccharides is an important aspect in the development of modern glycobiology. It not only provides oligosaccharides of definitive structure for elucidation of the biological roles, but also can provide new and unusual oligosaccharides with new functions. The current challenge in oligosaccharide synthesis is how to produce oligosaccharides in as large a quantity and at as low a cost as to be affordable as "glycodrugs". The combination of chemical and enzymatic synthesis offers the greatest promise in this direction. With more manageable quantities available, the clustering of oligosaccharides can also become feasible.

Acknowledgment

This work is supported in part by NIH Research Grant DK09970. The author is indebted to Dr. Reiko T. Lee for careful reading of the manuscript.

Literature Cited

1. Rademacher, T. W.; Parekh, R. B.; Dwek, R. A. *Ann. Rev. Biochem.* **1988**, *57*, 785-838.
2. Lasky, L. A. *Science* **1992**, *258*, 964-969.
3. Blithe, D. L. *Trends in Glycoscience & Glycotechnol.* **1993**, *5*, 81-95.
4. Rick, P. D.; Drewes, L. R.; Gander, J. E. *J. Biol. Chem.* **1974**, *249*, 2073-2078.
5. Lee, Y. C.; Lee, B. I.; Tomiya, N.; Takahashi, N. *Anal. Biochem.* **1990**, *188*, 259-266.
6. Brockhausen, I.; Matta, K. L.; Orr, J.; Schachter, H.; Koenderman, A. H.; van den Eijnden, H, D. *Eur. J. Biochem.* **1986**, *157*, 463-474.
7. Savage, A. V.; Donoghue, C. M.; D'Arcy, S. M.; Koeleman, C. A. M.; van den Eijnden, H, D. *Eur. J. Biochem.* **1990**, *192*, 427-432.
8. Wilson, I. B. H.; Gavel, Y.; Heijne, v. G. *Biochem. J.* **1991**, *275*, 529-534.
9. Paul, G. P.; Lottspeich, F.; Wieland, F. *J. Biol. Chem.* **1986**, *261*, 1020-1024.
10. Harris, R. J.; Halbeek, v.; H., G.; J., B.; J., L.; Ling, V. T.; Smith, K. J.; Spellman, M. W. *Glycoconj. J.* **1993**, *10*, 278-
11. Rodreiguez, I. R.; Whelan, W. J. *Biochem. Biophy. Res. Commun.* **1985**, *132*, 829-836.
12. Chen, L.-M.; Yet, M.-G.; Shao, M.-C. *FASEB J.* **1988**, *2*, 2819-2824.
13. Tai, T.; Yamashita, K.; Ito, S.; Kobata, A. *J. Biol. Chem.* **1977**, *252*, 6687-6694.
14. Suzuki, S.; Kakehi, K.; Honda, S. *Anal. Biochem.* **1992**, *205*, 227-236.

15. Jefferis, R.; Lund, J.; Mizutani, H.; Nakagawa, H.; Kawazoe, Y.; Arata, Y.; Takahashi, N. *Biochem. J.* **1990**, *268*, 529-537.
16. Sasaki, H.; Ochi, N. D.; Dell, A.; Fukuda, M. *Biochemistry* **1988**, *27*, 8618-8626.
17. Lee, Y. C.; Townsend, R. R.; Hardy, M. R.; Lönngren, J.; Arnarp, J.; Haraldsson, M.; Lönn, H. *J. Biol. Chem.* **1983**, *258*, 199-202.
18. DeFrees, S. A.; Gaeta, F. C. A.; Lin, Y.-C.; Ichikawa, Y.; Wong, C.-H. *J. Amer. Chem. Soc.* **1993**, *115*, 7349-7550.
19. Gottschalk, A. In "The chemistry and biology of sialic acids and related substances"; Cambridge Univ. Press: London, **1966**.
20. Itzstein, V.; M., W.; Y., W.-; Kok, G. B.; Pegg, M. S.; Dyason, J. C.; Jin, B.; Phan, T. V. et al. *Nature* **1993**, *363*, 418-423.
21. Glick, G. D.; Kowles, J. R. *J. Am. Chem. Soc.* **1991**, *113*, 4701-4703.
22. Sabesan, S.; Duus, J.; Neirs, S.; Domaille, P.; Kelm, S.; Paulson, J. C.; Bock, K. *J. Am. Chem. Soc.* **1992**, *114*, 8363-8375.
23. Weinhold, E. G.; Knowles, J. R. *J. Am. Chem. Soc.* **1992**, *114*, 9270-9275.
24. Mistry, P. K.; Davies, S.; Corfied, A.; Dixon, A. K.; Cox, T. M. *Q. J. Med.* **1992**, *83*, 5431-546.
25. Sato, Y.; Beutler, E. *J. Clin. Invest.* **1993**, *91*, 1909-1907.
26. Petitou, M.; Lormeau, J.-C.; Choay, J. *Supplement to Nature* **1991**, *350*, 30-33.
27. Duchaussoy, P.; Lei, P. S.; Petitou, M.; Sinay, P.; Lormeau, J. C.; Choay, J. *Bioorganic & Medical Chem. Lett.* **1991**, *1*, 99-102.
28. Lee, R. T. In "Liver Diseases. Targeted dianosis and therapy using specific receptors and ligands"; Wu, G. Y.; Wu, C., Eds; Marcel Dekker, Inc.: New York, **1991**, pp 65-86.
29. Kuhlenschmidt, T. B.; Lee, Y. C. *Biochemistry* **1984**, *23*, 3569-3575.
30. Lee, Y. C. In "Carbohydrate recognition in cellular function"; Ruoslati, E., Ed; CIBA Foundation Symposium; Wiley: Chichester, GB, **1989**, Vol. 145; pp 80-95.
31. Lee, Y. C. *FASEB J.* **1992**, *6*, 3193-3200.
32. Lee, Y. C.; Townsend, R. R.; Hardy, M. R.; Lonngren, J.; Bock, K. In "Biochemical and Biophysical Studies of Proteins and Nucleic Acids"; Lo, T.-P.; Liu, T.-Y.; Li, C.-H., Eds; Elsevier: New York, NY, **1984**, pp 349-360.
33. Bock, K.; Arnarp, J.; Loenngren, J. *Eur. J. Biochem.* **1982**, *129*, 171-178.
34. Lee, R. T.; Lee, Y. C. *Glycoconjugate J.* **1987**, *4*, 317-328.
35. Ohsumi, Y.; Hoppe, C. A.; Ogawa, T.; Lee, Y. C. *Archiv. Biochem. Biophys.* **1988**, *260*, 241-249.
36. Lee, R. T.; Ichikawa, Y.; Allen, H. J.; Lee, Y. C. *J. Biol. Chem.* **1990**, *265*, 7864-7871.
37. Crowley, J. F.; Goldstein, I. J.; Arnarp, J.; Lönngren, J. *Archiv. Biochem. Biophys.* **1984**, *231*, 524-533.
38. Lee, R. T.; Gabius, H.-J.; Lee, R. T. *J. Biol. Chem.* **1992**, *267*, 23722-23727.

39. Glaudemans, P. J.; Kovac, P. In "Fluorinated Carbohydrates"; Taylor, N. F., Ed.; ACS Symposium Series; Am. Chem. Soc., Washington, D.C.; **1988**, Vol. 374, pp 78-108.
40. Lemieux, R. U.; Bundle, D. R.; Baker, D. A. *J. Am. Chem. Soc.* **1975**, *97*, 4076-4083.
41. Lemieux, R. U.; Driguez, H. *J. Am. Chem. Soc.* **1975**, *97*, 4069-4075.
42. Lindh, I.; Hindsgaul, O. *J. Am. Chem. Soc.* **1991**, *113*, 216-223.
43. Moeller, G.; Reck, F.; Paulsen, H.; Kaur, K. J.; Sarkar, M.; Schachter, H.; Brockhausen, I. *Glycoconjugate J.* **1992**, *9*, 180-190.
44. Brockhausen, I.; Moeller, G.; Yang, J.-M.; Khan, S. H.; Matta, K. L.; Paulsen, H.; Grey, A. A.; Shah, R. N.; Schachter, H. *Carbohydr. Res.* **1992**, *236*, 281-299.
45. Paulsen, H. *Angew. Chem.* **1982**, *21*, 155-175.
46. Paulsen, H. *Angew. Chemie* **1990**, *29*, 823-839.
47. Yoshida, T.; Lasky, L.; Lee, Y. C. *Glycobiology* **1992**, *2*, 489-489.
48. Garegg, P. J. *Acc. Chem. Res.* **1992**, *25*, 575-580.
49. Mootoo, D. R.; Konradsson, P.; Udodong, U.; Frasier-Reid, B. *J. Am. Chem. Soc.* **1988**, *110*, 5583-5584.
50. Fugedi, P.; Garegg, P. J.; Lönn, H.; Norberg, T. *Glycoconjug. J.* **1987**, *4*, 97-108.
51. Zhang, Y.-M.; Mallet, J.-M.; Sinay, P. *Carbohydr. Res.* **1992**, *236*, 73-88.
52. Randall, J. L.; Nicolaou, K. C. In "Fluorinated Carbohydrates"; Taylor, N. F., Ed; ACS Symposium Series; Am. Chem. Soc.: Washington, D.C., **1988**, Vol. 373; pp 13-28.
53. Martin, T. J.; Schmidt, R. R. *Tetrahedron Lett.* **1992**, *33*, 6123-6126.
54. Ichikawa, Y.; Look, G. C.; Wong, C.-H. *Anal. Biochem.* **1992**, *202*, 215-238.
55. Okamotok, K.; Goto, T. *Tetrahedron* **1990**, *46*, 5835-5857.
56. Kanie, O.; Hindsgaul, O. *Current Opinion in Structural Biology* **1992**, *2*, 674-681.
57. Halcomb, R. L.; Wong, C.-H. *Current Opinion in Structural Biology* **1993**, *3*, 694-700.
58. David, S.; Hanessian, S. *Tetrahedron* **1985**, *41*, 643-663.
59. Lonn, H.; Stenvall, K. *Tetrahedron Lett.* **1992**, *33*, 115-116.
60. Nicolaou, K. C.; Hummel, C. W.; Iwabuchi, Y. *J. Am. Chem. Soc.* **1992**, *114*, 3126-3128.
61. Lee, Y. C.; Lee, R. T.; Rice, K.; Ichikawa, Y.; Wong, T.-C. *Pure & Appl. Chem.* **1991**, *63*, 499-506.
62. Lee, Y. C. *Carbohydr. Res.* **1978**, *67*, 509-514.
63. Vandekerckhove, F.; Schenkman, S.; Carvalho, P. d; L., T.; S., K.; M., Y.; M., H.; A., N.; , V. *Glycobiology* **1992**, *2*, 541-548.
64. Maru, I.; Ohta, Y.; Okamoto, K.; Suzuki, S.; Kakehi, K.; Tsukada, Y. *Biosci. Biotech. Biochem.* **1992**, *56*, 1557-1561.
65. Srivastava, G.; Kaur, K. J.; Hindsgaul, O.; Palcic, M. M. *J. Biol. Chem.* **1992**, *267*, 22356-22361.
66. Stowell, C. P.; Lee, Y. C. *Adv. in Carbohydr. Chem. and Biochem.* **1980**, *37*, 225-281.

67.   Lee, Y. C.; Lee, R. T. In "The Glycoconjugates"; Horowitz, M. I., Ed.; Academic Press, N.Y. **1982**, Vol. *IV*, pp 57-83.

68.   Orgeret, C.; Seiller, E.; Gautier, C.; Defaye, J.; Driquez, H. *Carbohydr. Res.* **1992**, *29*, 29-40.

69.   Martin, O. R.; Lai, W. *J. Org. Chem.* **1990**, *55*, 5188-5190.

70.   Bertpzzi, C.; Bednarski, M. *Carbohydr. Res.* **1992**, *223*, 243-253.

71.   Nagy, J. O.; Wang, P.; Gilbert, J. H.; Schaefer, M. E.; Hill, T. G.; Callstrom, M. R.; Benardski, M. *J. Med. Chem.* **1992**, *35*, 4501-4502.

72.   Lee, Y. C. *Biochem. Soc. Trans.* **1993**, *21*, 460-463.

RECEIVED March 30, 1994

Chapter 2

Drugs Based on Carbohydrates
Past and Future

Department of Carbohydrate Chemistry, Sanofi Recherche,
9, rue du Président Salvador Allende, F–94256 Gentilly, France

Several drugs currently available contain a carbohydrate structure. They are reviewed in the first section of the chapter. The second section is devoted to current research aimed at the improvement of existing drugs and the discovery of new leads. Minor improvements of existing drugs may consist of a slight modification of their biological profile, but major improvements, based upon the elucidation of the mechanism of action, may result in completely new drugs. Although biological screening of carbohydrate derivatives may provide new active principles, most of them are expected to come from research in glycobiology, a new, rapidly expanding, domain of biology.

Carbohydrate containing drugs have been used for a very long time: they can be traced back to 1600 BC to ancient Egyptian manuscripts, where a medicinal prescription of the squill bulb shows the use of cardiotonic glycosides. Prescription of the squill bulb was again reported two centuries later, in the *Corpus Hippocraticum*, to produce diuresis. Much later (1785) William Withering reported on the use of the foxglove, but it was only in 1869 that the different components and particularly digoxin, were purified by Nativelle. Digoxin and several other cardiac glycosides (17 International Nonproprietary Names -INN- could be counted in 1991 (*1*)) are still used nowadays and research in the field is going on.

Numerous other drugs, although with a shorter and less exciting history, contain a carbohydrate or a carbohydrate derivative, and altogether one can list more than 150 different INNs of carbohydrate-based active principles (*1*). The origin of all these different drugs is diverse: some, like the cardiac glycosides, have been extracted and identified in old medicines. Others derive from the known biological function of their parent carbohydrate. Finally, some others have been invented by screening carbohydrate derivatives as for any other chemicals. For the sake of completeness one

0097–6156/94/0560–0019$08.00/0
© 1994 American Chemical Society

should also mention those drugs (related to the latter category) that contain a carbohydrate as a carrier of pharmacophores: for example "mannitol hexanitrate" (vasodilator containing mannitol), "isosorbide dinitrate" (vasodilator containing 1,4:3,6-dianhydro-D-glucitol), "Calcium glucoheptonate" (mineral agent containing D-gluco-heptonic acid), "Ferrous Gluconate" (antianemic containing D-gluconic acid), in all of them the role of the carbohydrate-derived entity is probably not of prime importance.

Nowadays pharmaceutical research in the field of carbohydrates follows the same lines: improve old drugs; discover new leads based on recent breakthroughs in the biology of carbohydrates, or discovered by chance, by screening the hypothetical properties of any available chemical entity.

The first section of this article is devoted to existing carbohydrate containing drugs. The second explores the current areas of research.

Carbohydrate-Based Drugs of Today.

Cardiac Glycosides.   We have already mentioned this class of compounds (see above). Each cardiac glycoside results (Figure 1) from the combination of an aglycon (genin) linked to a glycon (a carbohydrate, mono- to tetrasaccharide). The mechanism of action, which is still being debated (2) involves inhibition of a membrane Na^+,K^+ ATPase. Pharmacological activity resides in the aglycon and the carbohydrate

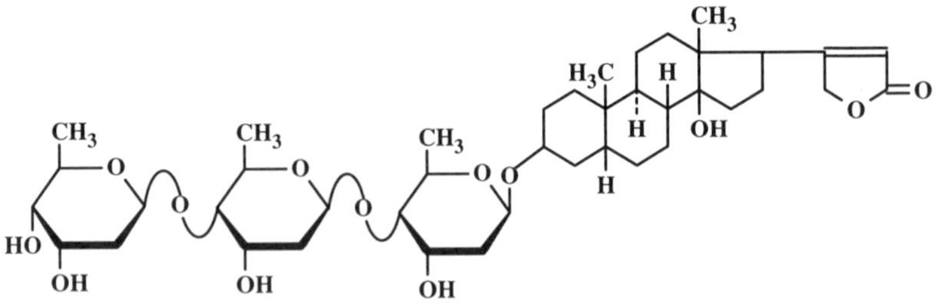

Figure 1. Digitoxin

component modifies the solubility, potency and pharmacokinetics (3). The most important therapeutic use of digitalis is to treat heart failure. Research is still going on to improve the monitoring of these very efficient drugs.

Aminoglycoside Antibiotics.   With about thirty members (1), the aminoglycoside antibiotics family is the largest family of carbohydrate containing drugs. The first member, streptomycin, was discovered in 1944 and its clinical efficiency proved in 1949. The aminoglycosides (Figure 2) consist of generally two (sometimes three as in neomycin) amino-deoxy monosaccharides joined in glycosidic linkages to an aminocyclitol (streptidine or 2-deoxy-streptamine) usually placed in a central position (although it may be terminal as in streptomycin) (4). Aminoglycosides interfere with protein synthesis mainly through binding to 30S ribosomal subunits in susceptible organisms. They are used primarily to treat infections by aerobic gram negative bacteria. The most widely used are gentamicin, tobramycin, amikacin, netilmicin, kanamycin, streptomycin and neomycin. Most of them are produced by fermentation

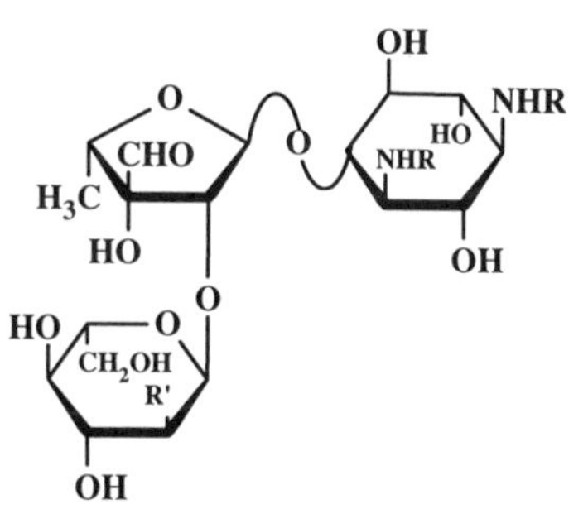

Streptomycin, R = CH₃NH R' = CH₃CN,

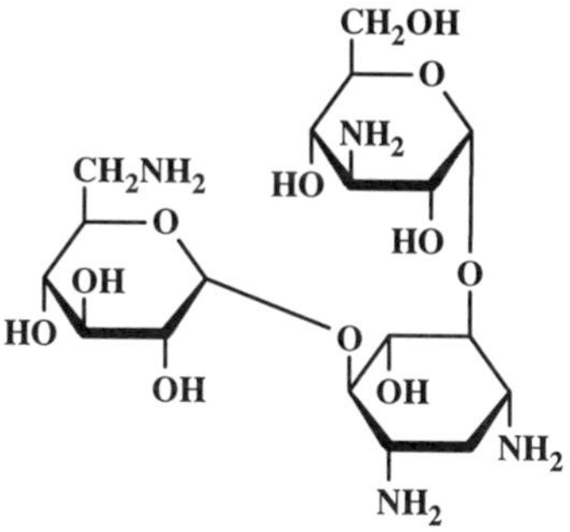

Kanamycin A

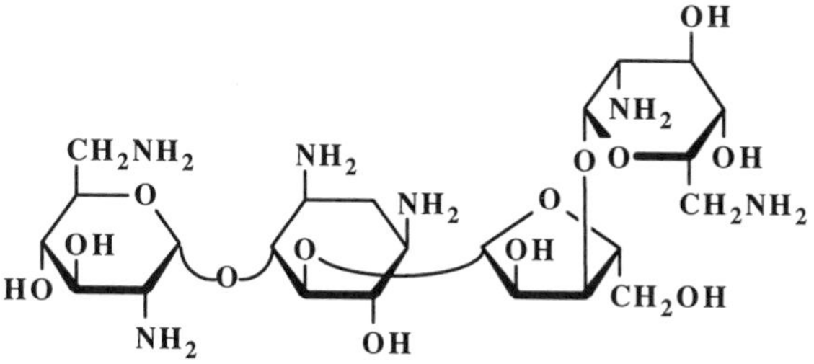

Neomycin B

Figure 2. Aminoglycoside antibiotics.

techniques, and sometimes semisynthesis is used to improve activity or bioavailability since they are poorly absorbed after oral administration. Thus amikacin derives from kanamycin and netilmicin derives from sisomicin.

The same pattern of toxicity is shared by all members of the group: ototoxicity mainly, and nephrotoxicity. This severely restricts the use of these essential antibiotics. Research is still going on, at a low rate, in this domain and several products are under clinical trial; isepamicin is among the most promising candidates.

Other Carbohydrate-Containing Antibiotics.

Macrolide Antibiotics.   These compounds (erythromycin etc...) contain one large lactone ring to which one or several deoxy-sugars are attached, the activity of which is not very well known. Of course the lipophilic or hydrophilic character of the molecule is modulated by the presence of the carbohydrate moiety. Macrolide antibiotics inhibit protein synthesis by binding to the 50s ribosomal subunits of sensitive microorganisms.

Clindamycin (Figure 3). This drug contains a substituted methylthio octoside, which can be esterified to produce a prodrug (clindamycin palmitate).

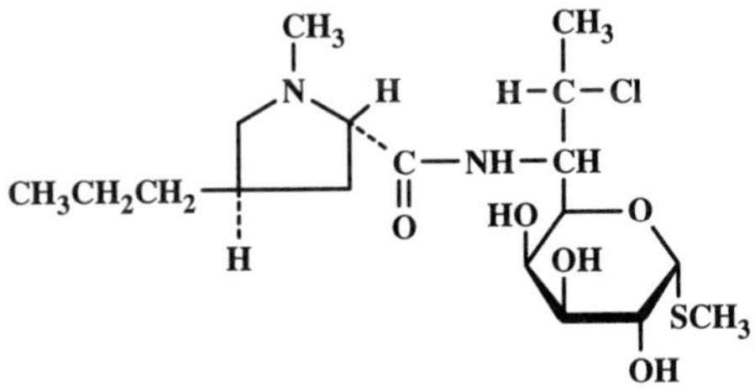

Figure 3. Clindamycin

Antiviral Agents. Most of the currently used antivirals are derivatives or analogs of purine or pyrimidine nucleosides. All these agents interfere with DNA synthesis or transcription, and some can be used as antimitotics and anticancer drugs. Those which have specific antiviral activity need to be phosphorylated by a viral kinase or act at the level of reverse transcriptase (in the case of retro-viruses).

Ribavirin is a pure nucleoside analog that inhibits the replication *in vivo* of a wide range of RNA and DNA viruses. Zidovudine (AZT) 3'azido-3'-deoxythymidine (HIV-1) is phosphorylated *in vivo* to the corresponding deoxynucleoside triphosphate which inhibits the viral reverse transcriptase. Vidarabine (arabinosyl adenine) is active against the herpes virus (see Figure 4).

Idoxuridine and Trifluridine are fluorinated pyrimidine which also inhibit viral DNA synthesis.

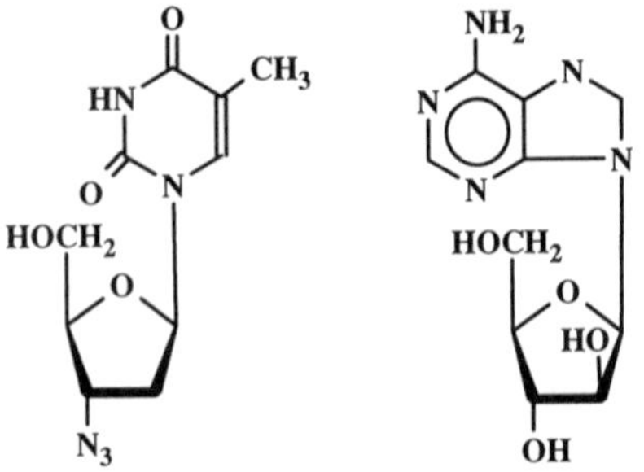

Figure 4. Zidovudine (left) and Vidarabine.

Anticancer Agents. Numerous anticancer agents, either isolated from natural sources or chemically synthesized, are glycosylated:

Pyrimidine nucleoside analogs inhibit the synthesis of pyrimidine nucleotides or they mimic the metabolites to such an extent that they interfere with vital cellular functions such as the synthesis of nucleic acids. Cytarabine (AraC) is an analog of 2'-deoxycytidine. It is the most important antimetabolite used in the therapy of acute myelocytic leukemia.

Purine nucleoside analogs are also used: pentostatine (2'-deoxycoformycin) is an inhibitor of adenosine deaminase (inhibits purine synthesis) which has a certain clinical efficacy against some leukemias and lymphomas.

Anthracycline antibiotics are an important class of anticancer drugs. Daunorubicin, doxorubicin and idarubicin are commonly used either alone or in combination, to treat acute leukemias and various types of solid tumors. They can damage DNA (via super-oxide radicals) and prevent its repair, intercalate with DNA, and/or inhibit RNA synthesis. They contain the monosaccharide L-daunosamine with other analogs containing di- and trisaccharides.

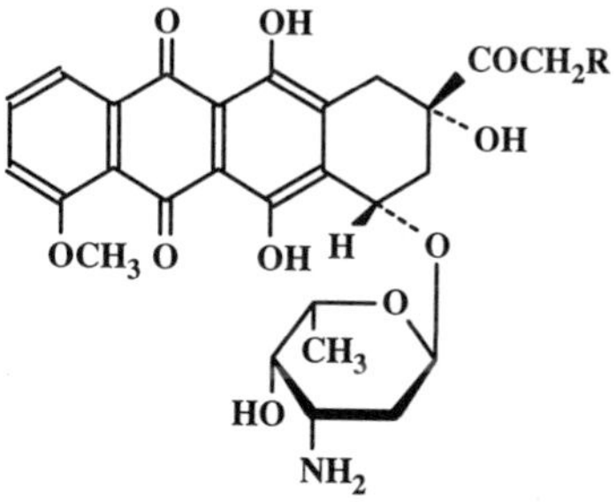

Figure 5. Daunorubicin, R = H; Doxorubicin, R = OH

In contrast, in the recently discovered calicheamicins group of antitumor agents, the aglycon is responsible for the cutting of the DNA molecule, whereas the specificity is directed by the complex carbohydrate-aromatic fragment which interact with the minor groove of the DNA (5).

Another cytotoxic antibiotic, Plicamycin, contains one disaccharide and one trisaccharide (6).

Some alkylating agents like nitrosoureas contain a monosaccharide (glucosamine in the case of streptozocin, (6)).

The carbohydrate moiety in most of these products plays a dual role in recognition (by metabolism enzymes or by nucleic acids) and to influence the hydrophilic/hydrophobic character and thereby the distribution of the drug. Thus, the carbohydrate moiety of daunorubicin (Figure 5) has been shown (7) to lay in the minor groove of a double DNA helix (without bonding to it) in a position where it could inhibit the polymerase. In the case of azaribine, the triacetyl and prodrug form of azauridine, the acetylated carbohydrate part allows better oral absorption and prevents metabolism of azauridin to azauracyl by intestinal microorganisms, a factor that contributes to CNS toxicity of azauridine.

Polysaccharides. Several polysaccharides are used in therapeutics. The most famous one is heparin, which was discovered in 1916 and underwent its first clinical trials in 1937. Heparin is used nowadays as an anticoagulant and venous antithrombotic. Its mode of action is now well known: it binds to the protein antihrombin III, a serine protease inhibitor present in plasma, and considerably reinforces the rate of inhibition of some blood coagulation factors, particularly thrombin and factor Xa. Although commercial heparin is of animal origin, it has been clearly established that it is also present in humans (8). In the past twenty years heparin has been the matter of an intense research effort. It was discovered that the biological properties of this structurally complex molecule (Figure 6) can be modified by changing the size of the molecule. This led to the so-called Low Molecular Weight Heparins, presently the best drug available for the prevention of venous thrombosis. More recent developments in the field will be reported in the section devoted to drugs of the future.

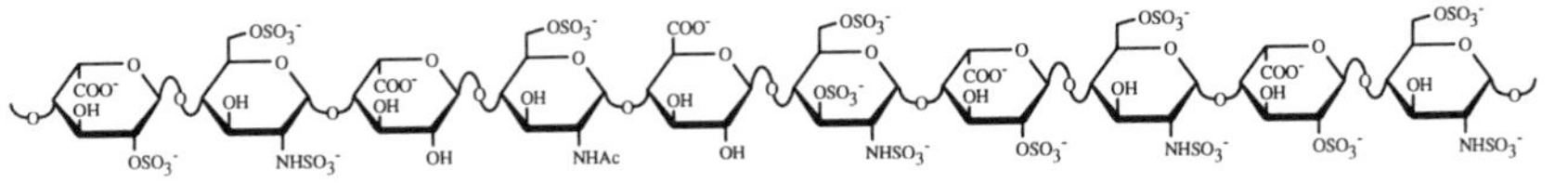

Figure 6. Structure of Heparin

Based on the idea that the action of heparin was due to its polyanionic character, several other sulfated polysaccharides were developed. Thus pentosan polysulfate, obtained by chemical sulfation of a polysaccharide extracted from beech wood is still being used in some indications (hematomas). Sulodexide, a mixture of dermatan sulfate and heparin-like compounds is another natural sulfated polysaccharide which is reported to lower plasma viscosity (9); and sodium amylosulfate, resulting from chemical sulfation of amylose, is a gastric secretory inhibitor (6).

Hyaluronic acid, another member of the glycosaminoglycan family also has a medical application as a dermatological/wound healing agent (6). Dextran or modified dextran (dextranomer) or modified starch (cadexomer) are also used as dermatological/wound healing agents (6).

Other polysaccharides, extracted from mushrooms are used as immunostimulants (sizofiran, 6).

Other Carbohydrate-based Drugs. Some other carbohydrate containing drugs do not belong to a family but are unique.

Acarbose, a pseudo tetrasaccharide (Figure 7) is an α-glucosidase inhibitor useful in the treatment of insulin dependent diabetes. This compound has a pronounced effect on intestinal glucosidases and decreases glucose absorption from food. The flattened cyclitol ring in acarbose mimics the transition state reached by the glucose ring in the normal glucosidase substrate, which results in a high affinity of these inhibitors for the glucosidases (10). Numerous other glycosidase inhibitors are currently under study.

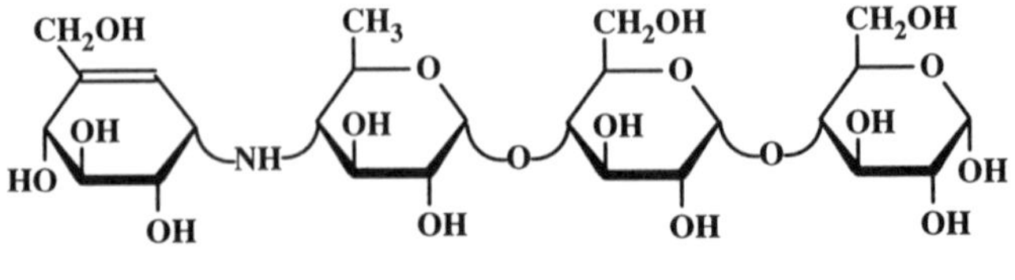

Figure 7. Acarbose

Aurothioglucose, also know as auranofin, is a glucose derivative that is used as a water soluble gold carrier. In auranofin, acetylated glucose is used allowing oral route administration. These drugs are very useful to treat rheumatoid diseases. Glucosamine has also been reported to have anti-rheumatoid properties (see Figure 8).

Figure 8. Auranofin, Ac = Acetyl

Sucralfate is a complex of aluminium hydroxide and sucrose octasulfate which has been shown to be effective for curing duodenal ulcers. However it is not clear whether this activity is related to its ability to bind basic fibroblast growth factor and protect it from degradation, as it has been suggested (*11*).

Gangliosides are a group of closely related lipids present in almost all the tissues of the body but particularly in the brain. Two preparations are currently available for therapeutical use, both obtained from bovine brains. One contains a mixture of the four major gangliosides and is used in the treatment of diabetic neuropathies, the other is a pure preparation of GM1 used in the treatment of stroke. There is evidence that gangliosides act in conjunction with nerve growth factor to repair the nervous tissue after injury (*12*).

Glucurolactone (D-glucuronic acid 3,6-lactone) and Aceglatone (2,5-di-O-acetyl-D-glucaric acid-1,4:6,3-di-γ-lactone) are β-glucuronidase inhibitors, useful as hepatic protectors. Their mode of action is to increase the clearance of carcinogens as glucuronides through inhibition of β-glucuronidase.

Most probably, in the following group of compounds, the carbohydrate part of the molecule has been initially introduced considering that carbohydrates are not only biologically active substances but also potential carriers for pharmacophores. This seems to be the case for glyconiazide (D-glucuronic acid 3,6-lactone-1-[(4-pyridinylcarbonyl)hydrazone]; antitubercular), tribenoside (ethyl 3,5,6-tri-O-benzyl-β-D-glucofuranoside; sclerosing agent), clobenoside (ethyl 3-O-propyl-5,6-di-O-p-chlorobenzyl-β-D-glucofuranoside; antiinflammatory agent) and chloralose (1,2-O-2,2,2-trichloroethylidene-α-D-glucofuranose; hypnotic sedative), see Figure 9. In the

Figure 9. Tribenoside (R = H, Bn = benzyl, Et = ethyl); Clobenoside (R = Cl); and Chloralose (right).

case of mitolactol (1,6-dibromo-1,6-dideoxy-D-galactitol; alkylating agent), and mitobronitol(1,6-dibromo-1,6-dideoxy-D-mannitol; alkylating agent), the hexitol chain was first considered to play only the role of a carrier, but later it was proved that the presence of the 2,5-OH groups is essential for the activity, as the actual metabolite is the 1,2:5,6-epoxide.

Current Research, the Drugs of Tomorrow?

Among the many active compounds mentioned in the first section, only the anticancer and antiviral agents were rationally designed from the known molecular biology of nucleic acids. The others were discovered by chance by people who were looking for something different from what they found. In the case of heparin, the most famous anticoagulant was isolated in place of the expected procoagulant substance (13). In contrast, current research in the biology and biochemistry of complex carbohydrates reveals new opportunities to rationally develop drugs and this is now the main pathway, together with improvements of old drugs and screening of new derivatives. We will briefly review these different domains.

Improvement of Existing Drugs. Several drugs mentioned in the first section of this article are the subject of constant study to improve their therapeutic use. Most of the time, new derivatives are sought with increased potency, decreased side-effects or both. Sometimes, however, major modifications are performed and finally result in completely new derivatives. We will develop the case of heparin to illustrate this course of action.

Heparin. Heparin has been used in clinics for half a century in the prevention and treatment of venous thrombosis. Standard heparin preparations, however, present undesired side effects like uncontrolled bleeding and thrombocytopenia. In an attempt to improve the therapeutic index of the drug, low molecular weight heparin preparations have been developped and their clinical efficacy is well proven (14). These compounds now tend to replace standard heparin in most of its indications. The antithrombotic properties of heparin are mediated by the serine protease inhibitor antithrombin III (AT III), a 432-amino-acid protein which is an endogenous inhibitor of blood coagulation. Binding of heparin to AT III induces a conformational change of the protein which considerably reinforces its anticoagulant activity. A specific pentasaccharide sequence in heparin is responsible for binding to AT III and the corresponding pentasaccharide has been synthesized (15). This compound binds to AT III and induces specific inhibition of blood coagulation factor Xa. Numerous

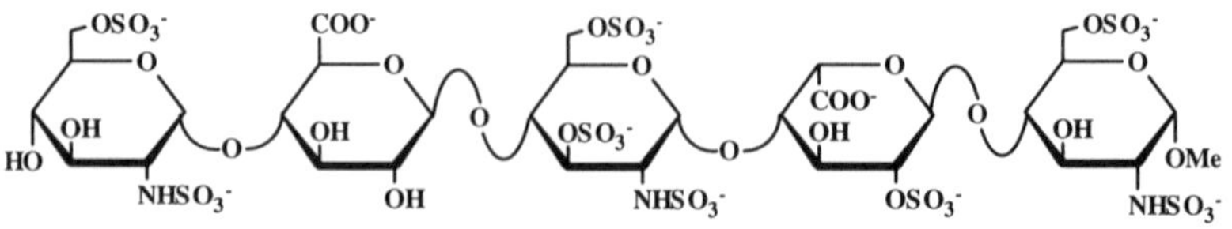

Figure 10. Pentasaccharide representing the binding site of heparin to AT III.

analogs of this lead compound (Figure 10) have been prepared (*16*) and are currently being evaluated for their actual antithrombotic effect.

Another glycosaminoglycan, dermatan sulfate, also possesses anticoagulant properties mediated by another serine protease inhibitor: heparin cofactor II. As for heparin, a specific sequence in the polymer has been identified which represents the active part of the compound. The corresponding hexasaccharide is another lead structure in the field of synthetic antithrombotic oligosaccharides.

In addition to its anticoagulant activity heparin has numerous pharmacological activities, among which is its action on the regulation of cell growth. This property is presently the matter of intensive investigations. Uncontrolled arterial smooth muscle cell proliferation is an important step in artery restenosis after angioplasty (restoration of blood flow in obstructed arteries). The antiproliferative effect of heparin (*17*) could be used to prevent restenosis. In this case, however, the anticoagulant activity represents a risk. To overcome this difficulty, heparin derivatives devoid of anticoagulant activity have been obtained (*18*) by periodate treatment of the polysaccharide. This treatment destroys the AT III binding site in heparin and abolishes the action on coagulation. Similar compounds will soon be tested in clinics. Another approach towards the same problem is to identify the oligosaccharide structures in heparin that are thought to be responsible for the antiproliferative activity, this approach is also under investigation.

To improve the therapeutic use of heparin and other glycosaminoglycans we have prepared O-acylated derivatives (*19*). Acylation allows the modulation of biological properties (prolongation of half-life in plasma) and reinforces the specific activity of the compounds (*20, 21*).

Some of the biological properties of glycosaminoglycans can also be simulated by synthetic polyanionic compounds consisting of a polysaccharide backbone substituted by anionic groups (sulfates, sulfonates or carboxylates) (*22-25*).

Biological Screening of Carbohydrate Derivatives.   A number of bioactive carbohydrate derivatives continue to be discovered by biological screening.

Systematic screening of the biological properties of the β-D-glucose derivatives shown in Figure 11 demonstrate that a hexose backbone can serve as a scaffold for molecules capable of binding to G-protein coupled receptors (*26*). Compound I was initially prepared as a non-peptide mimetic of a somatostatin agonist. It binds to the somatostatin receptor (IC_{50} 10 μM) and its analog II does the same (IC_{50} 1.3 μM). I and II also bind to the substance P receptor (IC_{50} 0.12 and 0.18 respectively). Surprisingly III binds to the substance P receptor (IC_{50} 0.06 μM) but in a very specific way since it does not interfere with some 50 other receptors at concentrations of 1 μM. In a functional assay, it inhibits Substance P-mediated inositol phosphate production.

Screening of the biological properties of sucrose derivatives led to the selection of   2,3:4,5-bis-*O*-(1-methylethylidene)-β-D-fructopyranose   sulfamate   as   an anticonvulsant (*27*). This compound (Topiramate, Figure 12) is currently undergoing clinical trials.

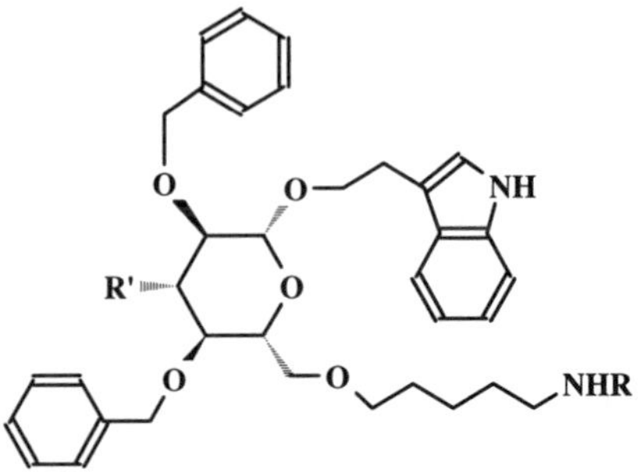

Figure 11. I) R = H, R' = benzyloxy, II) R = R' = H,
III) R = acetyl, R' = benzyloxy

Amiprilose (Figure 12) represents another example of a bioactive carbohydrate found by biological screening. This compound possesses anti-inflammatory properties and is now undergoing clinical trials for treatment of rheumatoid arthritis.

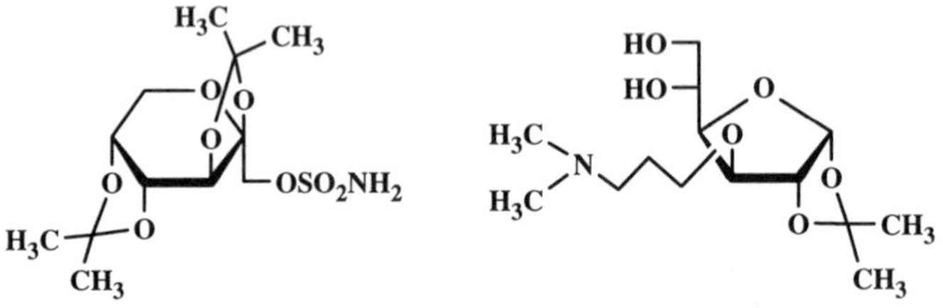

Figure 12. Topiramate, Amiprilose (right)

Other bioactive carbohydrate derivatives are prepared by analogy with natural active structures. In this case biological screening is limited to those expected properties. In this way sulfated lactobionic acid (Figure 13) was prepared by analogy with heparin and dermatan sulfate's ability to activate heparin cofactor II.

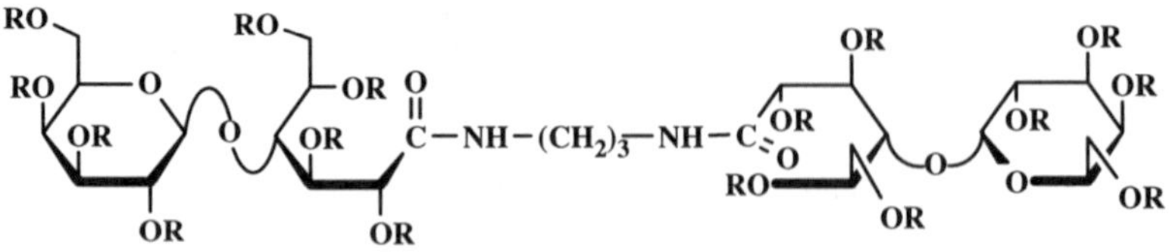

Figure 13. Sulfated bis-lactobionic acid amide (R=SO_3^-).

New Perspectives, from the Biological Function of Complex Carbohydrates. The increased knowledge about the key role carbohydrates play in living organisms opens up new perspectives for drug development.

Carbohydates are very often exposed to the surrounding media and thus they are very well placed to interact with invading bacterias or viruses, other cells, proteins or even other carbohydrates. Compounds that inhibit or modulate these interactions are drug candidates. This is illustrated in the following examples.

Bacterial Toxins. Some bacterial exotoxins (like the cholera toxin produced by *Vibrio cholerae* and the tetanus toxin of *Clostridium tetani* which blocks neurotransmitter release) bind to sialylated glycolipids, this binding can be inhibited by preparations containing gangliosides. However, in spite of several trials (*28*) no therapy so far has been developed based on this property.

Bacterial Adherence. Although bacterial infection does not require specific recognition, it is now established (*29*) that bacterial lectins, which are responsible for the recognition and binding of bacteria to cells (*30*), are involved in host adherence and the initiation of infection. Inhibition of this activity would result in the bacteria being washed away by a natural mechanism of elimination. Anti-adhesive drugs would provide substitutes or adjuvants to classical antibiotics. For example *Escherichia coli* strains specifically bind α-mannosides. This binding may be efficiently inhibited by mannose glycosides like *p*-methylumbelliferyl α-D-mannopyranoside which may be used to prevent bacterial adherence to epithelial cells. However the pharmacological interest of this interaction could be counteracted by the fact that the same kind of lectin-mediated binding to macrophages may be a way to facilitate phagocytosis of the pathogens (*31*). This type of approach is currently under investigation for the bacterium *Helicobacter pylori*, a cause of gastritis and ulcers (*32*). Recent studies indicate that, in addition to previously reported adhesins, bacteria could also display heparan sulfate-specific adhesins (*33*).

Viral infection. Several viruses bind to glycoconjugates (influenza viruses, rabies virus, Sendai virus, Newcastle disease virus, rotaviruses, coronaviruses, and reoviruses). The most studied in this respect are influenza viruses which bind to variously substituted sialic acids present at the cell surface (*34*). Numerous attempts have been reported to influence this mode of binding in order to prevent viral infection. Two proteins are involved in influenza virus attachment: the first one is a hemagglutinin which mediates attachment through binding to cell surface sialic acids, the second is a sialidase that is postulated to facilitate the elution of newly formed virus particules from the host cell and to help the movement of the virus in the mucus. Both proteins have been the target of drugs. Hemagglutinin attachment of the virus is a multivalent process and several polysialyl compounds have been prepared to inhibit this attachment (*35-38*). Recently, inhibitors of the sialidase have been designed from the crystal stucture of the enzyme. These compounds are effective inhibitors not only of the enzyme but also of the virus in cell culture and animal models (*39*).

Other viruses bind through different carbohydrate stuctures (*40*) and recently heparan sulfate type glycosaminoglycans have been shown to be involved in the case of Cytomegalovirus (*41*), Herpes simplex virus (*42, 43*), and HIV-1 (*44*). Several heparin, heparin derivatives and polyanionic compounds have been studied for their ability to prevent viral infection. Excellent results were obtained *in vitro* (*46*) but they need to be confirmed by *in vivo* trials. In fact other polyanionic polysaccharides, namely pentosan polysulfate and dextran sulfate, were tested on human HIV-infected patients. They were not found to be efficient in curing or preventing viral infections.

Another approach against viral infection is based on the observation that envelope-virus coat proteins are glycosylated (in HIV-1 N-glycosylation of gp 120 represents about 50% of the weight of the glycoprotein). This glycosylation is necessary for the virus to egress from the host cell (*47*), it also allows the virus to escape immune surveillance, and it provides ways to attach to host cells receptors then fuse to plasma membranes. This requirement for glycosylation seems to be a good target for pharmacological intervention. Since the host cell machinery is used for glycosylation, the resulting structures closely resemble those of the host. N-glycosylation involves the transfer of the Glc3Man9GlcNAc2 preformed oligosaccharide sequence from a dolichol carrier onto Asn-X-Ser/Thr sites in the protein (X is any amino acid except Pro and perhaps Asp). This triglucosylated precursor, from which the glucose residues need to be removed before introduction of other carbohydrate entities, is then processed by two glucosidases (I and II). Inhibition of these enzymes blocks the biosynthesis of the corresponding oligosaccharide and thereby alter the glycosylation of the glycoprotein of the virus. Many synthetic inhibitors of this enzyme like deoxynojirimycin and castanospermin (Figure 14) are available.

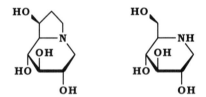

Figure 14. Castanospermin (left) and Deoxynojirimycin.

Surprisingly, they affect the glycosylation pattern of the viral glycoproteins (*48*) while those of the host are not always altered, probably because there exist another glycosylation pathway which is not available to viral proteins. The current search for better inhibitors of viral glycoprotein synthesis is directed towards greater antiviral potency and improved selectivity (*49,50*).

Inflammation. The key reaction in inflammatory reactions is the recruitment and extravasation of leukocytes at the site of inflammation. However excessive recruitment of leukocytes can also cause damage to healthy tissues, particularly in certain settings such as adult respiratory distress syndrome, ischemic reperfusion injury, and arthritis. The control of the inflammatory reaction is thus of crucial importance. Recent research has shown that the intervention of leukocytes occurs in two steps, the first involves oligosaccharide recognition: in response to inflammation mediators, endothelial cells expose a protein, E-selectin, that specifically recognizes the sialyl-Lewis x determinant (a complex tetrasaccharide, Figure 15) borne by leukocytes. As a result the cells are attracted by the blood vessel and roll on this vessel. In the second step, the rolling cells firmly bind to another adhesion molecule, an integrin, present at the site of inflammation. In the presence of the sialyl-Lewis x tetrasaccharide, the inflammation reaction is practically stopped since the rolling step cannot occur. It is worth mentioning that the lack of an antiinflammatory response is not a serious problem (particularly for a short period), since individuals born with an inability to produce sialyl-Lewis x only get occasional bacterial infections, manageable with antibiotics.

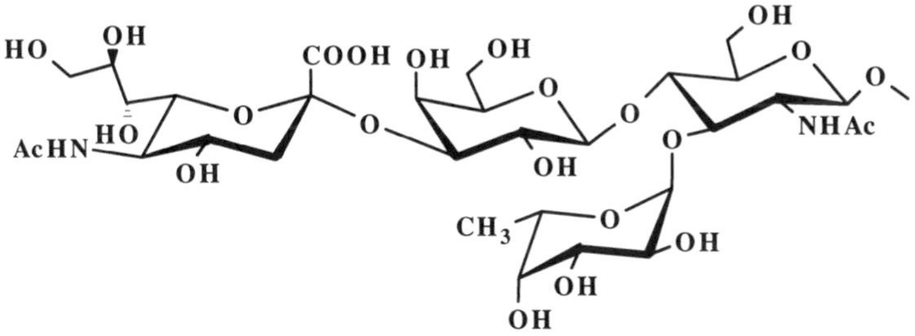

Figure 15. The sialyl-Lewis x tetrasaccharide

Several compounds, the sialyl-Lewis x tetrasaccharide itself, analogs of this tetrasaccharide (*51*), and several other potential new antiinflammatory compounds are now available for pharmacological and, ultimately, clinical testing. Large scale production of such a complex carbohydrate structures seems more probable, thanks to chemoenzymatic synthesis (*52*). Another approach is to look for simpler analogs having the same biological properties. Scientists have identified much simpler disaccharide structures endowed with the same inhibitory properties (*53, 54*).

Thrombosis.   We have already mentioned the therapeutic use of glycosaminoglycans (particularly heparin) in the prevention of venous thrombosis. These compounds require parenteral administration and home nursing when patients are discharged from hospital. In spite of many attempts, oral administration remains impossible. Since these glycosaminoglycans are also endogenous components, one can imagine overproducing them in the body to obtain the desired therapeutic effect. The biosynthesis of proteoglycans begins by synthesis of the core protein which is afterwards glycosylated. The monosaccharides are then added stepwise, the first one, xylose, is introduced by a xylosyl transferase then a galactosyl transferase introduces galactose etc... It has been known for a long time that the introduction of β-xylosides in cell culture results in the inhibition of proteoglycan synthesis while free glycosaminoglycan chains are formed and released into the culture medium. It was only recently, however, that *in vivo* experiments revealed the antithrombotic properties of orally administered xylosides (*55*). Most probably this antithrombotic activity is mediated by the newly formed glycosaminoglycans. The clinical relevance of this approach is currently being investigated. Since glycosaminoglycans are most probably involved in several biological processes, this type of approach might find other applications. The difficulty then would be the specificity of the compounds.

Immunology. Carbohydrate containing immuno-adjuvants have been studied for a long time, since the first chemical synthesis of the muramyl dipeptide lead compound (56). Their efficiency in therapeutics is however still to be established (57,58).

Lipopolysaccharides present in the outer membrane of gram-negative bacteria contain the so-called Lipid A (a phosphorylated derivative of β-1,6-glucosamine disacccharide substituted by various fatty acids) which possesses immunoadjuvant and antitumor properties. A large number of lipid A analogs have been synthesized but their clinical interest is not clearly proven.

Active specific immunotherapy of cancer seems possible through the use of artificial antigens containing the Thomsen-Friedenreich determinant (βGal1-3α GalNAc) or the sialyl-Tn antigen (αNANA2-6αGalNAc) which show increased expression on a variety of human carcinomas with little expression on normal tissues. The oligosaccharides have been conjugated to a protein and injected to ovarian or breast cancer patients, together with an immuno-adjuvant. The safety of the vaccine and the generation of an immune response was demonstrated (59), and phase II clinical trials are being contemplated.

Fertilization. The first interaction in mammalian development is between the egg and the sperm. This interaction more precisely occurs between one glycoprotein of the egg extracellular coat (glycoprotein ZP3 of the zona pellucida) and the sperm cell. It has been clearly established that O-linked carbohydrate chains of ZP3, particularly (in mice) those containing a terminal galactose residue, are responsible for this interaction (60, 61). In the future, this approach could be followed to control the implantation process.

An Original Aspect of Carbohydrate Medicinal Chemistry: Recombinant Proteins. Several glycoproteins produced by genetic engineering are used as drugs. In general, these glycoproteins are mixtures of glycoforms sharing the same polypeptide backbone but glycosylated to different extents. This pattern of glycosylation is not random, it depends on the cell line used for production of the glycoprotein. Studies on recombinant tissue plasminogen activator (a drug used to treat thromboembolic diseases) have shown that different glycoforms may exhibit different biological properties (62). The same observation holds for erythropoietin (63), a glycoprotein used to treat anemia in the case of renal failure. It is therfore another aspect of carbohydrate medicinal chemistry to deal with this problem of protein glycosylation.

Conclusion.

After the era of nucleic acids and proteins, understanding the role of other cell constituents, among which carbohydrates seem crucial to achieve further significant progress in biology. It is now acknowledged that carbohydrates play a key role in many biological processes, but, due to their frequent occurence as glycoconjugates (in glycoproteins, glycolipids and proteoglycans) this role is difficult to appraise. In this article, we have briefly reviewed currently existing carbohydrate-based drugs and we have outlined new directions of applied research to convert into drugs the most recent discovery in the field of glycobiology.

Complex carbohydrates are sometimes assimilated to oligosaccharides found in glycoproteins or at the cell surface, and made up of neutral sugars and sialic acids. Numerous research teams devote their effort to this important class of compounds (like the sialyl Lewis x tetrasaccharide) which may lead to new drugs. One should keep in mind, however, that another type of complex carbohydrates, the glycosaminoglycans, made up of highly charged sugar units, might play important biological roles as well. Thus, heparan sulfate, which is evenly distributed in living organisms and possesses a highly complex structure, is involved in the regulation of many cellular functions. For this class of compounds, the examples of heparin and dermatan sulfate show that, here also, unique structural features are responsible for the biological effect. The results of ongoing research in these directions will probably also provide opportunities for the development of new carbohydrate-based therapeutics.

Literature Cited.

1. *Index Nominum: International Drug Directory;* Medpharm Scientific Publishers: Stuttgart, Germany, 1990, Vol. 1990/91.
2. Heller, M. *Biochem. Pharmacol.* **1990**, *40*, 919-925.
3. Marschall, P. G. In *Rodd's Chemistry of Carbon Compounds;* Coffey, S., Ed.; Elsevier Publishing Co: Amsterdam, NL, 1970, 2nd ed Vol. 2D; pp 121-200.
4. *Aminoglycoside Antibiotics;* Umezawa, H.; Hooper, I. R.; Eds.; Springer Verlag: Berlin, 1982.
5. Ellestad, M. D.; Boredrs, D. B. *Acc. Chem. Res.* **1991**, *24*, 235-243.
6. Umezawa, H. In *Cancer Drug Development;* de Vita, V.T. Jr.; Busch, H., Ed.; Methods in Cancer Research; Academic Press Inc.: New-York, 1979, Vol. 16, pp 43-72.
7. Quigley, G. J.; Wang, A.-H. J.; Ughetto, G.; van der Marel, G.; van Boom, J. H.; Rich, A. *Proc. Natl. Acad. Sci. U.S.A.* **1980**, *77*, 7204-7208.
8. Linhardt, R. J.; Ampofo, S. A.; Fareed, J.; Hoppensteadt, D.; Mulliken, J. B.; Folkman, J. *Biochemistry* 1992, *31*, 12441-12445.
9. Lunetta, M.; Salatrini, T. *J. Int. Med. Res.* **1992**, *20*, 45-53.
10. Hanozet, G.; Pircher, H. P.; Vanni, P.; Oesch, B.; Semenza, G. *J. Biol. Chem.* **1981**, *265*, 3703-2711.
11. Folkman, J.; Szabo, S.; Stovroff, M.; McNeil, P.; Li, W.; Shing, Y. *Ann. Surg.* **1991**, *214*, 414-426.
12. Jack, D. B. *J. Clin. Pharm. Therapeut.* **1990**, *15*, 233-239.
13. MacLean, J. *Circulation* **1959**, *19*, 75-78.
14. Verstraete, M. *Drugs* **1990**, *40*, 498-530.
15. Petitou, M. In *Heparin;* Lane, D. A.; Lindahl, U. Eds.; Edward Arnold: London, 1989; pp 65-79.
16. Petitou, M.; van Boeckel, C. A. A. In *Chemical Synthesis of Heparin Fragments and Analogues;* Herz, W.; Kirby, G. W.; Moore, R. E.; Steglich, W.; Tamm, Ch., Eds.; Progr. Chem. Org. Nat. Prod.; Springer-Verlag: New-York, 1992, Vol. 60; pp 143-210.
17. Clowes, A. W.; Karnovsky, M. J. *Nature* **1977**, *265*, 625-626.
18. Lormeau, J. -C.; Petitou, M.; Choay, J. *Chem. Abstr.* **1989**, *111*, 66.

19. Petitou, M. In *Heparin and related polysaccharides;* Lane, D. A.; Björk, I.; Lindahl, U. Eds.; Advan. Exp. Med. Biol.; Plenum Press: New York, N.Y., 1992, Vol. 313; pp 21-30.

20. Saivin, S.; Petitou, M.; Lormeau, J.-C.; Dupouy, D.; Sié, P.; Caranobe, C.; Houin, G.; Boneu, B. *J. Lab. Clin. Med.* **1992,** *119,* 189-196.

21. Pukac, L. A.; Hirsch, G. M.; Lormeau, J.-C.; Petitou, M.; Choay, J.; Karnovsky, M. *J. Am. J. Pathol.* **1991,** *139,* 1501-1509.

22. Fischer, A. M.; Mauzac, M.; Tapon-Brétaudière, J.; Josefonvicz, J. *Biomaterials* **1985,** *6,* 198-202.

23. Sun, X. J.; Chang, J. Y. *Eur. J. Biochem.* **1989,** *185,* 225-230.

24. Sugidachi, A.; Asai, F.; Koike, H. *Thromb. Res.* **1993,** *69,* 71-80.

25. Giedrojc, J.; Radziwon, P.; Chen, T. J.; Breddin, H. K. *Thromb. Haemostas.* **1993,** *69,* 673.

26. Hirschmann, R.; Nicolaou, K. C.; Pietranico, S.; Salvino, J.; Leahy, E. M.; Sprengeler, P. A.; Furst, G.; Smith, A. B. III; Strader, C. D.; Cascieri, M. A.; Candelore, M. R.; Donaldson, C.; Vale, W.; Maechler, L. *J. Am. Chem. Soc.* **1992,** *114,* 9217-9218.

27. Maryanoff, B. E.; Nortey, S. O.; Gardocki, J. F.; Shank, R. P.; Dodgson, S. P. *J. Med. Chem.* **1987,** *30,* 880-887.

28. Stoll, B. J.; Holmgren, J.; Bardhan, P. K.; Huq, I.; Greenough, W. B. III; Fredman, P.; Svennerholm, L. *Lancet* **1980,** ii 888-891.

29. Karlsson, K. A. *Annu Rev. Biochem.* **1989,** *58,* 309-350.

30. Sharon, N.; Lis, H. *Science* **1989,** *246,* 227-234.

31. Ofek, I.; Sharon, N. *Infect. Immun.* **1988,** *56,* 539-547.

32. Lingwood, C. A.; Law, H.; Pellizzari, A.; Sherman, P.; Drumm, B. *Lancet* **1989,** *2,* 238-241.

33. Ascencio, F.; Fransson, L. A.; Wadström, T. *J. Med. Microbiol.* **1993,** *38,* 240-244.

34. Higa, H. H.; Rogers, G. N.; Paulson, J. C. *Virology* **1985,** *144,* 279-282.

35 Sparks, M. A.; Williams, K. W; Whitesides, G. M. *J. Med. Chem.* **1993,** *36,* 778-783.

36. Spevak, W.; Nagy, J. O.; Charych, D. H.; Schaefer, M. E.; Gilbert, J. H.; Bednarski, M. D. *J. Am. Chem. Soc.* **1993,** *115,* 1146-1147.

37. Sabesan, S.; Duus, J. Φ.; Domaille, P.; Kelm, S.; Paulson, J. C. *J. Am. Chem. Soc.* **1991,** *113,* 5865-5866.

38. Glick, G. D.; Knowles, J. R. *J. Am. Chem. Soc.* **1991,** *113,* 4701-4703.

39. von Itzstein, M.; Wu, W.-Y.; Kok, G. B.; Pegg, M. S.; Dyason, J. C.; Jin, B.; Phan, T. V.; Smythe, M. L.; White, H. F.; Oliver, S. W.; Colman, P. M.; Varghese, J. N.; Ryan, D. M.; Woods, J. M.; Bethell, R. C.; Hotham, V. J.; Cameron, J. M.; Penn, C. R. *Nature* **1993,** *363,* 418-423.

40. Srnka, C. A.; Tiemeyer, M.; Gilbert, J. H.; Moreland, M.; Schweingruber, H.; de Lappe, B. W.; James, P. G.; Gant, T.; Willoughby, R. E.; Yolken, R. H.; Nashed, M. A.; Abbas, S. A.; Laine, R. A. *Virology* **1992,** *190,* 794-805.

41. Compton, T.; Nowlin, D. M.; Cooper, N. R. *Virology* **1993,** *193,* 834-841.

42. Shieh, M. -T.; WuDunn, D.; Montgomery, R.; Esko, J. D.; Spear, P. G. *J. Cell Biol.* **1992,** *116,* 1273-1281.

43. Gruenheid, S.; Gatzke, L.; Meadows, H.; Tufaro, F. *J. Virol.* **1993**, *67*, 93-100.
44. Patel, M.; Yanagishita, M.; Roderiquez, G.; Bou-Habib, D. C.; Oravecz, T.; Hascall, V. C.; Norcross, M. A. *Aids Res. Human Retrovir.* **1993**, *9*, 167-174.
45. Baba, M; Schols, D.; Pauwels, R.; Nakashima, H.; De Clercq, E. *J. Acquir. Immunodef. Synd.* **1990**, *3*, 493-499.
46. Barzu, T.; Level, M.; Petitou, M.; Lormeau, J.-C.; Choay, J.; Schols, D.; Baba, M.; Pauwels, R.; Witrouw, M.; de Clercq, E. *J. Med. Chem.*, in press.
47. Gallagher, P. J.; Henneberry, J. M.; Sambrook, J. F.; Gething, M. -J. H. *J. Virol.* **1992**, *66*, 7136-7145.
48. Karlsson, G. B.; Butters, T. D.; Dwek, R. A.; Platt, F. M. *J. Biol. Chem.* **1993**, *268*, 570-576.
49. Taylor, D. L.; Sunkara, P. S.; Liu, P. S.; Kang, M. S.; Bowlin, T. L.; Tyms, A. S. *Aids* **1991**, *5*, 693-698.
50. van den Broek, L. A. G. M.; Vermaas, D. J.; Heskamp, B. M.; van Boeckel, C. A. A.; Tan, M. C. A. A.; Bolscher, J. G. M.; Ploegh, H. L.; van Kemenade, F. J.; de Goede, R. E. Y.; Miedema, F. *Recl. Trav. Chim. Pays-Bas* **1993**, *112*, 82-94.
51. Nelson, R. M.; Dolich, S.; Aruffo, A.; Cecconi, O.; Bevilacqua, M. P. *J. Clin. Invest.* **1993**, *91*, 1157-1166.
52. Ichikawa, Y.; Lin, Y.-C.; Dumas, D. P.; Shen, G.-J.; Garcia-Junceda, E.; Williams, M. A.; Bayer, R.; Ketcham, C.; Walker, L. E.; Paulson, J. C.; Wong, C.-H. *J. Amer. Chem. Soc.* **1992**, *114*, 9283-9287.
53. Tyrrell, D.; James, P.; Rao, N.; Foxall, C.; Abbas, F.; Dasgupta, F.; Nashed, M.; Hasegawa, A.; Kiso, M.; Asa, D.; Kidd, J.; Brandley, B. K. *Proc. Natl. Acad. Sci. U.S.A* **1991**, *88*, 10372-10376.
54. Allanson, N. M.; Davidson, H. M.; Martin, F. M. *Tetrahedron Lett.* **1993**, *34*, 3945-3948.
55. Bellamy, F.; Horton, D.; Millet, J.; Picart, F.; Samreth, S.; Chazan, J. B. *J. Med. Chem.* **1993**, *36*, 898-903.
56. Merser, C.; Sinaÿ, P.; Adam, A. *Biochem. Biophys. Res. Commun.* **1975**, *66*, 1316-1322.
57. Ichihara, N.; Kanazawa, R.; Sasaki, S.; Ono, K.; Otani, T.; Yamaguchi, F.; Une, T. *Arzneim.-Forsch./Drug Res.* **1988**, *38(II)*, 1043-1069.
58. *Drugs Fut.* **1993**, *18*, 282-283.
59. MacLean, G. D.; Reddish, M.; Koganty, R. R.; Wong, T.; Gandhi, S.; Smolenski, M.; Samuel, J.; Nabholtz, J. M.; Longenecker, B. M. *Cancer Immunol. Immunother.* **1993**, *36*, 215-222.
60. Bleil, J. D.; Wassarman, P. M. *Proc. Natl. Acad. Sci. U.S.A.* **1988**, *85*, 6778-6782.
61. Miller, D. J.; Macek, M. B.; Shur, B. D. *Nature* **1992**, *357*, 589-592.
62. Parekh, R. B.; Dwek, R. A.; Rudd, P. M.; Thomas, J. R.; Rademacher, T. W.; Warren, T.; Wun, T. C.; Hebert, B.; Reitz, B.; Palmier, M.; Ramabhadran, T.; Tiemeier, T. W. *Biochemistry* **1989**, *28*, 7670-7679.
63. Yamaguchi, K.; Akai, K.; Kawanishi, G.; Ueda, M.; Masuda, S.; Sasaki, R. *J. Biol. Chem.* **1991**, *266*, 20434-20439.970, 2nd ed Vol. 2D; pp 121-200.

RECEIVED April 19, 1994

Chapter 3

Sugar Cyanoethylidene Derivatives

Useful Tools for the Chemical Synthesis of Oligosaccharides and Regular Polysaccharides

L. V. Backinowsky

N. D. Zelinsky Institute of Organic Chemistry, Russian Academy of Sciences, 117913 Moscow, Russia

Condensation of sugar 1,2-*O*-(1-cyano)ethylidene derivatives (CED's) with sugar trityl ethers catalyzed by triphenylmethylium salts (tritylcyanoethylidene condensation) is an efficient method of glycosylation. Methods are elaborated to make the starting materials readily accessible. The reaction is highly 1,2-*trans*-stereoselective. CED's of pyranose sugars with *manno*-configuration and glycofuranose CED's react stereospecifically; stereospecificity is also observed for reactions of *gluco*- and *galacto*-configurated CED's with primary trityl ethers. The role of various factors on the reaction stereoselectivity is examined. Synthesis of regular homo- and heteropolysaccharides can be performed by condensation polymerization of tritylated CED's as monomers.

The roles played by carbohydrates in life processes are extremely diverse, cell recognition and immunological activity being among the most important. Terminal fragments of polymeric carbohydrate structures, polysaccharides and glycoconjugates can produce specific serological activity and the polysaccharide chain itself (or its internal oligosaccharide sequences) is also involved in the antigen – antibody interactions (*1, 2*). It is not surprising therefore, that numerous studies aimed at synthesizing specific terminal oligosaccharide structures, which are known, or presumed, to be the immunodominant part of glycoconjugates, are complemented by the synthesis of internal fragments of immunologically active polymeric carbohydrates. Modified analogs of these oligosaccharides have also been the subject of extensive synthetic studies (*3*).

The progress in achieving these goals to a great extent has depended on creation of novel, effective glycosylation methods, and contributions by Canadian (*4*), German (*5, 6*), Russian (*7*), and Swedish (*8*) groups cannot be overestimated.

The other facet of the glycoside synthesis is the challenging problem of synthetic polysaccharides. Undoubtedly, polysaccharides exhibit broader spectrum of biological

0097–6156/94/0560–0036$08.00/0

activity than the constituent oligosaccharides. Only few of the existing methods of glycosylation (such as orthoester (*7*) or thioglycoside (*9*) approaches) could be adopted to the synthesis of polysaccharides or higher regular oligosaccharides. Ring-opening polymerization of anhydro sugars (*10*), which is only applicable to the synthesis of polysaccharides, is another approach. A method of glycosylation based on the use of 1,2-*O*-(1-cyano)ethylidene derivatives (CED's) of sugars as glycosyl donors and sugar trityl ethers as glycosyl acceptors has already demonstrated its effectiveness in both the hypostases of the glycoside synthesis since the first publications in 1975 (*11, 12*).

Prior to 1975, the only known reaction, which made use of sugar trityl ethers as glycosyl acceptors, was introduced by Bredereck (*13*), wherein an acylglycosyl halide served as the glycosyl donor and silver perchlorate (or tetrafluoroborate) as the glycosylation promoter. The organic solvent-soluble silver salt generates the actual glycosylation species (cations **A** and/or **B**, scheme 1) by detachment of the halide from C(1).

Scheme 1

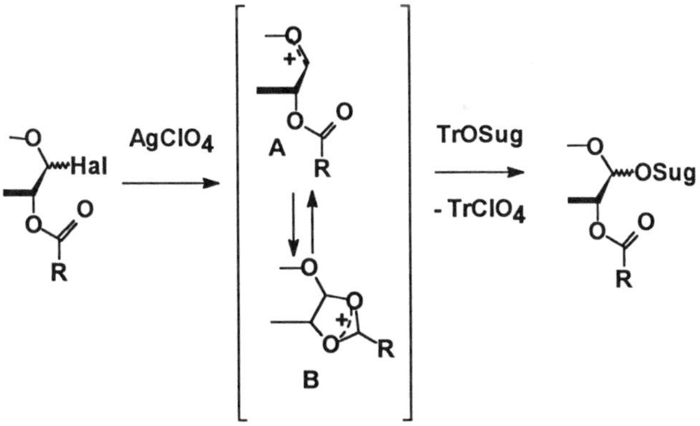

Meerwein's data (*14*) on the formation of dioxolenium ions from 2-cyanodioxolanes and cyanophilic triphenylmethylium salts (Scheme 2) led to the idea (*11*) of employing sugar CED's as the precursors of glycosylation species type **B**:

Scheme 2

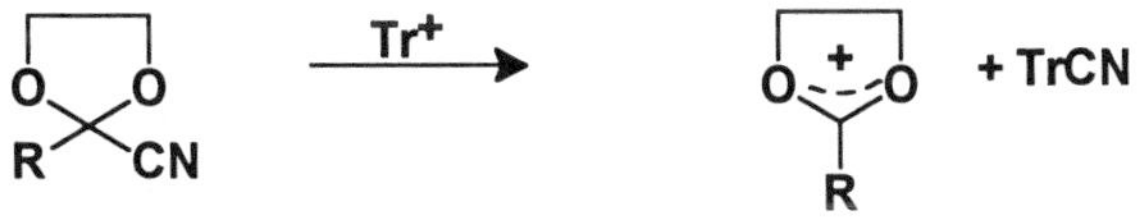

Thus, a combination of two previously unrelated reactions gave a novel result, the so-called trityl-cyanoethylidene condensation (Scheme 3).

A characteristic feature of this reaction is that the glycosylation promoter (triphenylmethylium salt) is formed *in situ* and thus it can be used in catalytic amounts.

Scheme 3

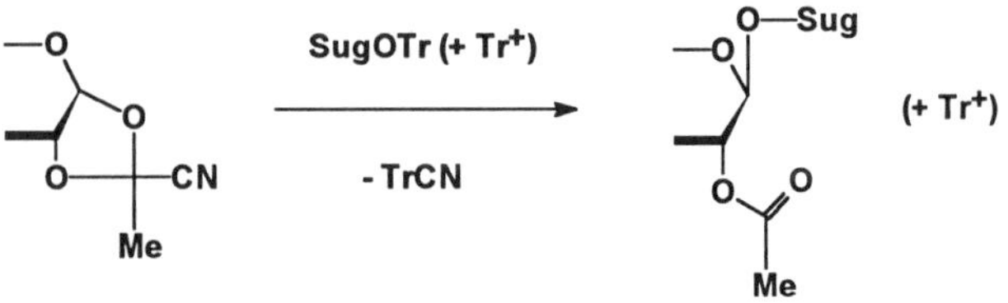

Furthermore, in the absence of the promoter, the 1,2-O-(1-cyano)ethylidene and trityl ether groups can be combined in one sugar moiety, thereby converting it into a monomer suitable for Tr$^+$-catalyzed condensation polymerization (*12*) (Scheme 4).

Scheme 4

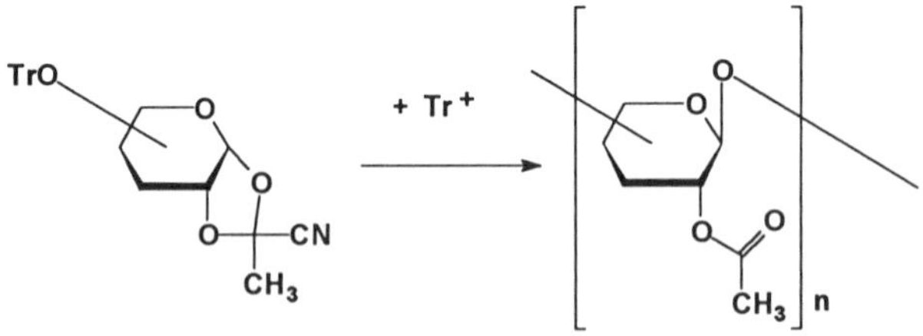

These two conceptually novel condensations were exemplified by syntheses of a (1→6)-β-linked disaccharide gentiobiose (*11*) and a (1→6)-β-D-glucan with a number-average degree of polymerization (DP) of 15 (*12*).

To ascertain whether this glycosylation is of general character, the following problems had to be solved: (i) synthesis of 1,2-O-(1-cyano)ethylidene derivatives of different sugars, (ii) synthesis of specifically tritylated sugars to be used as glycosyl acceptors, (iii) estimation of the steric outcome and preparative efficiency of the reaction. The compatibility of different protective groups with the glycosylation conditions had also to be evaluated. Our studies were aimed at the solution of these problems.

Synthesis of 1,2-O-(1-Cyano)ethylidene Derivatives

The first representative of compounds of this class, D-glucose CED **1** (Scheme 5),

Scheme 5

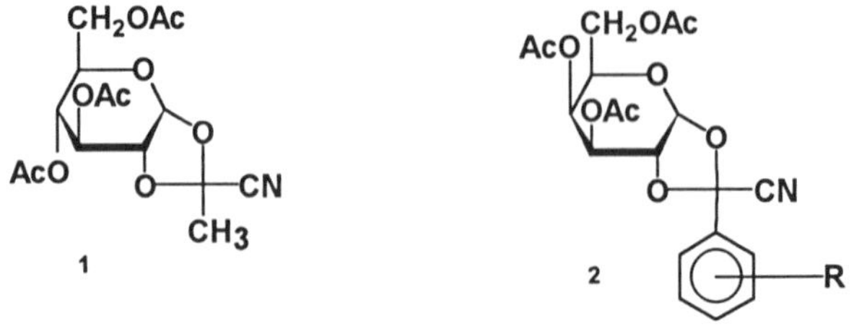

which was used as the glycosyl donor in the disaccharide synthesis (*11*) and as the precursor of the monomer for condensation polymerization (*12*), has been synthesized by treating 2,3,4,6-tetra-*O*-acetyl-α-D-glucopyranosyl bromide with silver cyanide in boiling xylene (*15*). Similar procedure has been applied to the synthesis of D-glucuronate and D-galacturonate CED's (*16*) and the maltose CED (*17*).

The presumed (*15*) intermediacy of a dioxolenium derivative of the **B** type (Scheme 1) in the formation of **1** implied that other routes to this intermediate are conceivable and an analogy with the synthesis of sugar 1,2-orthoesters (cf. *7* and references cited therein) suggested itself. Indeed, interaction of peracetylated 1,2-*trans*-glycosyl bromides (L-rhamnose and D-mannose derivatives) with potassium or sodium cyanide in acetonitrile at room temperature resulted in high yields of the respective CED's. With 1,2-*cis*-glycosyl bromides (D-glucose and D-galactose derivatives), the reaction only proceeded in the presence of tetrabutylammonium bromide (*18*). This procedure was equally applicable to benzoylated glycosyl bromides, whereby 1,2-*O*-(α-cyano)benzylidene derivatives, e.g. of the type **2** (Scheme 5) are formed (*18,19*).

In a similar fashion, the disaccharide (*18, 20, 21*) and trisaccharide (*22*) CED's were obtained, also in consistently high yields. The trisaccharide CED's were further elaborated for the synthesis of a monomer, which corresponds to a biological repeating unit of the O-specific polysaccharide of *Salmonella newington* (-6-D-Manβ 1-4-L-Rhaα1-3-D-Galβ1-) and its α-D-Man-analog. Condensation polymerization of the former resulted for the first time in a regular, natural heteropolysaccharide, its DP was about 12 (referred to trisaccharide units). It was shown to possess high inhibitory properties in passive hemagglutination reaction in the *Salmonella* O-factor 3 – anti-3 system, while the polysaccharide derived from the monomer containing α-D-Man (analog) showed no such activity (*22*).

Glycofuranose CED's could be prepared by treating the respective peracetates with trimethylsilyl cyanide in the presence of Lewis acid catalysts [stannous chloride (*23*) or trimethylsilyl trifluoromethanesulfonate (*24*)].

All CED's contain an additional chiral center, C(2) of the dioxolane ring, and a diastereomeric mixture is produced in most cases. The diastereomers with *exo*- (major) and *endo*-cyano group differed, as a rule, in chromatographic mobility and could be separated; their structure was ascertained by spectroscopic methods (NMR, X-ray).

In addition to its role in the coupling reaction, the cyanoethylidene group acts as a 1,2-*O*-protective group. It is stable to a wide range of protective group manipulations, including *O*-deacylation by base or acid methanolysis (*12, 25*), acylation, acetalization, and deacetalization (*26*). The CED's can also be benzylated with benzyl trichloroacetimidate (*20*) and glycosylated under conditions of Helferich reaction (*27*).

The latter transformation has opened up new possibilities for the synthesis of CED's of oligosaccharides. Thus, the assembly of a tetrasaccharide CED and a tritylated tetrasaccharide CED, both units corresponding to a chemical repeating unit of the basic chain of *Shigella flexneri* O-specific lipopolysaccharide (-3-D-GlcNAcβ 1-2-L-Rhaα1-2-L-Rhaα1-3-L-Rhaα1-) and which were employed in the synthesis of an octasaccharide (*28*) and a polysaccharide (*29*), involved several deprotection – glycosylation operations. It is relevant to mention that the synthetic polysaccharide exhibited notably higher inhibitory activity in passive hemagglutination test in a homologous antibody – antiserum system than the octasaccharide (*28*).

We made use of a combination of O-acetyl and O-benzoyl protective groups in monosaccharide synthons, which enabled selective removal of the former with retention of the latter by mild acid-catalyzed methanolysis (25), whereupon the cyano group was partly converted into methoxycarbonyl group.

This disadvantageous property could be reverted to the benefit by applying the following reaction sequence (Scheme 6) (30):

Scheme 6

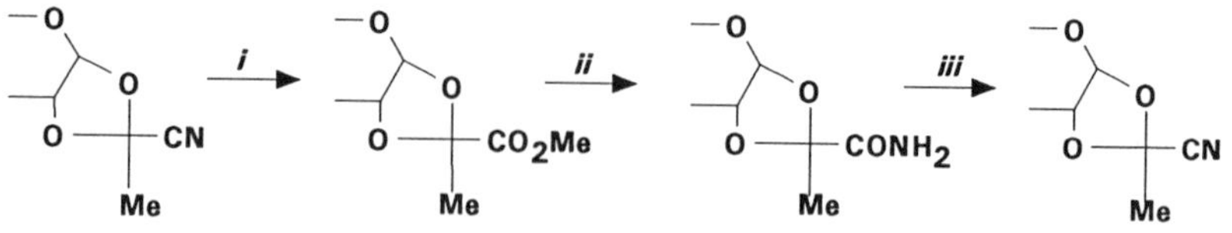

i, MeOH/HCl or MeONa; ii, NH₃/MeOH; iii, BzCl/Py

In this scheme, the aforementioned partial conversion has been brought to completion and following all the required manipulations the cyano group was recovered by successive ammonolysis and dehydration. The latter was effected by treatment with benzoyl chloride in pyridine at room temperature and all the O-acyl groups initially present in the starting CED's were replaced by O-benzoyl groups. This "by-pass" sequence is especially advantageous in the synthesis of oligosaccharide monomers to be used in condensation polymerization.

Thus, the existing approaches to the synthesis of mono- and oligosaccharide CED's are rather flexible to make them accessible compounds.

Synthesis of Trityl Ethers

Tritylation with trityl chloride under mild conditions is known to be selective with respect to primary hydroxyl groups in carbohydrates. At high temperatures, ditritylation occurs and the substitution pattern is structure- and temperature-dependent (31). Thus, selective introduction of a trityl group at a position other than primary might be problematic. Fortunately, it is not the case. The use of triphenylmethylium perchlorate as a tritylating agent and a sterically hindered pyridine base as acid scavenger [2,4,6-tri-*tert*-butylpyridine (32), 2,6-di-*tert*-butyl-4-methylpyridine, 2,4,6-collidine, or 2,6-lutidine (33)] has opened the way to secondary trityl ethers of different structure. Sugar derivatives containing a free hydroxyl group were tritylated at this position. With diols, tritylation was selective in certain cases. The reaction proceeds smoothly at room temperature and an equimolar amount of TrClO₄ is usually required, though exceptions are documented (33). Sometimes, difficulties were encountered during the tritylation of certain CED's (29). These were shown to be associated with an elimination (dehydrocyanation) reaction (34). The use of the 1-(methoxycarbonyl)ethylidene group as a substitute for the cyanoethylidene group eliminated this possibility and the CED could be regenerated following tritylation.

Thus, both the glycosyl donor and glycosyl acceptor counterparts of the trityl-cyanoethylidene condensation are readily accessible.

Synthesis of Oligosaccharides by Trityl-Cyanoethylidene Condensation

Since the first publication in 1975 (*11*), the conditions of the title condensation have basically not been changed. The reaction is usually performed in dichloromethane with equimolar amounts of the components [or, more often, with small excess of a trityl ether (*35*)] and *ca.* 10 mol.% of a freshly prepared [or reprecipitated (*22*)] triphenylmethylium perchlorate. A tuning fork-shaped tube (人) attached to a high-vacuum system (*ca.* 10^{-3} Torr) was used as a reaction vessel. Benzene solution of a CED and a trityl ether (TrOR) was placed in one limb of the tube, a solution of $TrClO_4$ in dichloromethane or nitromethane, in the other limb, and the solvents were removed *in vacuo* (benzene was freeze-dried). Dry benzene (distilled from calcium hydride) was introduced and then removed by freeze-drying from the limb with the reagents, whereafter dry dichloromethane was distilled (from calcium hydride) into the reaction tube, and the solutions were mixed. This procedure ensures strictly anhydrous conditions, which is essential in order to avoid both, the hydrolysis of $TrClO_4$ and acid-mediated cleavage of TrOR. Similar technique was applied in some Helferich-type glycosylations (*29*). The reaction mixtures were usually kept overnight at room temperature, then methanol and pyridine or aqueous pyridine were added to destroy the catalyst, and the products were isolated conventionally.

Examination of diastereomeric (*exo-* and *endo*-cyano) CED's of D-glucopyranose, D-galactopyranose, D-mannopyranose, L-rhamnopyranose (*35*), D-xylopyranose (*36*), and L-arabinofuranose (*23*) as glycosyl donors and methyl 2,3-*O*-isopropylidene-4-*O*-trityl-α-L-rhamnopyranoside as the glycosyl acceptor revealed no difference between the diastereomers in terms of the yield (normally, around 90%) of the respective disaccharides. However, when ethylene glycol ditrityl ether, 1,2;3,4-di-*O*-isopropylidene-6-*O*-trityl-α-D-galactopyranose,   or   1,6-anhydro-3,4-*O*-isopropylidene-2-*O*-trityl-β-D-galactopyranose were used as glycosyl acceptors (*37*), the reactions studied were shown to proceed faster with *endo*-cyano isomers of D-gluco-, D-galacto-, and D-xylopyranose CED's. Although this is promising from a synthetic point of view, it deserves additional kinetic studies. Of note are relatively low yields attained in these syntheses, which can be attributed to the "on-the-bench" operation (without recourse to vacuum technique) rather than to the use of $TrBF_4$ as a catalyst instead of $TrClO_4$ (cf. *11*).

Our conclusion on equal effectiveness of *exo-* and *endo*-CED's was of crucial importance for the prospective synthesis of polysaccharides from the corresponding tritylated CED's of mono- and oligosaccharides as monomers. All the glycosylations were carried out with the goal of developing the approaches to, and performing the synthesis of regular polysaccharides built of repeating units. Thus, the synthetic scheme can be so designed that the formation of the cyanoethylidene function followed the proper oligosaccharide chain assembly, and a mixture of diastereomeric CED's is used at final steps (*22, 38*). On the other hand, the opposite order of an oligosaccharide CED assembly with a chain elongation from the "reducing" terminus already bearing the 1,2-O-(1-cyano)ethylidene function (or its equivalent, i.e. 1,2-O-(1-methoxycarbonyl)ethylidene group) may require the use of an individual isomer as a starting material to facilitate chromatographic and spectroscopic control (*26, 28, 29, 39, 40*). Silver trifluoromethanesulfonate was used as an initiator of trityl-cyano-ethylidene condensation in the latter two cases.

The reactivity of various tritylated derivatives and the stability of different protective groups such as ester groups (acetates and benzoates), acetals (benzylidene, ethylidene, and isopropylidene), and benzyl ethers was also estimated in model oligosaccharide (mainly, disaccharide) syntheses (*vide infra*).

With the aim of easy deprotection of the polysaccharides produced on condensation polymerization of the corresponding tritylated CED's as the monomers, ester protection was thought to be the most suitable to this end. Acetates and benzoates are equally stable under glycosylation conditions and are appropriate for use in both the glycosyl donor and acceptor: no migration of these protective groups has been observed and the glycosylations were always regiospecific. In favor of the preferred use of benzoate was the observation that 3-O-benzoylated 4-O-trityl-L-rhamnose CED was a more efficient monomer for the synthesis of (1→4)-α-L-rhamnan than the 3-O-acetyl analog (*41*). The trend toward wider utilization of O-benzoyl protection stems also from its resistance to mild acid-catalyzed methanolysis of O-acetates (*25*). This permits flexible synthetic schemes for preparation of oligosaccharide monomers based on a selective protection – deprotection strategy where O-acetates are selectively removed in the presence of O-benzoates (cf., e.g. *29*). Oligosaccharide monomers with O-benzoates as the only O-acyl protective groups are produced if their preparation involves a "by-pass" approach (see Scheme 6).

In order to determine whether acetal groups can be utilized in the monomers, a series of model oligosaccharide syntheses has been undertaken. The following acetals (compounds **3–9**, Scheme 7) were tested for their compatibility with the conditions of the trityl-cyanoethylidene condensation.

Scheme 7

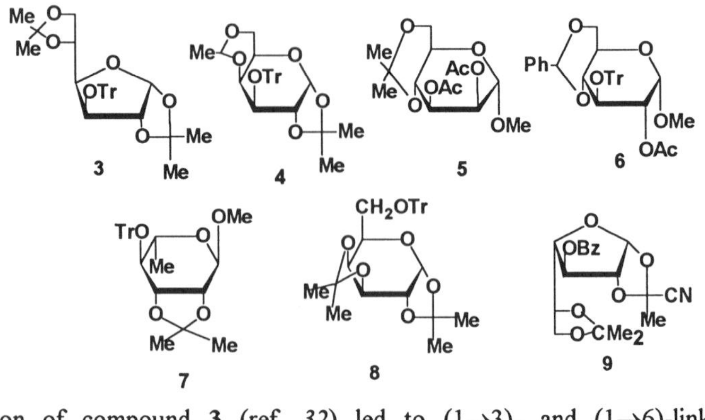

Glycosylation of compound **3** (ref. *32*) led to (1→3)- and (1→6)-linked disaccharides (*42*). The 5,6-O-isopropylidene group in the galactofuranose CED **9,**

however, was stable to the glycosylation conditions (*39, 40*). The formation of two disaccharides from **3** is the only exception of the otherwise regiospecific glycosylation, wherein the position of the trityl group unambiguously predetermined the glycosylation site. Decomposition was observed for compounds **4** and **5** (*42, 43*), while the 4,6-*O*-benzylidene derivative **6** gave a disaccharide in *ca.* 80% yield (*43*). The rhamnose derivative **7** was extensively used as a "test-compound" for evaluation of different CED's and high yields of the respective disaccharides were consistently obtained (*vide supra*). Also suitable for glycosylation is galactose derivative **8**, with which some features of detritylation – acylation were also studied. This occurs as a side reaction under glycosylation conditions and results in replacement of the *O*-trityl group in a glycosyl acceptor by an acyl group that is incorporated into the 1,2-*O*-cyanoalkylidene fragment of a glycosyl donor molecule (*44, 45*) (Scheme 8).

Scheme 8

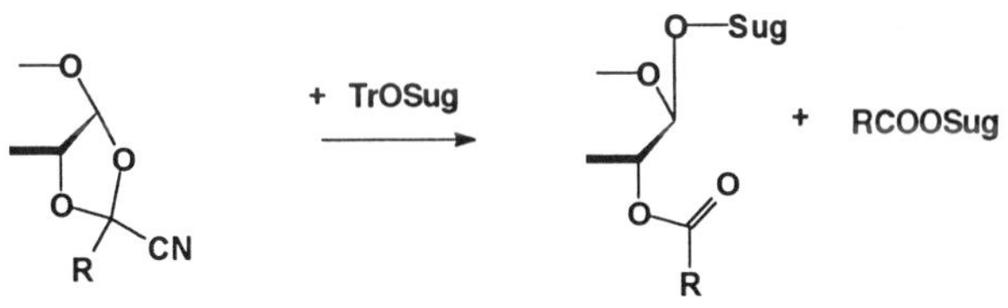

$R = Me, Me_3C, Ph, p\text{-}BrC_6H_4, p\text{-}MeOC_6H_4$

This process is similar to transacetylation observed with sugar 1,2-orthoacetates (*46*) and 2-*O*-acetylated glycosyl halides (*47*) as glycosyl donors and resulted in the formation of *O*-acetyl derivatives of the nucleophiles used.

As a persistent protective group, benzyl ethers are extensively employed in oligosaccharide synthesis. Our studies (*43*) have shown that this group is also compatible with the glycosylation conditions and may be present both in the donor and acceptor molecules, unless the primary hydroxyl is protected with a benzyl group. This observation was taken into account in designing disaccharide monomers with a 4-*O*-benzyl-L-rhamnose CED moiety (*20*).

The need for involvement of amino sugars in trityl-cyanoethylidene condensation, ultimately aimed at the synthesis of regular glycosaminoglycans, necessitated the choice of an appropriate *N*-protective group. A study of tritylated *N*-acetyl- and *N*-phthaloyl-glucosamine derivatives as acceptors revealed the former to be much less reactive: the yields of 80 - 88% were obtained for acceptors of both types after 80 and 20 h, respectively (*48*). The results of a condensation polymerization of glucos-aminyl-rhamnose monomers (regarded as simplified models for a tetrasaccharide repeating unit of *Shigella flexneri* O-specific polysaccharide) with *N*-phthaloyl- and *N*-acetyl-glucosamine moieties (*20*), corroborated this finding. *N*-Acetylglucosamino-rhamnan with an average DP of 40 – 45 disaccharide units was obtained from the former (following *O,N*-deprotection by hydrazinolysis and *N*-acetylation) in *ca.* 60% yield after 40 h, while the latter afforded only octa- to decasaccharides in 16% yield after 13 days (*49*). Thus, *N*-phthaloyl was the *N*-protective group of choice in tetrasaccharide monomers, which were used for the synthesis of *Sh. flexneri* O-specific, linear polysaccharide (*29*) and *Streptococcus pneumoniae* type 14 capsular, branched polysaccharide (*50*) by condensation polymerization. Success of this particular strategy was clearly demonstrated by the high yield synthesis of the tetrasaccharide repeating unit [D-Galβ1-4-D-Glcβ1-6(D-Galβ1-4)-D-GlcNAc] from lactose CED and 6-*O*- tritylated lactosamine acceptor (*21*) and finally by synthesizing the polysaccharide itself. Of note is the observation that whereas the synthetic and the natural polysaccharides of *Str. pneumoniae* type 14 possessed almost the same inhibitory properties in *Str. pneumoniae* type 14 – anti-14 system (ELISA), the tetrasaccharide was practically inactive.

In addition to phthaloyl, trifluoroacetyl (Backinowsky, L.V., unpublished results) or azide (*38*) can also be used for protecting the aminogroup. This was demonstrated by the synthesis of the disaccharide (L-Rhaα1-3-D-Man2N₃α1-OMe) in virtually quantitative yield from a reaction between 2-azido-2-deoxy-D-mannoside **10** (Scheme 9) and a rhamnose CED. This spectacular result paved the way for the synthesis of disaccharide **11**, which could be used as a monomer for the synthesis of the O-antigenic polysaccharide of *Pseudomonas aeruginosa* X (Meitert), consisting of (-3-D-ManNAcβ1-4-L-Rhaα1-) repeating units.

Scheme 9

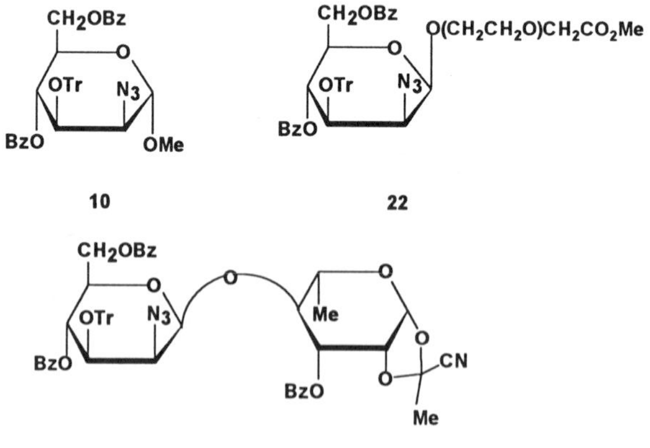

10   **22**

11

Stereochemical Aspects

The most essential aspect of the use of trityl-cyanoethylidene condensation is the stereochemistry of glycosylation. It is of critical importance in the synthesis of polysaccharides, since only stereospecificity of the reaction ensures their regular structure. Therefore, steric outcome of glycosylation has continued to be a matter of especial concern. This was addressed through oligosaccharide synthesis using the appropriate donor and acceptor that will mimic the polymerization in hand. NMR spectroscopy is the main analytical method for determination of the α:β ratio, although it is not the only tool available. For example, the structure of a synthetic (1→6)-D-glucan, which was the first polysaccharide prepared by trityl-cyanoethylidene condensation polymerization, was deduced from optical rotation data (*12*). In the case of the first synthetic heteropolysaccharide, O-specific polysaccharide of *Salmonella newington*, Smith oxidation and GLC analysis of a glycosylalditol produced supported the spectroscopic evidence (*22*). The newly formed glycosidic bonds in this and many other syntheses were established to possess exclusively or nearly exclusively 1,2-*trans*-configuration.

An unexpected violation of stereospecificity was observed (*51*) in the xylosylation of a tritylated xyloside 12 (Scheme 10) as a model for the synthesis of natural (1→4)-β-D-xylan. The ratio of the isolated (1→4)-β- and (1→4)-α-linked xylobiosides was 2.1:1 and 2.3:1 for *endo*- and *exo*-CED's 13, respectively, with overall yields of *ca.* 90%.

Scheme 10

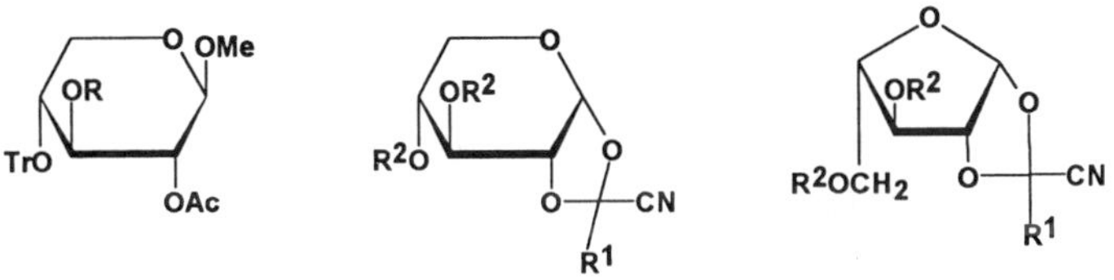

12 R = Ac

21 R = Bn

13 R^1 = Me, R^2=Ac

14 R^1 = Me, R^2=Bz

15 R^1 = Me, R^2=Bn

16 R^1 = Ph, R^2=Bz

17 R^1 = Me, R^2=Ac

18 R^1 = Ph, R^2=Bz

Other D-xylose CED's (14, 15) gave β- and α-linked disaccharides in *ca.* 80% overall yield and β:α-ratio of 7.7:1 and 2.1:1, respectively. In the xylosylation of 12 by cyanobenzylidene derivative 16, the lowest amount of the α-linked disaccharide (*ca.* 10%) was observed .

This lack of specificity was not unique to D-xylose CED's [note that with many other acceptors it behaved as highly stereoselective, 1,2-*trans*-glycosylating agent (*36*)], but D-glucose, D-galactose, and L-rhamnose CED's also gave mixtures of 1,2-*trans*- and 1,2-*cis*-linked disaccharides (*51*). The only exceptions were L-arabinofuranose cyanoethylidene and cyanobenzylidene derivatives 17 and 18, both reacted virtually stereospecifically (*52*).

Nonstereospecificity of the glycosylation of xyloside 12 by xylose CED 13 manifested itself in the condensation polymerization of the corresponding 4-O-tritylated xylose CED as the monomer (*53*). The resulting (1→4)-D-xylan contained up to 26% of 1,2-*cis*-linked xylose units. Of note also is the nonstereospecific polycondensation of isomeric, 3-O-tritylated xylose CED (*ca.* 11% of α-xylosidic units), whereas the proportion of (1→3)-α-linked xylobioside in a model disaccharide synthesis amounted to only *ca.* 6% (*36*).

The formation of 1,2-*cis*- and 1,2-*trans*-linked disaccharides is not unique to the xyloside 12. Some other O-tritylated monosaccharides also behaved similarly (*54*). Nor were irregular (1→4)- and (1→3)-xylans the only exceptions, since other monomers also produced anomerically mixed polysaccharides (*55*, *56*).

The results obtained allow one to make following conclusions on stereoselectivity of trityl-cyanoethylidene condensation. It is exceptionally high for mannopyranose, rhamnopyranose, and glycofuranose CED's and almost any type of acceptors. Pyranose CED's with *gluco-* and *galacto*-configuration are stereospecific glycosylating agents, as a rule, with respect to acceptors with primary trityloxy groups. This follows from model disaccharide syntheses and is supported by syntheses of regular polysaccharides.

Attempts were made to correct the deviation from the desired 1,2-*trans*-stereospecificity. The replacement of a perchlorate anion in the catalyst (triphenylmethylium salt) by a non-nucleophilic tetrafluoroborate (*53, 54, 57*) or a less nucleophilic trifluoromethanesulfonate (*57, 58*) decreased or even eliminated 1,2-*cis*-linkages in some synthetic disaccharides and polysaccharides. Conducting the reaction under high pressure (14 kbar) also resulted in stereospecific syntheses of disaccharides (*59*) and polysaccharides (*60*). Introduction of an *O*-benzyl group instead of an *O*-acetyl group at a position vicinal to *O*-trityl in an acceptor molecule greatly diminished the content of 1,2-*cis*-linked disaccharides (*61*; Kitov, P.I.; Tsvetkov, Yu.E.; Backinowsky, L.V., unpublished results). 1,2-*O*-(α-Cyano)benzylidene derivatives with electron-donating groups in the aromatic nucleus (*58*) can be regarded as favorable substitutes for the cyanoethylidene derivatives.

The knowledge of the scope and limitations of trityl-cyanoethylidene condensation and promising results on correction of the deviations from stereospecificity allow one to make realistic prognosis in regards to the design and synthesis of complex, regular polysaccharides. The potential and value of this approach has not been fully exploited.

Examples of a novel application of the condensation polymerization are the syntheses of regular polysaccharides and block-polysaccharides as glycosides bearing a functionalized spacer-arm (*62, 63*). The former involved the condensation of a tritylated CED as a monomer in the presence of a tritylated aglycon-acceptor, which served as a "primer" for a polysaccharide chain growth. When following the condensation of one monomer, another monomer was introduced, it resulted in the formation of a block-polysaccharide. Thus, initial model syntheses with monosaccharide monomers were followed with oligosaccharide monomers, which correspond to repeating units of a group A-variant streptococcal polysaccharide (-2-L-Rhaα1-3-L-Rhaα1-) (*64*) or a common polysaccharide antigen of *Ps. aeruginosa* (-2-D-Rhaα1-3-D-Rhaα1-3-D-Rhaα1-) (*65*). In the latter case, the polysaccharide produced was identical in its NMR spectral parameters to the natural one. This was coupled to bovine serum albumin by virtue of a 6-aminohexyl spacer at the "reducing" terminus and the neoglycoprotein thus obtained was used to raise polysaccharide-specific antibodies (Makarenko, T.A., *et al.*, *FEMS Immunol. Med. Microbiol.*, 1993, in press).

With new targets, new problems have arisen, and preferential coupling of a monomer to the aglycon-acceptor as opposed to its self-condensation, became the problem of primary importance. Therefore, the relative reactivity of acceptors toward CED's had to be evaluated. It was shown, for example, that the tritylated rhamnoside **19** (Scheme 11) is preferentially selected by a rhamnose CED from an equimolar mixture of **19** and 6-*O*-trityl-mannoside **20**, the yield of a rhamnosyl-rhamnoside being *ca.* 90% and the recovery of **20** being 96.5%. This accounted for a low degree of (poly)rhamnosylation of a mannan anchored to a 6-phthalimidohexyl glycoside in an attempted synthesis of rhamnomannan, a block-polysaccharide (*63*). Benzylated xyloside **21** (Scheme 10) not only was almost stereospecifically glycosylated by xylose CED, it was also much more reactive than the diacetate **12**. The recovery of the latter was 70% and the yield of a mono-*O*-benzylated xylobioside was 76% when an equimolar mixture of these two acceptors was subjected to glycosylation by xylose CED **13** (*61*).

Scheme 11

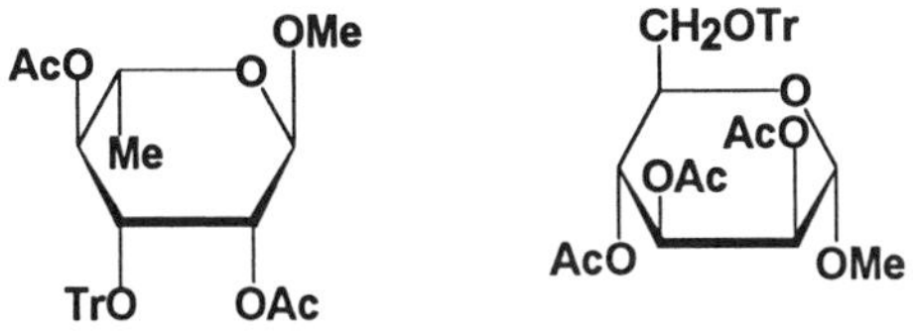

19   **20**

Some features of the synthesis of polysaccharide glycosides still remain unexplained. For example, despite obvious similarity in the acceptor sites in the monomer **11** and the tritylated acceptor glycoside **22** (Scheme 9), the yield of a polysaccharide glycoside is extremely low (*38*) with overall high efficiency of condensation polymerization of the monomer **11**. Analogously, polymerization of a monomer **23** (Scheme 12), which represents the synthon of a repeating unit of a group A-variant streptococcal polysaccharide, in the presence of a 2-*O*-tritylated acceptor **24** resulted in polysaccharide glycoside in lower yield and lesser DP as compared to the polycondensation of the same monomer in the presence of an isomeric acceptor **25** (Backinowsky, L.V., unpublished results).

Scheme 12

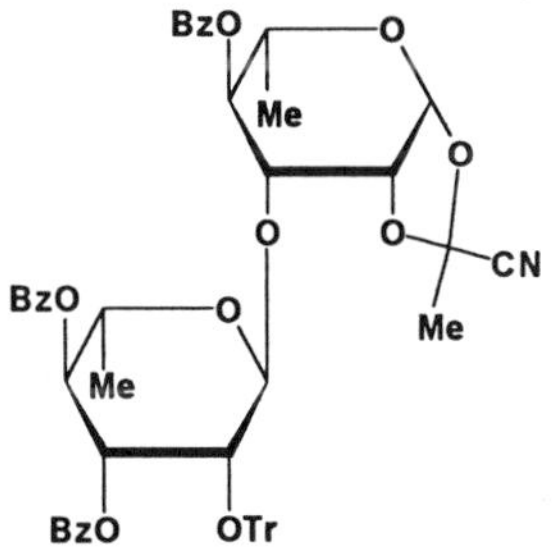

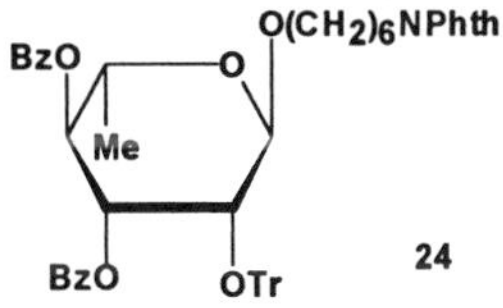

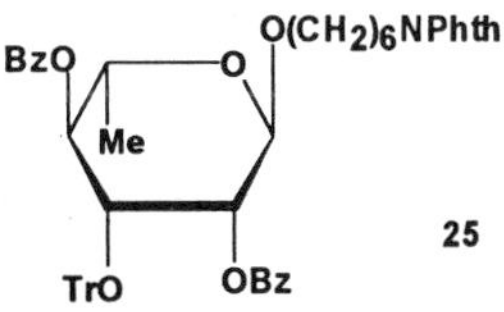

23

Recently we have begun the study of kinetics of this glycosylation aimed at solving the mechanistic problems and evaluation of the relative reactivity of CED's as glycosyl donors and trityl ethers as acceptors. Our preliminary results show that of several ordinary CED's studied, none exhibited exceptionally high or low activity (Kitov, P.I.; Tsvetkov, Yu.E.; Backinowsky, L.V.; Kochetkov, N.K., *Izv. Akad. Nauk. Ser. Khim. [Russ. Chem. Bull.]*, 1993, in press). At the same time, the higher reactivity of the galactose *endo*-cyano-CED vs. its *exo*-cyano isomer was confirmed (cf. *37*). Relatively great difference in reactivities of various acceptors was observed, and the pattern found for one CED does not necessarily hold for another CED.

Quite unexpected was the finding that the secondary trityl ethers are more reactive than the primary ones (though experimental data were documented, cf. *63*). This allowed us to perform selective monoglycosylation of some monosaccharide primary – secondary ditrityl ethers at the secondary sites (Tsvetkov, Yu.E; Kitov, P.I.; Backinowsky, L.V.; Kochetkov, N.K., *Tetrahedron Lett.*, in press). This order of reactivity contrasts that of primary – secondary diols, and may open new possibilities, although additional studies are required to reveal characteristic features of this version of the synthesis of oligosaccharides by trityl-cyanoethylidene condensation.

Literature Cited

1. Glaudemans, C.P.J. *Chem. Rev.* **1991**, *91*, 25.
2. Schuerch, C. in *Polymer and Fiber Science: Recent Advances*; Fornes, R.E. and Gilbert, R.D., Eds.; VCH: Weinheim, 1992; pp.9 - 16.
3. Glaudemans, C.P.J.; Kovác, P. *ACS Symp. Ser.* **1988**, *374*, 78.
4. Lemieux, R.U. *Chem. Soc. Rev.* **1978**, *7*, 423.
5. Paulsen, H. *Angew. Chem.* **1982**, *94*, 184.
6. Schmidt, R.R. *Angew. Chem.* **1986**, *98*, 213.
7. Kochetkov, N.K.; Bochkov, A.F. in *Recent Developments in the Chemistry of Natural Carbon Compounds*; Bognár, R.; Bruckner, V., and Szántay, Cs., Eds.; Akadémiai Kiadó: Budapest, 1971, Vol. 4; pp. 75 - 191.
8. Fügedi, P.; Garegg, P.; Lönn, H.; Norberg, T. *Glycoconjugate J.* **1987**, *4*, 97.
9. Hashimoto, H.; Abe, Y.; Horito, S.; Yoshimura, J. *J. Carbohydr. Chem.* **1988**, *8*, 307.
10. Schuerch, C. *Adv. Carbohydr. Chem. Biochem.* **1981**, *39*, 157.
11. Bochkov, A.F.; Kochetkov, N.K. *Carbohydr. Res.* **1975**, *39*, 355.
12. Bochkov, A.F.; Obruchnikov, I.V.; Kalinevich, V.M.; Kochetkov, N.K. *Tetrahedron Lett.* **1975**, 3403.
13. Bredereck, H.; Wagner, A.; Faber, G. *Angew. Chem.* **1957**, *69*, 438.
14. Meerwein, H.; Hederich, V.; Morschel, H.; Wunderlich, K. *Liebigs Ann. Chem.* **1960**, *635*, 1.
15. Coxon, B.; Fletcher, H.G., Jr. *J. Am. Chem. Soc.* **1963**, *85*, 2637.
16. Betaneli, V.I.; Litvak, M.M.; Backinowsky, L.V.; Kochetkov, N.K. *Carbohydr. Res.* **1981**, *94*, C1.
17. Obruchnikov, I.V.; Kochetkov, N.K. *Izv. Akad. Nauk SSSR. Ser. Khim.* **1977**, 2571 [*Bull. Acad. Sci. USSR. Div. Chem. Sci.*].
18. Betaneli, V.I.; Ovchinnikov, M.V.; Backinowsky, L.V.; Kochetkov, N.K. *Izv. Akad. Nauk SSSR. Ser. Khim.* **1979**, 2751 [*Bull. Acad. Sci. USSR. Div. Chem. Sci.*].

19. Betaneli, V.I.; Kryazhevskikh, I.A.; Ott, A.Ya.; Kochetkov, N.K. *Bioorg. Khim.* **1988**, *14*, 664 [*Sov. J. Bioorg. Chem.*].
20. Backinowsky, L.V.; Tsvetkov, Yu.E.; Ovchinnikov, M.V.; Byramova, N.E.; Kochetkov, N.K. *Bioorg. Khim.* **1985**, *11*, 66 [*Sov. J. Bioorg. Chem.*].
21. Nifant'ev, N.E.; Backinowsky, L.V.; Kochetkov, N.K. *Bioorg. Khim.* **1987**, *13*, 967 [*Sov. J. Bioorg. Chem.*].
22. Kochetkov, N.K.; Betaneli, V.I.; Ovchinnikov, M.V.; Backinowsky, L.V. *Tetrahedron*, **1981**, *37*, Suppl. 9, 149.
23. Backinowsky, L.V.; Nepogod'ev, S.A.; Shashkov, A.S.; Kochetkov, N.K. *Carbohydr. Res.* **1985**, *138*, 41.
24. Backinowsky, L.V.; Nepogod'ev, S.A.; Kochetkov, N.K. *Bioorg. Khim.* **1988**, *14*, 1234 [*Sov. J. Bioorg. Chem.*].
25. Byramova, N.E.; Ovchinnikov, M.V.; Backinowsky, L.V.; Kochetkov, N.K. *Carbohydr. Res.* **1983**, *124*, C8.
26. Nifant'ev, N.E.; Backinowsky, L.V.; Kochetkov, N.K. *Bioorg. Khim.* **1987**, *13*, 1093 [*Sov. J. Bioorg. Chem.*].
27. Betaneli, V.I.; Backinowsky, L.V.; Byramova, N.E.; Ovchinnikov, M.V.; Litvak, M.M.; Kochetkov, N.K. *Carbohydr. Res.* **1983**, *113*, C1.
28. Tsvetkov, Yu.E.; Byramova, N.E.; Backinowsky, L.V.; Kochetkov, N.K.; Yankina, N.F. *Bioorg. Khim.* **1986**, *12*, 1213 [*Sov. J. Bioorg. Chem.*].
29. Kochetkov, N.K.; Byramova, N.E.; Tsvetkov, Yu.E.; Backinowsky, L.V. *Tetrahedron* **1985**, *41*, 3363.
30. Byramova, N.E.; Backinowsky, L.V.; Kochetkov, N.K. *Izv. Akad. Nauk SSSR. Ser. Khim.* **1987**, 1120 [*Bull. Acad. Sci. USSR. Div. Chem. Sci.*].
31. Koto, S.; Morishima, N.; Yoshida, T.; Uchino, M.; Zen, S. *Bull. Chem. Soc. Japan* **1983**, *56*, 1171.
32. Wozney, Ya.V.; Kochetkov, N.K. *Carbohydr. Res.* **1977**, *54*, 300.
33. Backinowsky, L.V.; Tsvetkov, Yu.E.; Balan, N.F.; Byramova, N.E.; Kochetkov, N.K. *Carbohydr. Res.* **1980**, *85*, 209.
34. Tsvetkov, Yu.E.; Backinowsky, L.V. *Bioorg. Khim.* **1988**, *14*, 1589 [*Sov. J. Bioorg. Chem.*].
35. Betaneli, V.I., Ovchinnikov, M.V.; Backinowsky, L.V.; Kochetkov, N.K. *Carbohydr. Res.* **1979**, *76*, 252.
36. Backinowsky, L.V.; Nifant'ev, N.E.; Betaneli, V.I.; Struchkova, M.I.; Kochetkov, N.K. *Bioorg. Khim.* **1983**, *9*, 74 [*Sov. J. Bioorg. Chem.*].
37. Vicent, C.; Coteron, J.-M.; Jimenez-Barbero, J.; Martin-Lomas, M.; Penades, S. *Carbohydr. Res.* **1989**, *194*, 163.
38. Tsvetkov, Yu.E.; Backinowsky, L.V.; Kochetkov, N.K. *Bioorg. Khim.* **1991**, *17*, 1534 [*Sov. J. Bioorg. Chem.*].
39. Nepogod'ev, S.A.; Backinowsky, L.V.; Kochetkov, N.K. *Bioorg. Khim.* **1989**, *15*, 1555 [*Sov. J. Bioorg. Chem.*].
40. Kochetkov, N.K.; Nepogod'ev, S.A.; Backinowsky, L.V. *Tetrahedron* **1990**, *46*, 139.
41. Malysheva, N.N.; Kochetkov, N.K. *Carbohydr. Res.* **1982**, *105*, 173.
42. Wozney, Ya.V.; Backinowsky, L.V.; Kochetkov, N.K. *Carbohydr. Res.* **1979**, *73*, 282.

50 SYNTHETIC OLIGOSACCHARIDES

43. Ovchinnikov, M.V.; Byramova, N.E.; Backinowsky, L.V.; Kochetkov, N.K. *Bioorg. Khim.* **1983**, *9*, 391 [*Sov. J. Bioorg. Chem.*].
44. Tsvetkov, Yu.E.; Kitov, P.I.; Backinowsky, L.V.; Kochetkov, N.K. *Bioorg. Khim.* **1990**, *16*, 98 [*Sov. J. Bioorg. Chem.*].
45. Betaneli, V.I.; Kryazhevskikh, I.A.; Ott, A.Ya.; Kochetkov, N.K. *Bioorg. Khim.* **1989**, *15*, 217 [*Sov. J. Bioorg. Chem.*].
46. Bochkov, A.F.; Betaneli, V.I.; Kochetkov, N.K. *Bioorg. Khim.* **1976**, *2*, 927 [*Sov. J. Bioorg. Chem.*].
47. Ziegler, T.; Kovac, P.; Glaudemans, C.P.J. *Liebigs Ann. Chem.* **1990**, 613.
48. Ovchinnikov, M.V.; Byramova, N.E.; Backinowsky, L.V.; Kochetkov, N.K. *Bioorg. Khim.* **1983**, *9*, 401 [*Sov. J. Bioorg. Chem.*].
49. Tsvetkov, Yu.E.; Backinowsky, L.V.; Kochetkov, N.K. *Bioorg. Khim.* **1985**, *11*, 77 [*Sov. J. Bioorg. Chem.*].
50. Kochetkov, N.K.; Nifant'ev, N.E.; Backinowsky, L.V. *Tetrahedron* **1987**, *43*, 3109.
51. Backinowsky, L.V.; Nifant'ev, N.E.; Kochetkov, N.K. *Bioorg. Khim.* **1983**, *9*, 1089 [*Sov. J. Bioorg. Chem.*].
52. Nepogod'ev, S.A.; Backinowsky, L.V.; Kochetkov, N.K. *Bioorg. Khim.* **1986**, *12*, 1139 [*Sov. J. Bioorg. Chem.*].
53. Backinowsky, L.V.; Nifant'ev, N.E.; Shashkov, A.S.; Kochetkov, N.K. *Bioorg. Khim.* **1984**, *10*, 1212 [*Sov. J. Bioorg. Chem.*].
54. Kochetkov, N.K.; Malysheva, N.N.; Struchkova, M.I.; Klimov, E.M. *Bioorg. Khim.* **1985**, *11*, 391 [*Sov. J. Bioorg. Chem.*].
55. Kochetkov, N.K.; Malysheva, N.N.; Klimov, E.M. *Izv. Akad. Nauk SSSR. Ser. Khim.* **1983**, 1170 [*Bull. Acad. Sci. USSR. Div. Chem. Sci.*].
56. Kochetkov, N.K.; Ott, A.Ya. *Izv. Akad. Nauk SSSR. Ser. Khim.* **1983**, 1177 [*Bull. Acad. Sci. USSR. Div. Chem. Sci.*].
57. Backinowsky, L.V.; Nifant'ev, N.E.; Kochetkov, N.K. *Bioorg. Khim.* **1984**, *10*, 226 [*Sov. J. Bioorg. Chem.*].
58. Betaneli, V.I.; Kryazhevskikh, I.A.; Kochetkov, N.K. *Dokl. Akad. Nauk* **1992**, *322*, 540 [*Dokl. Chem.*].
59. Zhulin, V.M.; Klimov, E.M.; Makarova, Z.G.; Malysheva, N.N.; Kochetkov, N.K. *Dokl. Akad. Nauk SSSR* **1986**, *289*, 105 [*Dokl. Chem.*].
60. Zhulin, V.M.; Klimov, E.M.; Malysheva, N.N.; Makarova, Z.G.;Kochetkov, N.K. *Dokl. Akad. Nauk SSSR* **1987**, *295*, 873 [*Dokl. Chem.*].
61. Nifant'ev, N.E.; Backinowsky, L.V.; Kochetkov, N.K. *Carbohydr. Res.* **1989**, *191*, 13.
62. Tsvetkov, Yu.E.; Backinowsky, L.V.; Kochetkov, N.K. *Bioorg. Khim.* **1986**, *12*, 1144 [*Sov. J. Bioorg. Chem.*].
63. Tsvetkov, Yu.E.; Bukharov, A.V.; Backinowsky, L.V.; Kochetkov, N.K. *Carbohydr. Res.* **1988**, *175*, C1.
64. Tsvetkov, Yu.E.; Bukharov, A.V.; Backinowsky, L.V.; Kochetkov, N.K. *Bioorg. Khim.* **1988**, *14*, 1428 [*Sov. J. Bioorg. Chem.*].
65. Tsvetkov, Yu.E.; Backinowsky, L.V.; Kochetkov, N.K. *Carbohydr. Res.* **1989**, *193*, 75.

RECEIVED October 18, 1993

Chapter 4

Applications of Stannyl Ethers and Stannylene Acetals to Oligosaccharide Synthesis

T. Bruce Grindley

Department of Chemistry, Dalhousie University, Halifax, Nova Scotia B3H 4J3, Canada

The chemistry of trialkylstannyl ethers and dialkylstannylene acetals is reviewed with emphasis on those aspects that are useful for oligosaccharide synthesis. The causes of the remarkable regio-selectivity obtained with these intermediates are considered and the trends observed with particular types of carbohydrates and reaction conditions are summarized.

Since the discovery of their utility in 1974 (*1*),   trialkylstannyl ethers and dialkylstannylene acetals have become widely used intermediates in the synthesis of carbohydrate derivatives. A number of advances have been made since this subject was last reviewed (*2,3,4*).   In only a few publications (*5,6*) has this type of intermediate been used directly in oligosaccharide synthesis; however,   methods involving these intermediates are among the most important of those employed to prepare synthons for oligosaccharide synthesis. The major reason for their extensive use is that they react with electrophiles to give monosubstituted products, often with high regioselectivity. In addition, the reactions occur under much milder conditions or at rates that are much faster than those of the parent alcohols. As a result, these reactions have become part of the arsenal of techniques commonly used by organic chemists and large numbers of publications using them appear each year. This report will summarize the results that are considered to be useful for oligosaccharide synthesis and will highlight those publications that have introduced new concepts or types of applications.

The most widely employed intermediates are dibutylstannylene acetals and tributylstannyl ethers. Dibutylstannylene acetals are prepared by reaction of diols with dibutyltin oxide in methanol with heating or in benzene or toluene with azeotropic removal of water. Reaction in methanol, where dibutyldimethoxytin is an intermediate for the preparation, is more rapid; however, it was observed recently that yields were lower and that starting material remained after reaction workup if the dibutylstannylene acetal was formed by this method (*7*). The technique used for

0097–6156/94/0560–0051$09.08/0

preparation may also have been a factor in the higher yields reported by Kováč and Edgar (8), who formed the stannylene acetal in toluene, than reported for identical reactions by earlier workers, who formed it in methanol (9). The question as to why lower yields are obtained from at least some preparations using the more convenient procedure in methanol remains open; possibilities include moisture in the methanol, which is very hygroscopic, or incomplete formation of the dibutylstannylene acetal because of incomplete removal of methanol.

Tributylstannyl ethers are prepared in the same manner by reaction of alcohols with hexabutyldistannoxane, more commonly known as bistributyltin oxide. Holzapfel *et al.* noted that the reaction in benzene requires only 0.5 molar equivalents of bistributyltin oxide to go to completion but takes 16 h at reflux. This is probably

$$2\ ROH\ +\ (Bu_3Sn)_2O\ \longrightarrow\ 2\ ROSnBu_3\ +\ H_2O$$

because the tin-containing by-product of the first half of the reaction, tributyltin hydroxide, reacts much slower than the initial reagent (10).

Structures

In the solid state, trimethyltin methoxide (11) and trimethyltin hydroxide (12,13) are linear polymers. The tin atoms are pentacoordinate with distorted trigonal bipyramidal geometries having apical oxygen atoms. In solution, simple and more complex trialkyltin alkoxides exist predominantly as monomers with tetrahedral tetracoordinate tin atoms. This was determined by molecular weight measurements and from ^{119}Sn NMR chemical shifts, which are diagnostic for coordination status (14,15,16). Formation of tributylstannyl ethers from a polyol using less than a stoichiometric amount of bistributyltin oxide yields a mixture that contains all possible tributylstannyl ethers (15,16). These tributyltin ethers do not interconvert rapidly on the NMR timescale at room temperature, but do interconvert rapidly under reaction conditions (15,16).

Acyclic dialkyltin dialkoxides exist as dimers in solution unless the alkyl or alkoxy groups are large, *e.g.*, *t*-butoxy groups, where they are present either as monomers or mixtures (14,17). Dialkyltin dialkoxides, in which the dialkoxides form a ring are called dialkylstannylene acetals, or more properly, 2,2-dialkyl-1,3,2-dioxastannolanes, if the ring is five-membered, and 2,2-dialkyl-1,3,2-dioxastannanes if the ring is six-membered. Compounds having primary alkyl substitution exist as infinite polymers in the solid state with octahedral hexacoordinate tin atoms (18,19,20). Compounds having larger substituents, as in 2,2-di-*t*-butyl-1,3,2-dioxastannolane (21), methyl 4,6-O-benzylidene-2,3-O-dibutylstannylene-α-**D**-glucopyranoside (22,23), or methyl 4,6-O-benzylidene-2,3-O-dibutylstannylene-α-**D**-mannopyranoside (24) are present in the solid state as dimers or oligomers.

As with the trialkyltin ethers, these compounds are less aggregated in solution. For instance, 2,2-dibutyl-1,3,2-dioxastannolane, a polymer in the solid-state (*18*), has been shown by variable temperature ^{119}Sn NMR spectroscopy to be a mixture of dimers, trimers, and tetramers in solution, with dimers predominating at room temperature and above (*25,26*). This technique has also indicated that most carbohydrate-derived stannylene acetals are present predominantly as dimers in solution (*27,28,29*). Supporting evidence has been obtained from mass spectral studies (*27*) and by comparison of solid-state NMR spectra with those of solutions (*28*).

Cyclic 1,3,2-dioxastannolanes or 1,3,2-dioxastannanes show a much increased tendency to exist as species containing pentacoordinate or hexacoordinate tin atoms than do acyclic dialkyltin dialkoxides. This tendency has been attributed to the bond angles imposed by ring formation on the tin atom (*22*). The O-Sn-O bond angles are about 78 - 80° for both five- and six-coordinate tin atoms in five-membered stannylene acetals (*18,20-24*), and 93.2° for the six-coordinate tin atom in the one six-membered stannylene acetal studied by X-ray crystallography (*19*). These bond angles are close to the 90° bond angles needed for trigonal bipyramidal geometry or octahedral geometry of five and six-coordinate tin but are much smaller than those in the tetrahedral geometry favored by tetracoordinate tin. The small O-Sn-O bond angles in the five- and six-membered stannylene acetal rings are imposed on the rings because the Sn-O bond lengths are much longer than the other bond lengths (*22*). In dimers, oligomers, or polymers, the Sn-O bonds inside the monomer units are shorter, 1.98 to 2.13 Å, than those between monomer units, which are 2.23 to 2.27 Å if the tin atom is pentacoordinate, or 2.43 to 2.60 Å if the tin atom is hexacoordinate (*18-24*).

The geometries of dimers of the stannylene acetals of 1,2-propanediol shown below (Figure 1) (*22,23*) can be related to the distorted trigonal bipyramidal geometry at tin. The tricoordinate oxygens are equatorial to one tin atom but apical to the other. The dicoordinate oxygen atoms are apical and the alkyl groups on tin are equatorial.

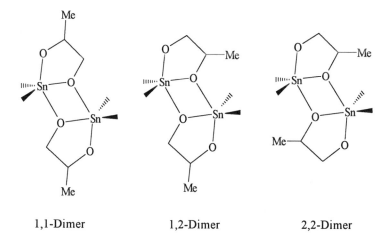

1,1-Dimer   1,2-Dimer   2,2-Dimer

Figure 1. Dimers of the stannylene acetal of 1,2-propanediol

When the two oxygen atoms involved in the stannylene acetal are diastereotopic, three dimers, two with C_2 symmetry, can be formed as shown in Figure 1 for those obtained from 1,2-propanediol. Dimers are named by means of the numbers of the tricoordinate oxygen atoms (28). Steric effects appear to be the most important factor in determining the relative populations of the three dimers. In particular, stannylene acetals derived from *trans*-diols with one adjacent axial substituent exist in solution to the level of detection of ^{119}Sn NMR spectroscopy as the symmetric dimer in which the tricoordinate oxygen atom is not adjacent to the axial substituent (28). Similarly, dialkylstannylene acetals from carbohydrate-derived terminal diols exist predominantly as symmetric dimers with the primary oxygen atoms tricoordinate (Figure 2) (29).

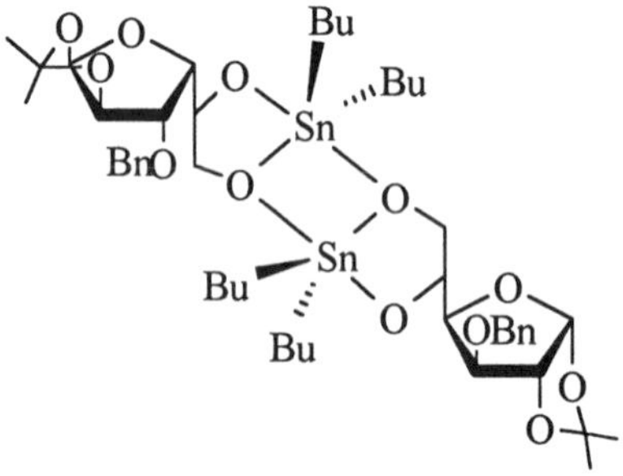

Figure 2. The populated dimer of 3-O-benzyl-5,6-O-dibutylstannylene-1,2-O-isopropylidene-α-**D**-glucofuranose

Simple 2,2-dibutyl-1,3,2-dioxastannolanes form 1:1 crystalline complexes of monomer units with nucleophiles such as pyridine and dimethyl sulfoxide (30). Such complexes are less stable for more substituted stannylene acetals, *e.g.*, those derived from carbohydrates (30,31). Addition of nucleophiles to solutions of stannylene acetals in non-polar solvents has been found to markedly increase the rates of reaction with electrophiles (32) and transient complexes of this type are likely intermediates. Similar rate enhancements were observed in reactions of tributylstannyl ethers (33). Tetrabutylammonium iodide was the nucleophile used first but a wide variety of nucleophiles have been used subsequently; tetraalkylammonium halides, *N*-methylimidazole (10) and cesium fluoride (34,35) are used most. Nucleophilic solvents, such as *N*,*N*-dimethylformamide and probably ethers, also must act as added nucleophiles. As well as increasing the rates of reaction, in certain cases the added nucleophiles reverse the regioselectivity from that observed in non-polar solvents (10,29).

Cause of the regioselectivity

Tributylstannyl ethers. When a polyol is reacted with bis(tributyltin) oxide, all tributylstannyl ethers are formed and these interconvert readily at the reaction temperatures (15,16). Ogawa and Matsui suggested that oxygen atoms of tributyltin ethers that have adjacent oxygen atoms in appropriate orientations to coordinate with the tin atom are activated toward electrophiles (*36,37*), specifically, equatorial oxygen atoms that have an adjacent axial oxygen atom, or primary oxygen atoms where the tin atom of the tributylstannyl ether can coordinate to the ring oxygen as shown in Figure 3. Evidence from ^{119}Sn NMR spectroscopy indicates that tributylstannyl ethers

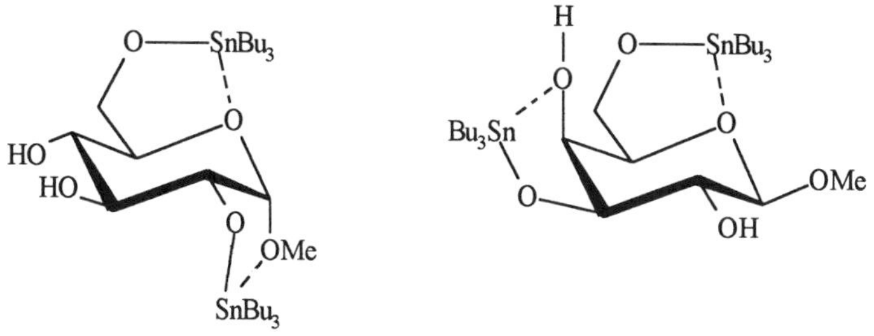

Figure 3. Coordination of tributylstannyl ethers

exist in solution predominantly as monomers with tetracoordinate tin atoms.   The additional coordination pictured above (*15,16*) does not occur to any great extent but such transient species are likely reaction intermediates.

The same reactivity preferences are observed in the presence of added nucleophiles (*33,38*). In one suggested mechanism (Figure 4), the nucleophile adds

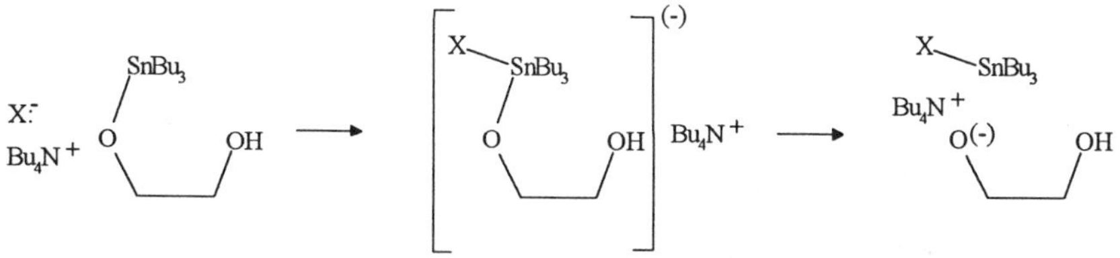

Figure 4.   A mechanism proposed for activation of tributylstannyl ethers by nucleophiles (*33*)

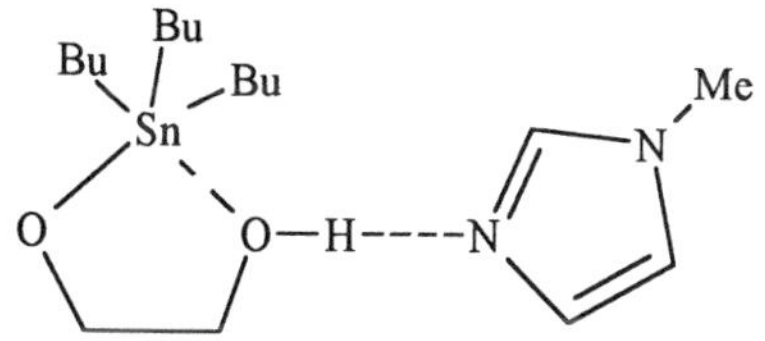

Figure 5.   An intermediate proposed for tributylstannyl ether reactions in the presence of *N*-methylimidazole (*16*)

to the tetracoordinate tin atom to give a species which dissociates to the reactive species, an oxyanion coordinated to the counterion of the nucleophile (*33*). In another suggested mechanism (Figure 5), the added nucleophile acts as an acceptor of a hydrogen bond from a hydroxyl group, enhancing the coordination of the hydroxyl oxygen atom to a tin atom attached to an adjacent carbon (*16*) and also increasing its nucleophilicity. Evidence for the latter proposal was obtained from the effects of adding N-methylimidazole (NMI) on ^{119}Sn NMR chemical shifts. With one equivalent added, signals were shielded by up to 26 ppm with the largest shieldings being observed if adjacent OH groups were in a *cis*-relationship. A change in coordination state of tin from four to five would be expected to induce a shielding of 100 to 200 ppm for the ^{119}Sn NMR signals (*39,40,41*). The smaller changes in shifts observed with the addition of one equivalent of NMI (0 to 26 ppm) and the observation that changes in shifts of up to 58 ppm were obtained when four equivalents were added indicate that equilibria occur between species having coordinated and non-coordinated tin atoms. The position of the equilibria must lie very far towards the non-coordinated species with one equivalent or less of NMI added. Both suggested mechanisms appear plausible for the nucleophile considered; however the first requires a counterion that is not present for neutral nucleophiles to stabilize the oxyanion, the second is not reasonable for non-basic nucleophiles like iodide ions. Yet, all types of nucleophiles cause similar regioselectivity. The similarity of results obtained for both basic and non-basic nucleophiles suggests that the role of the nucleophile is to add to tin so as to produce a five-coordinate tin atom that enhances the nucleophilicity of the stannyl ether oxygen atom. Species incorporating six-coordinate tin atoms appear to have only slightly less stability than those with only five-coordinate tin atoms (*25*). Thus, a possible explanation of the regioselectivity is that it results from additional coordination of the five-coordinate tin atom to appropriately oriented oxygen atoms.

In recent systematic studies, Tsuda *et al.* observed that oxidation of tributylstannyl ethers of methyl glycosides with bromine gave different regioselectivities than those found for substitution reactions. Oxidation of secondary carbons bearing axial oxygen atoms of *cis*-diols is preferred over reaction either at secondary carbons bearing equatorial oxygen atoms or at primary carbon atoms (*42,43*). An additional but lesser preference is for O-3 for equatorial glycosides and O-4 for axial glycosides. No explanation for this pattern of regioselectivity is currently available.

Dialkylstannylene Acetals. A wider range of regioselectivities has been observed with dialkylstannylene acetals than with tributylstannyl ethers although there are some similarities in the regioselectivities observed. When added nucleophiles are present, reaction always occurs at the site that is preferred for reaction of the parent diols (*44*). Thus, it seems likely that the reactive intermediates in this case are coordinated monomers, and that the tin-containing intermediates simply magnify the inherent reactivity differences between pairs of oxygen atoms.

In non-polar solvents without added nucleophiles, different selectivities are often observed. Oxidation reactions using bromine or N-bromosuccinimide usually give the product of reaction at the inherently less reactive oxygen atom, *e.g.*, dibutylstannylene acetals of *cis*-diols on pyranose rings react at the axial oxygen atom (*45,46*), and those of terminal 1,2-diols react at the secondary oxygen atom (*45,47*). In certain cases, benzoylation reactions also occur preferentially on axial oxygen atoms from dibutylstannylene acetals of *cis*-diols on pyranose rings (*10,48*) and on secondary or tertiary oxygen atoms of dibutylstannylene acetals from terminal 1,2-

diols (*49*). Toluenesulfonation reactions on dialkylstannylene acetals of terminal 1,2-diols give mainly the product of reaction at the primary oxygen atom if the alkyl groups are *n*-butyl groups but the regioselectivity can be reversed if the two alkyl groups are replaced by a cyclic hexamethylene group (*29*).

These results suggest that the regioselectivity under these conditions is related to the structure and reactivity of the dominant species present in solution, the dimers (*27,28*). Monomers are unlikely to be intermediates in these reactions because they are much less populated than dimers and because their oxygen atoms would not be expected to be more reactive than those of the dimers. Any gain in reactivity conveyed by decreased steric hindrance to electrophile approach would be offset by the increased electron density on tin that presumably accompanies dimer formation. Since the same considerations apply to oligomer reactions as apply to dimer reactions and oligomers are usually much less populated, reactions conducted in the absence of added nucleophiles will be discussed only in terms of dimers.

Results obtained for *trans*-diols on pyranose rings are consistent with this conclusion. If only a single symmetric dimer is present in solution as determined by ^{119}Sn NMR spectroscopy, highly regioselective acylation reactions are obtained in which reaction occurs on the dicoordinate oxygen atoms of the dimer. If mixtures of dimers are present, mixtures of products are observed (*28*).

When the oxygen atoms of a diol have very different reactivities, as in the primary and secondary oxygens of terminal 1,2-diols, more complicated results are obtained that can only be explained in terms of dimer reactivity. The stannylene acetals of these diols are present primarily as symmetric dimers with the secondary oxygen atoms dicoordinate. Benzoylation of stannylene acetals of carbohydrate-derived terminal diols occurs preferentially on the secondary oxygen atoms but the initial products rearrange to the primary benzoates at about the same rate as they are formed (*50*). This is in contrast to the situation for terminal diols not bearing other electronegative groups (*49*). Toluenesulfonate groups do not rearrange under these conditions. Toluenesulfonation reactions of dibutylstannylene acetals of terminal diols give mainly terminal substitution, whereas reactions with hexamethylenestannylene acetals give mainly secondary toluenesulfonates (*29*).

The results of the toluenesulfonation reactions can be explained by means of the following kinetic scheme which will apply if one symmetric dimer (Dimer 1) is considerably more populated than the other symmetric dimer (Dimer 3), with the unsymmetrical dimer (Dimer 2) intermediate in concentration. The scheme is simplified by omission of Dimer 3 as explained below.

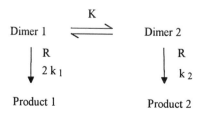

Figure 6. Kinetic scheme for reactions of unequally populated dimers

In Figure 6, K is the dimer equilibrium constant, k_1 is the rate constant for reaction at one of the two identical dicoordinate oxygens in the more populated symmetric dimer (Dimer 1) with the reagent R, and k_2 is the rate constant for reaction at the one different oxygen atom in the non-symmetric dimer, Dimer 2. The rate constant for reaction of the other dicoordinate oxygen atom in Dimer 2 (the same oxygen atom that is dicoordinate in Dimer 1) can be neglected. This rate constant will be about the same size as k_1 and Dimer 2 is much less populated than Dimer 1. Similarly, reaction from Dimer 3 can be neglected because it is much less populated than Dimer 2 and the rate constant for reaction from its two identical oxygen atoms will be about the same size as k_2.

If equilibration of the dimers is faster than reaction, the following equation is obtained,

$$(d[\text{Product 2}]/dt) \, / \, (d[\text{Product 1}]/dt) \; = \; K \, k_2 \, / \, (2 \, k_1) \qquad (1)$$

that is, the ratio of the rates of product formation, which will equal the final product ratio, is equal to the product of equilibrium constant times the ratio of the rate constants. Where the rate constants are not very different, the equilibrium constant will dominate, as in benzoylation of the *trans*-diols (28). Where the oxygen atoms involved have very different reactivities as for terminal diols where primary and secondary oxygen atoms are competing, either the equilibrium constants or the ratio of the rate constants can dominate (29).

When the reaction becomes very fast, equilibration may become slower than reaction. If $K = k_f \, / \, k_b$, the following equations can be derived using the approximation that the [Dimer 2] is at a steady-state concentration.

$$(d[\text{Product 1}]/dt) \, / \, (d[\text{Product 2}]/dt) \; = \; 2 \, k_1 \, (\, k_b + k_2 \, [R] \,) \, / \, (k_f \, k_2) \qquad (2)$$

Since reaction is now faster than equilibration, $k_b \ll k_2 \, [R]$, and equation 2 becomes

$$(d[\text{Product 1}]/dt) \, / \, (d[\text{Product 2}]/dt) \; = \; 2 \, k_1 \, [R] \,) \, / \, k_f \qquad (3)$$

that is, the product ratio is determined by whether the major dimer prefers to go to product or to the other dimer. Because product formation is assumed to be fast, the reagent essentially traps the major dimer under these conditions. These are probably the conditions under which oxidations of stannylene acetals by bromine occur.

Trends in Regioselectivity

Because only acylation and alkylation reactions of pyranosides are likely to be most important in oligosaccharide synthesis, only these will be surveyed. Results obtained since the last reviews (2,3) will be emphasized.

Tributylstannyl ethers.   Acylation reactions of tributyltin ethers formed from compounds having both primary and secondary alcohols occur fastest on primary oxygen atoms. If a secondary equatorial oxygen atom has an adjacent axial oxygen atom, reaction on that atom occurs at a slightly slower rate than on the primary atom so that diacyl or monoacyl derivatives can be obtained (Figure 7) (*36*).

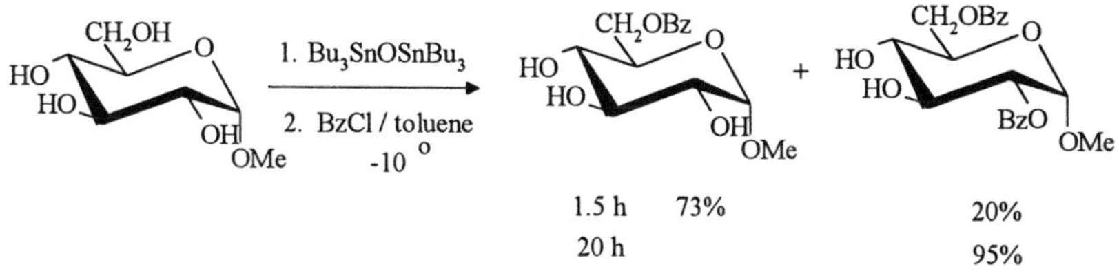

Figure 7. Regioselectivity in benzoylation of tributylstannyl ethers (36)

Benzoylation of the tris(tributylstannyl ether) of 2,2,2-trichloroethyl 2-deoxy-2-phthalimido-β-D-glucopyranoside with three equivalents of benzoyl chloride gave the 3,6-di-*O*-benzoate in 82% yield and the 6-*O*-benzoate in 5% yield (*51*). When only secondary alcohols are present, acylation occurs on the equatorial oxygen of a *cis*-diol unit or next to an axial oxygen-containing substituent (*10,36,52*).

Similar trends in regioselectivity were noted in alkylation reactions although reaction with benzyl bromide alone is very slow (*53,54*). Alkylation occurs much faster with the same regiochemistry in the presence of added nucleophiles (*33*).

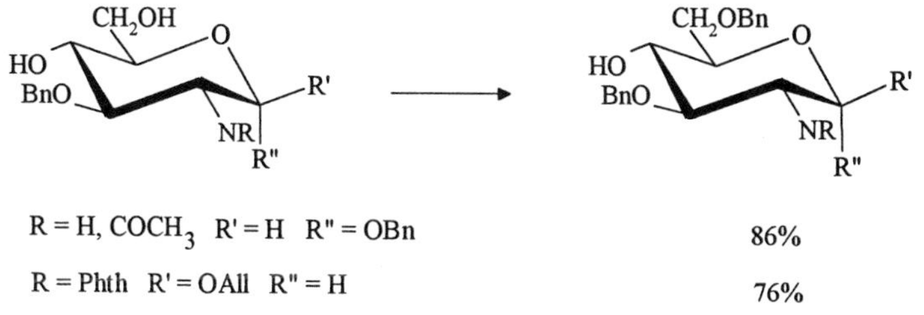

Figure 8. Regioselectivity in benzylation of tributylstannyl ethers (55,56)

Reactions can be performed preferentially at primary centers as shown in Figure 8 if no secondary oxygen atom has an adjacent axial oxygen atom (*55,56,57*). Liu and Danishefsky have achieved very good regioselectivity for the primary oxygen atom in the reaction of the tributylstannyl ether of a terminal 1,3-diol with a glycosidic oxirane in the presence of a Lewis acid catalyst, $Zn(OTf)_2$ (*58*). For other examples, see ref. *51, 59*, and *60*. When a secondary oxygen atom has an adjacent axial oxygen atom and a primary oxygen atom is available, mixtures of the products of substitution at the two sites are obtained (*51,54,61*). If an adjacent oxygen atom is axial, compounds bearing two or more secondary hydroxyl groups on a pyranose ring give regioselective reactions at an equatorial oxygen atom in some cases (Figure 9) (*38*) but mixtures in others (*37*). Surprisingly, Paulsen *et al.* have obtained regioselective alkylation at the secondary oxygen atom adjacent to an axial oxygen substituent in a triol containing a primary oxygen atom (Figure 10) (*62*). Occasionally somewhat selective reactions are obtained from *trans*-diols (Figure 11) (*63*).

In an interesting recent application, Danishefsky *et al.* (*5*) have reacted

Figure 9. Regioselectivity obtained in benzylation of tributylstannyl ethers (38)

Figure 10. Regioselective alkylation of a tributylstannyl ether of a triol (62)

Figure 11. Regioselective benzylation of a lactose derivative (63)

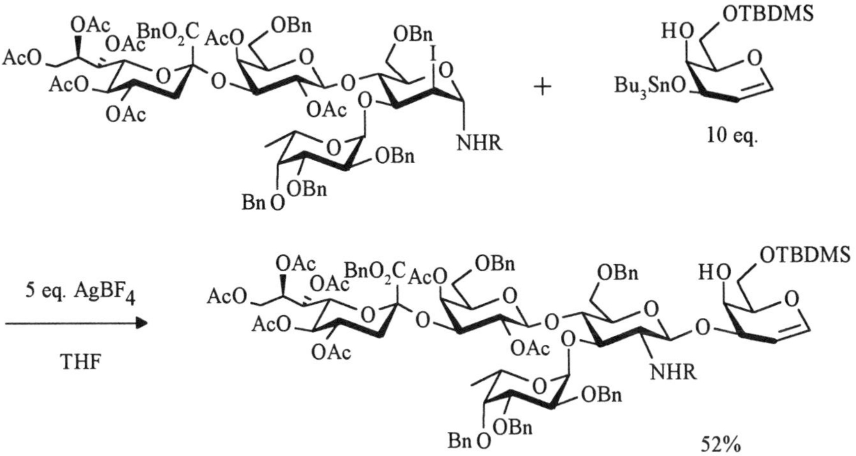

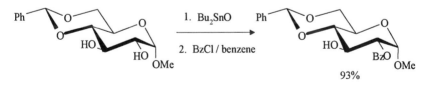

Figure 12. Reaction of tributylstannyl ethers with iodosulfonamides (5)

tributylstannyl ethers derived from 6-O-*t*-butyldimethylsilyl-**D**-galactal and 6,6'-di-O-*t*-butyldimethylsilyl-**D**-lactal with glycal-derived iodosulfonamides to give *trans*-2-sulfonylamino-2-deoxy glycosides in yields of 52 and 42%, respectively (Figure 12).

Dialkylstannylene Acetals. Considerably more studies have appeared that employ dialkylstannylene acetals than tributylstannyl ethers. The regioselectivity obtained in reactions bears many similarities to that observed for tributylstannyl ethers but there are several important differences. In addition, reactions have been performed on dialkylstannylene acetals of types of molecules that have not as yet been explored through tributylstannyl ethers.

Benzoylations performed in the presence or absence of added nucleophiles on dibutylstannylene acetals derived from *trans*-diequatorial secondary diols on pyranose rings give monobenzoates in a highly regioselective manner if one adjacent oxygen atom is axial (*10,45,64,65*). Substitution occurs adjacent to the axial oxygen atom as shown in Figure 13. Even when the anomeric substituent is large, regioselective

Figure 13. Benzoylation of a dibutylstannylene acetal of a *trans*-diol (*45*)

reactions can be obtained; *e.g.*, Baer and Wu (*66*) reacted the dibutylstannylene acetal of 4,6:4',6'-di-O-benzylidene-α,α-trehalose with one equivalent of palmitoyl chloride in benzene in the presence of triethylamine and obtained the 2-O-palmitate in 66% yield. When the *trans*-diols have both adjacent substituents equatorial (*28,45,67*) or both are axial (*28,68*), mixtures are normally obtained. However, methyl 4-O-benzoyl-β-**D**-xylopyranoside is regioselectively benzoylated and chloroacetylated at O-3 (*69*), suggesting that electronic factors also play a role. Dibutylstannylene acetals of acyclic derivatives are also acylated adjacent to ester groups under some conditions (Figure 14) (*70*).

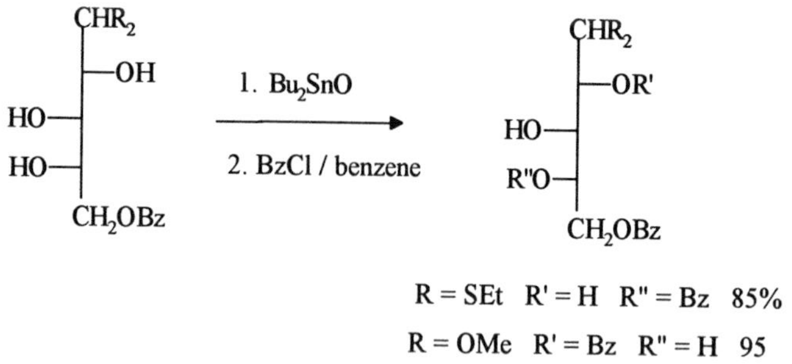

CHR₂ —OH
HO—
HO—
CH₂OBz

1. Bu₂SnO
⟶
2. BzCl / benzene

CHR₂ —OR'
HO—
R"O—
CH₂OBz

R = SEt R' = H R" = Bz 85%

R = OMe R' = Bz R" = H 95

Figure 14. Benzoylation of dibutylstannylene acetals of acyclic triols (*70*)

Cis-diols give varying results. The dibutylstannylene acetal of methyl 4,6-*O*-benzylidene-α-**D**-mannopyranoside reacted with benzoyl chloride in benzene in the presence of *N*-methylimidazole only at the equatorial oxygen atom, O-3, (*10*), but with benzoyl chloride in dioxane in the presence of triethylamine gave approximately equal amounts of the two monoesters (*48,64*).   When benzoylation was performed in benzene without added nucleophiles, the ratio of axial to equatorial esters was 85 to 15. It has been observed that if the stannylene acetal is prepared in methanol, it is essential to remove all methanol in order to obtain good regioselectivity (*10*). When the dibutylstannylene acetal of methyl 4,6-di-*O*-benzyl-α-**D**-mannopyranoside was prepared in methanol, and reacted with benzoyl chloride in benzene, the 2-*O*- and 3-*O*-benzoates were obtained 41 and 59% yields, respectively (*71*).   The dibutyl-stannylene acetal of methyl 4,6-*O*-benzylidene-α-**D**-allopyranoside reacted with benzoyl chloride in dioxane in the presence of triethylamine only at the equatorial oxygen atom, O-2 (*64*).   Galactopyranose derivatives having only O-3 and O-4 free react exclusively at O-3 in the presence (*72,73*) or absence of added nucleophiles (*45*).

Reactions of dibutylstannylene acetals from terminal 1,2-diols with benzoyl chloride or *p*-toluenesulfonyl chloride strongly favor substitution at the primary center in the presence of added nucleophiles and to a lesser extent in their absence (*29,74,75,76,77*). Similarly, the less hindered 4-oxygen atom is benzoylated (*69,78*) and tosylated (*77*) regioselectively in reactions of the dibutylstannylene acetal of methyl β-**D**-xylopyranoside (Figure 15).

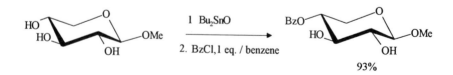

HO
HO
OH
OMe

1 Bu₂SnO
⟶
2. BzCl,1 eq. / benzene

BzO
HO
OH
OMe

93%

Figure 15. Benzoylation of the dibutylstannylene acetal of methyl β-**D**-
xylopyranoside (*69*)

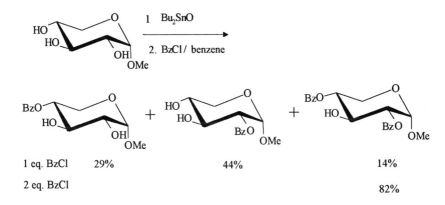

1 eq. BzCl	29%	44%	14%
2 eq. BzCl			82%

Figure 16. Benzoylation of the dibutylstannylene acetal of methyl α-D-xylopyranoside (*69*)

The α-anomer, which would be expected to give the 2-O-benzoate, based on results for methyl α-**D**-glucopyranoside derivatives, yielded a mixture of 2- and 4-substitution in agreement with this idea (Figure 16) (*69,78*). Benzoylation of the dibutylstannylene acetal of methyl 2,6-dideoxy-α-**L**-*arabino*-hexopyranoside similarly occurs regioselectively adjacent to the deoxy center (Figure 17) (*79*) and benzoylation

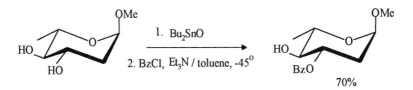

Figure 17. Benzoylation of a dibutylstannylene acetal of a deoxy derivative (*79*)

and tosylation of 4,6-O-benzylidene-N-carbobenzoxy-1-deoxy-norjiromycin was also accomplished regioselectively adjacent to the unsubstituted center on O-2 (*80*). Probably for the same reasons, benzoylation or acetylation of dibutylstannylene acetals of glycals containing *trans*-diols occurred, in the presence of added nucleophiles, adjacent to the double bond if the primary oxygen atom is protected but gave mixtures when the primary oxygen was not protected (Figure 18) (*81*). A 2-ene derivative with only O-4 and O-6 free was tosylated regioselectively at the primary oxygen atom (Figure 19) (*82*).

The selectivity obtained in fast reactions in the absence of added nucleophiles, where the least hindered oxygen atom reacts, is probably due to the greater reactivity of this oxygen atom. If the dimer having the more hindered oxygen atom dicoordinate is more populated, reversed regioselectivity can be achieved as in the tosylation of hexamethylene-stannylene acetals of terminal diols (Figure 20) (*29*) or the benzoylation of dibutylstannylene acetals of non-carbohydrate terminal diols (*49*).

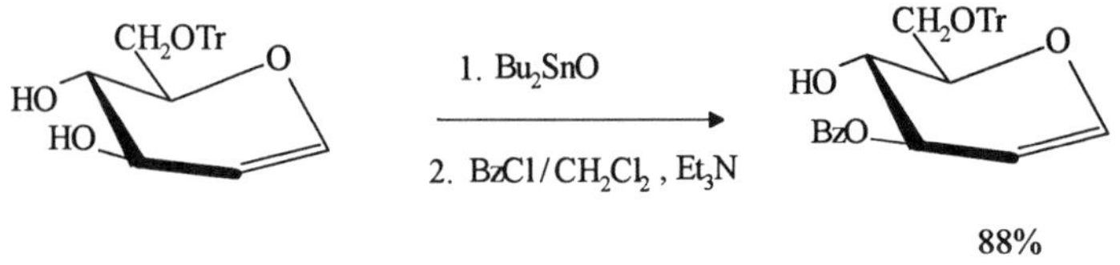

Figure 18. Benzoylation of a dibutylstannylene derivative of a glycal (*81*)

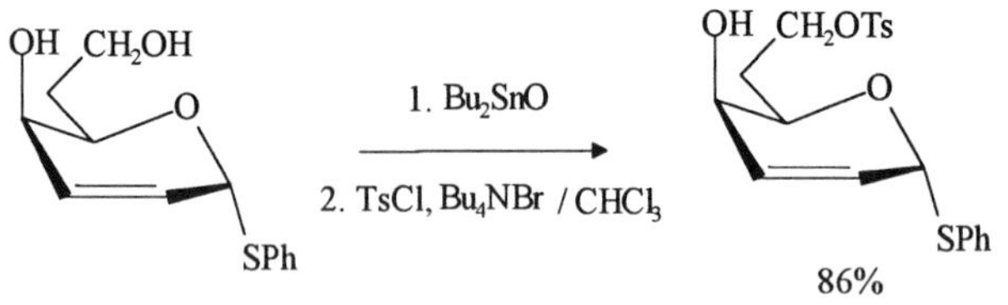

Figure 19. Tosylation of the dibutylstannylene acetal of a 2-ene derivative (*82*)

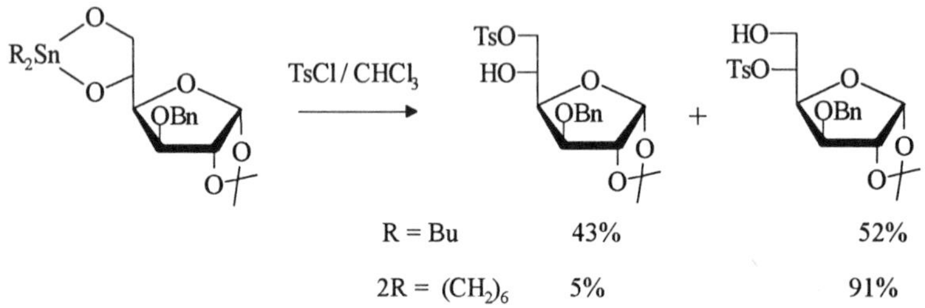

R = Bu	43%	52%
2R = (CH₂)₆	5%	91%

Figure 20. Tosylation of stannylene acetals of 3-*O*-benzyl-1,2-*O*-isopropylidene-
α-**D**-glucofuranose (*29*)

Hexamethylenetin oxide, the starting material for hexamethylenestannylene acetals, can be prepared conveniently by cleavage of the phenyl groups from hexamethylenediphenyltin by chloroacetic acid followed by hydrolysis (*83*).

When both primary and a number of secondary oxygen atoms are available, benzoylation takes place with excellent regioselectivity on a secondary equatorial oxygen atom if an adjacent oxygen is axial, e.g. at O-2 on methyl α-**D**-glucopyranoside, or at O-3 on methyl α-**D**-mannopyranoside and methyl β-**D**-galactopyranoside (*64,78*). If no adjacent axial oxygen is present, benzoylation takes place on the primary oxygen atom, e.g. in the reaction of the dibutylstannylene acetal of methyl β-**D**-glucopyranoside. Mixtures are obtained (*78*) if there is more than one axial group present that can activate different oxygen atoms.   When the slower reaction, tosylation, was performed in dioxane in the presence of the added nucleophile, 4-dimethylaminopyridine, the same trends were observed except that there was a greater preference for reaction at the less hindered oxygen atom and at equatorial oxygen atoms of *cis*-diols (*77*) (see Figure 21).

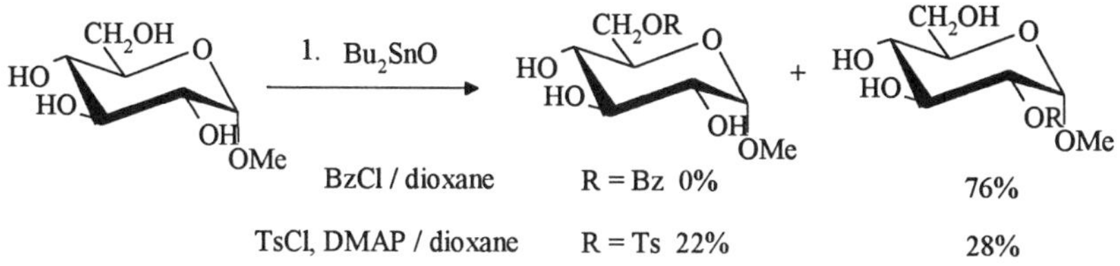

Figure 21. Acylation of the dibutylstannylene acetal of methyl α-D-glucopyranoside (77,78)

Probably the most valuable reaction for oligosaccharide synthesis that involves organotin intermediates is regioselective alkylation. Benzylation has been studied most, but substituted benzyl ethers, allyl ethers, methyl ethers, and a variety of others have been produced by means of this technique.   This reaction was performed initially in N,N-dimethylformamide without added nucleophiles (84), but reaction in benzene in the presence of added tetrabutylammonium iodide or bromide was found to give more selective reactions (32) and most researchers now employ added nucleophiles in alkylation reactions. Cesium fluoride (34,35) has become one of the most commonly employed nucleophiles, usually in N,N-dimethylformamide, and it has been observed for specific compounds that these conditions give better yields than previous methods (8,85,86).   However, the yields reported for most compounds seem comparable when either the tetraalkylammonium halides or cesium fluoride are used.

By far the most common use of dibutylstannylene acetals has been to alkylate regioselectively the equatorial oxygen of a *cis*-diol on a pyranose ring that is isolated or part of a triol of all secondary hydroxyl groups. This can be achieved with yields in the range of 75 to 98% reliably.   Most applications have been to 6-protected or 6-deoxy derivatives of mannopyranosides (7,9,87,88,89,90,91,92,93,94,95,96,97,98,99, 100,101) or galactopyranosides (9,68,73,86,102,103,104,105,106,107,108,109,110,111, 112,113,114), but other carbohydrates (79,115,116,117,118,119,120,121,122,123,124, 125,126,127), inositols (128,129,130,131,132,133,134,135), and non-carbohydrate molecules (136) also have been studied. In a recent example reported by Vogel *et al.* for a methyl D-galactopyranosyluronate derivative, partial transesterification accompanied regioselective benzylation (Figure 22) (137). In addition, there are a number of examples known where the equatorial oxygen atom of a *cis*-diol on a

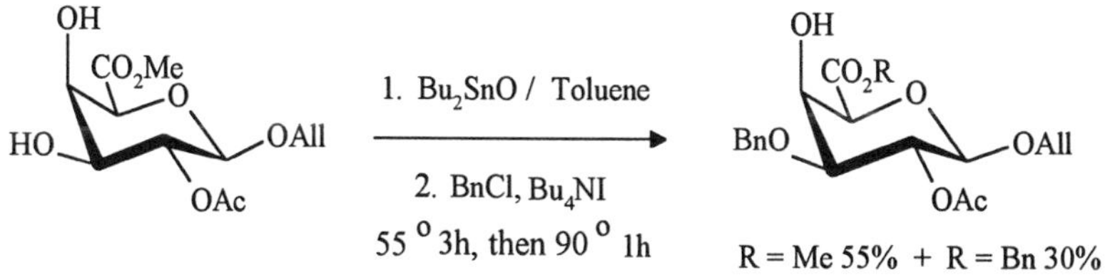

Figure 22. Benzylation of the dibutylstannylene acetal of methyl (allyl α-D-galactopyranosyluronate (137)

pyranose ring reacts with good regioselectivity even when the substrates contain unprotected primary oxygen atoms (*9,32,87,106,116,138,139,140*).

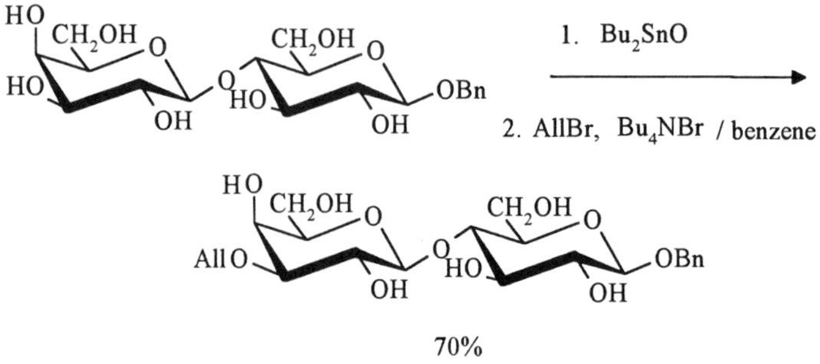

70%

Figure 23.   Allylation of the dibutylstannylene acetal of benzyl β-**D**-lactose (*138*)

A spectacular example is shown in Figure 23 where the product was isolated without chromatography (*138*) and a similar yield was obtained with the 2-(trimethylsilyl)ethyl glycoside (*140*).   Reactions that give this type of regioselectivity allow preparation of oligosaccharide synthons by monosubstitution, blocking the remainder of the free hydroxyls, then removal of the original substituent.   Di-*O*-benzylation can also be achieved with reasonable regioselectivity for some substrates containing *cis*-diols (Figure 24) (*141*).

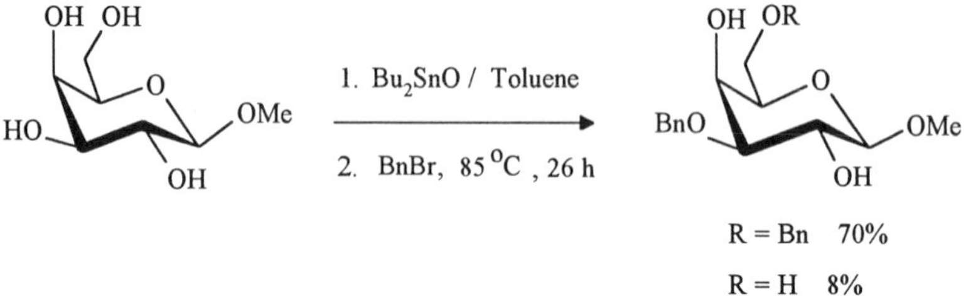

R = Bn   70%

R = H   8%

Figure 24.   Di-*O*-benzylation of the dibutylstannylene acetal of methyl β-**D**-galactopyranoside (*141*)

Dibutylstannylene acetals of isolated diequatorial *trans*-diols usually can be alkylated with reasonable selectivity if one and only one adjacent oxygen atom is axial, although the regioselectivity attained is lower than that in acylation reactions (*32,45,142,143*).   If the adjacent substituents are only equatorial, mixtures are usually obtained (*28,67,144,145,146*).   Moderate yields of single 3-*O*-substituted products have been obtained in reactions on ethyl 4,6-*O*-benzylidene-1-thio-β-**D**-glucopyranoside (Figure 25) (*147,148*).   In an unusual example shown in Figure 26, a glucopyranose derivative with a bulky β-glycosyl substituent reacted regioselectively at O-3 (*149,150*) and in another case, a bulky substituent at O-4 caused regioselective reaction at O-2 (*151*).

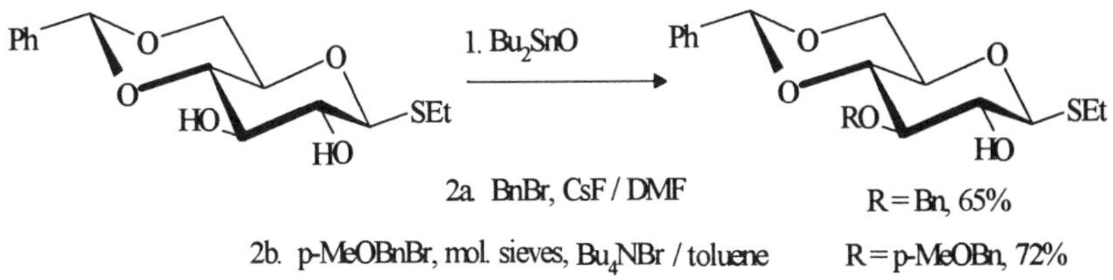

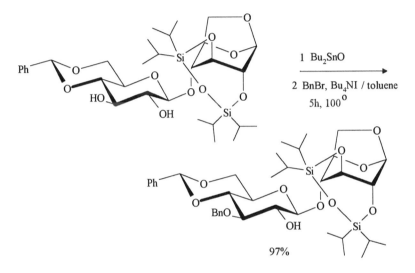

Figure 25. Benzylation of the *trans*-diol in ethyl 4,6-*O*-benzylidene-1-thio-β-**D**-glucopyranoside (*147,148*)

Figure 26. Benzylation of a dibutylstannylene acetal derived from a *trans*-diol with two adjacent equatorial substituents, one of which is large (*150*)

If a primary and several equatorial secondary hydroxyl groups are present, alkylation via the dibutylstannylene acetal intermediates can be achieved with reasonable selectivity adjacent to a substituted axial oxygen atom if one is present (Figure 27) (*116*). If no axial oxygen atom is available, mixtures of primary and secondary substitution are obtained (*98*). The 2,6-di-*O*-benzyl derivative of methyl-α-**D**-glucopyranoside can be obtained in 82% yield by heating the dibutylstannylene acetal of the unsubstituted glycoside in benzyl bromide (*141*).

When one of the sites adjacent to the *trans*-diol is a deoxy center, reactions occur with good regioselectivity on the oxygen atom adjacent to the deoxy center (Figure 28) (*79,152,153*). The preference for reaction with the oxygen atoms on stannylene acetals adjacent to deoxy centers is sufficient to reverse the normal specificity of *cis*-diols to react at the equatorial oxygen atoms, either partially (*154*), or totally, albeit with a bulky electrophile (Figure 29) (*155*).

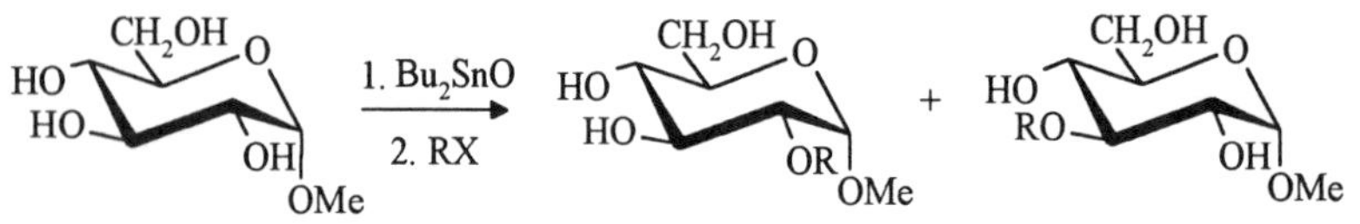

in dioxane, R = Bn	55%	18%
in acetonitrile, R = Bn	56%	6%
in dioxane, R = All	59%	16%
in dioxane, R = All	46%	trace

Figure 27. Alkylation of dibutylstannylene acetals derived from unsubstituted methyl α-**D**-glucopyranosides (*116*)

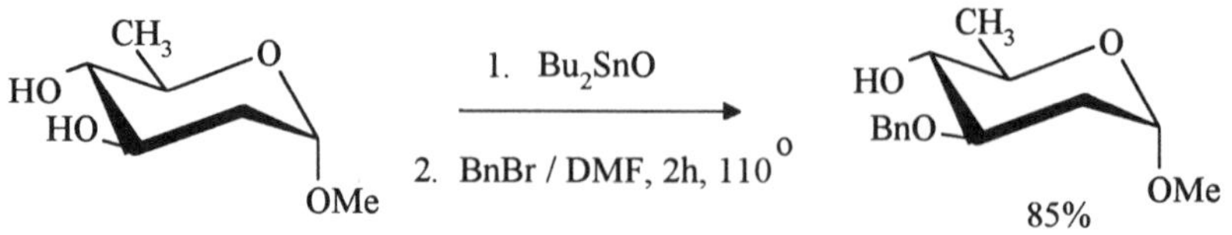

Figure 28. Benzylation of a dibutylstannylene acetal of a deoxy derivative (*152*)

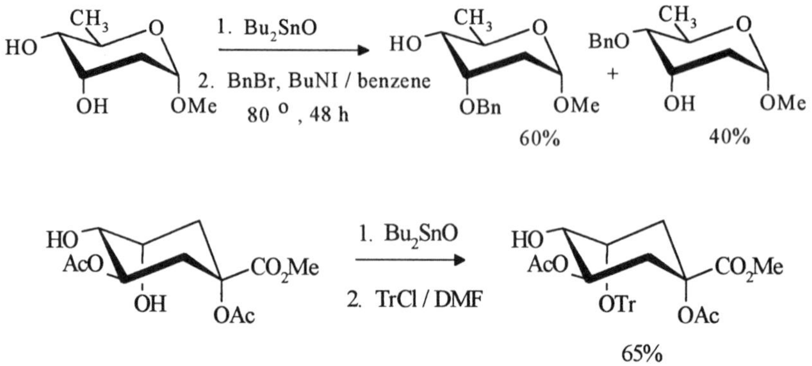

Figure 29. Benzylation of dibutylstannylene acetals of *cis*-diols adjacent to deoxy centers (*154,155*)

This preferred reaction adjacent to unsubstituted centers is also manifested in a tendency for pentopyranosides to react at O-4. Tsuda *et al.* have shown that methyl β-**D**-xylopyranoside reacts with benzyl bromide in dioxane only at O-4 (*116*). A similar, but slightly decreased, preference was obtained for a methyl 2-azido-2-deoxy-α-**D**-xylopyranoside (*156*). In reactions of the dibutylstannylene acetal of methyl α-**D**-xylopyranoside, the preference for reaction adjacent to unsubstituted centers is opposed by the preference for reaction adjacent to the axial glycosyl oxygen; mixtures were obtained with a variety of alkyl halides in dioxane that slightly favor substitution at O-4 (*116*). It seems likely that, as for acylation reactions (*69*), 2,4-di-*O*-alkyl derivatives could be prepared in good yield. The dibutylstannylene acetal of phenyl α-L-arabinopyranoside follows the normal *cis*-diol preference and reacts only at O-3 (*116*) but those of alkyl β-arabinopyranosides give mixtures of O-3 substitution and substitution at the nominally axial oxygen atom, O-4 (*116,123*).   In contrast, the dibutylstannylene acetal of methyl 2-*O*-benzyl-β-D-arabinopyranoside was reported to react in *N,N*-dimethylformamide containing cesium fluoride preferentially at O-3 in unspecified yield (*157*).

Dibutylstannylene acetals of glycals with the primary oxygen atom protected react regioselectively on the oxygen atom adjacent to the double bond (*82,155,158,159,160*).   Figure 30 shows a typical example (*158*).

Dibutylstannylene acetals formed from the 1,3-diols of glycosides of 2-enopyranosides react on the primary oxygen in preference to the site adjacent to the alkene (Figure 31) (*161,162*).

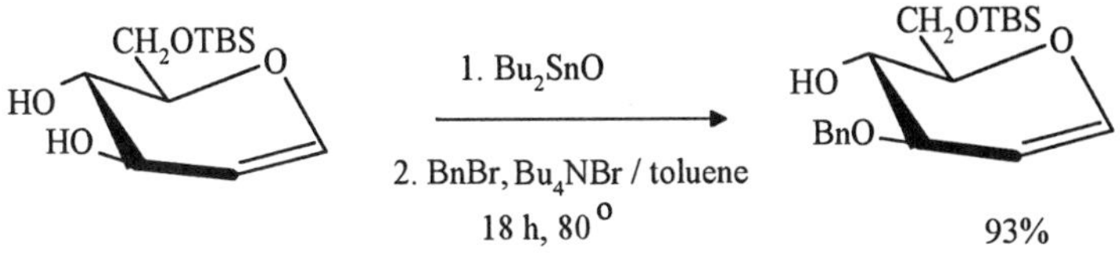

Figure 30. Benzylation of the dibutylstannylene acetal of 6-*O-t*-butyldimethylsilyl-**D**-glucal (*158*)

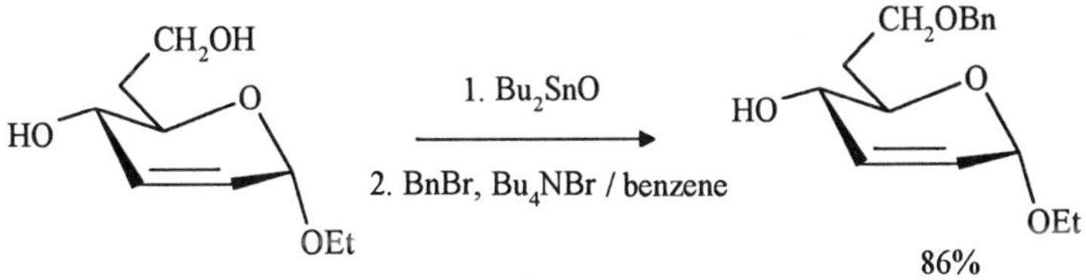

Figure 31. Benzylation of the dibutylstannylene acetal of ethyl 2,3-dideoxy-α-**D**-2-enopyranoside (*162*)

A number of alkylation reactions have been performed on dibutylstannylene acetals derived from terminal 1,2-diols (*85,163,164,165,166,167*). Reaction occurs on the primary oxygen atom both in the presence and absence of added nucleophiles. Most of the examples that have been studied involve reactions on glycerol or mannitol derivatives. A particularly interesting one is shown in Figure 32 where reaction takes place on a primary oxygen atom in preference to reaction on a secondary oxygen atom adjacent to a deoxy center (*164*).

SYNTHETIC OLIGOSACCHARIDES

70

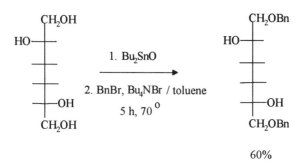

60%

Figure 32. Benzylation of the dibutylstannylene acetal of 3,4-dideoxy-**D**-*threo*-
hexitol (*164, 167*)

Glycoside Formation

Alkyl glycosides and alkyl and phenyl thioglycosides can be prepared in useful yields
by reaction of glycosyl acetates with tributyltin ethers and thioethers in the presence
of Lewis acids, such as trimethylsilyl trifluoromethanesulfonate (*168*) or tin
tetrachloride (*51,54,59,65, 169,170,171*). A recent application was to the synthesis
of *S*-glycosylated peptides (*171*). Dibutyltin derivatives (Bu$_2$Sn(SR)$_2$, R=Ph,Me,
cyclohexyl, *t*-Bu; Bu$_2$Sn(SePh)$_2$) have also been found to react in the same way,
particularly using Bu$_2$Sn(OTf)$_2$ as the Lewis acid catalyst (*172*). These methods only
give good control of stereochemistry if a participating group is present on C-2.
    Allyl glycosides and ethers can also be prepared by reaction of tributylstannyl
ethers or dibutylstannylene acetals of alcohols with allyl acetates in the presence of
palladium(0) compounds. *Cis*-glycosides appear to be preferred even in the presence
of participating groups (*173*).

Acknowledgement

I would like to thank NSERC for support and my former graduate students, Dr. R.
Thangarasa and Dr. X. Kong, for their important contributions to the development of
the concepts presented here.

References

1. Wagner, D.; Verheyden, J. P. H.; Moffatt, J. G. *J. Org. Chem.* **1974**, *39*,
    24-30.
2. David, S.; Hanessian, S. *Tetrahedron* **1985**, *41*, 643-663.
3. Pereyre, M.; Quintard, J. P.; Rahm, A. *Tin in Organic Synthesis*;
    Butterworths: London, 1987; pp 261-323.
4. Blunden, S. J.; Cusack, P. A.; Smith, P. J. *J. Organomet. Chem.* **1987**,
    325, 141-152.

5. Danishefsky, S. J.; Koseki, K.; Griffith, D. A.; Gervay, J.; Peterson, J. M.; McDonald, F. E.; Oriyama, T. *J. Am. Chem. Soc.* **1992**, *114*, 8331-8333.
6. Danishefsky, S. J.; Gervay, J.; Peterson, J. M.; McDonald, F. E.; Koseki, K.; Oriyama, T.; Wong, C.-H.; Dumas, D. P. *J. Am. Chem. Soc.* **1992**, *114*, 8329-8331.
7. Kjølberg, O.; Neumann, K. *Acta Chem. Scand.* **1993**, *47*, 721-727.
8. Kováč, P.; Edgar, K. J. *J. Org. Chem.* **1992**, *57*, 2455-2467.
9. Yang, G.; Kong, F.; Zhou, S. *Carbohydr. Res.* **1991**, *211*, 179-182.
10. Holzapfel, C. W.; Koekemoer, J. M.; Marais, C. F. *S. Afr. J. Chem.* **1984**, *37*, 19-26.
11. Domingos, A. M.; Sheldrick, G. M. *Acta Cryst., Section B* **1974**, *30*, 519-521.
12. Kasa, N.; Yasuda, K.; Okawara, R. *J. Organomet. Chem.* **1965**, *3*, 172-173.
13. Harris, R. K.; Packer, K. J.; Reams, P.; Sebald, A. *J. Magn. Reson.* **1987**, *72*, 385-387.
14. Smith, P. J.; White, R. F. M.; Smith, L. *J. Organomet. Chem.* **1972**, *40*, 341-353.
15. Blunden, S. J.; Smith, P. J.; Beynon, P. J.; Gillies, D. G. *Carbohydr. Res.* **1981**, *88*, 9-18.
16. Cruzado, C.; Bernabe, M.; Martin-Lomas, M. *J. Org. Chem.* **1989**, *54*, 465-469.
17. Kennedy, J. D. *J. Chem. Soc., Perkin Trans. II* **1977**, 242-248.
18. Davies, A. G.; Price, A. J.; Dawes, H. M.; Hursthouse, M. B. *J. Chem. Soc., Dalton Trans.* **1986**, 297-302.
19. Grindley, T. B.; Thangarasa, R.; Bakshi, P. K.; Cameron, T. S. *Can. J. Chem.* **1992**, *70*, 197-204.
20. Grindley, T. B.; Wasylishen, R. E.; Thangarasa, R.; Power, W. P.; Curtis, R. E. *Can. J. Chem.* **1992**, *70*, 205-217.
21. Bates, P. A.; Hursthouse, M. B.; Davies, A. G.; Slater, S. D. *J. Organomet. Chem.* **1989**, *363*, 45-60.
22. David, S.; Pascard, C.; Cesario, M. *Nouv. J. Chim.* **1979**, *3*, 63-68.
23. Cameron, T. S.; Bakshi, P. K.; Thangarasa, R.; Grindley, T. B. *Can. J. Chem.* **1992**, *70*, 1623-1630.
24. Holzapfel, C. W.; Koekemoer, J. M.; Marais, C. F.; Kruger, G. J.; Pretorius, J. A. *S. Afr. J. Chem.* **1982**, *35*, 80-88.
25. Grindley, T. B.; Thangarasa, R. *J. Am. Chem. Soc.* **1990**, *112*, 1364-1373.
26. Roelens, S.; Taddei, M. *J. Chem. Soc., Perkin Trans. II* **1985**, 799-804.
27. David, S.; Thiéffry, A.; Forchioni, A. *Tetrahedron Lett.* **1981**, *22*, 2647-2650.
28. Grindley, T. B.; Thangarasa, R. *Can. J. Chem.* **1990**, *68*, 1007-1019.
29. Grindley, T. B.; Kong, X. *Tetrahedron Lett.* **1993**, *34*, 5231-5234.
30. Davies, A. G.; Price, A. J. *J. Organomet. Chem.* **1983**, *258*, 7-13.
31. Thangarasa, R. *PhD Thesis*, Dalhousie University, 1990.
32. David, S.; Thiéffry, A.; Veyrières, A. *J. Chem. Soc., Perkin Trans. I* **1981**, 1796-1801.

33. Veyrières, A.; Alais, J. *J. Chem. Soc., Perkin Trans. 1* **1981**, 377-381.
34. Nagashima, N.; Ohno, M. *Chem. Lett.* **1987**, 141-144.
35. Danishefsky, S. J.; Hungate, R. *J. Am. Chem. Soc.* **1986**, *108*, 2486-2487.
36. Ogawa, T.; Matsui, M. *Tetrahedron* **1981**, *37*, 2363-2369.
37. Ogawa, T.; Takahashi, Y.; Matsui, M. *Carbohydr. Res.* **1982**, *102*, 207-215.
38. Cruzado, C.; Martin-Lomas, M. *Carbohydr. Res.* **1988**, *175*, 193-199.
39. Smith, P. J.; Tupčiauskas, A. P. *Ann. Rep. NMR Spectrosc.* **1978**, *8*, 291-370.
40. Wrackmeyer, B. *Ann. Rep. NMR Spectrosc.* **1985**, *16*, 73-186.
41. Kennedy, J. D.; McFarlane, W. In *Multinuclear NMR*; Mason, J., Ed.; Pergamon Press: London, 1988; pp 305-333.
42. Tsuda, Y.; Hanajima, M.; Matsuhira, N.; Okuno, Y.; Kanemitsu, K. *Chem. Pharm. Bull.* **1989**, *37*, 2344-2350.
43. Liu, H.-M.; Sato, Y.; Tsuda, Y. *Chem. Pharm. Bull.* **1993**, *41*, 491-501.
44. For a summary of reaction preferences of carbohydrate diols and polyols, see Haines, A.H. *Adv. Carbohydr. Chem. Biochem.* **1976**, *33*, 11-109.
45. David, S.; Thieffry, A. *J. Chem. Soc., Perkin Trans. 1* **1979**, 1568-1573.
46. Van Drijver, L.; Holzapfel, C. W.; Koekemoer, J. M.; Kruger, G. J.; van Dyk, M. S. *Carbohydr. Res.* **1986**, *155*, 141-150.
47. Kong, X.; Grindley, T. B. *J. Carbohydr. Chem.* **1993**, *12*, 557-571.
48. Nashed, M. A.; Anderson, L. *Tetrahedron Lett.* **1976**, *39*, 3503-3506.
49. Reginato, G.; Ricci, A.; Roelens, S.; Scapecchi, S. *J. Org. Chem.* **1990**, *55*, 5132-5139.
50. Kong, X.; Grindley, T.B., unpublished results.
51. Ogawa, T.; Nakabayashi, S.; Sasajima, K. *Carbohydr. Res.* **1981**, *96*, 29-39.
52. Hanessian, S.; Roy, R. *J. Am. Chem. Soc.* **1979**, *101*, 5839-5841.
53. Ogawa, T.; Katano, K.; Matsui, M. *Carbohydr. Res.* **1978**, *64*, C3-C9.
54. Ogawa, T.; Katano, K.; Sasajima, K.; Matsui, M. *Tetrahedron* **1981**, *37*, 2779-2786.
55. Ogawa, T.; Kitajima, T.; Nukada, T. *Carbohydr. Res.* **1983**, *123*, C5-C7.
56. Veyrières, A. *J. Chem. Soc., Perkin Trans. 1* **1981**, 1626-1629.
57. Toepfer, A.; Schmidt, R. R. *J. Carbohydr. Chem.* **1993**, *12*, 809-822.
58. Liu, K. K.-C.; Danishefsky, S. J. *J. Am. Chem. Soc.* **1993**, *115*, 4933-4934.
59. Ogawa, T.; Nakabayashi, S. *Carbohydr. Res.* **1981**, *97*, 81-86.
60. Ogawa, T.; Katano, K.; Sasajima, K.; Matsui, M. *Tetrahedron* **1981**, *37*, 2779-2786.
61. Ogawa, T.; Nukada, T.; Matsui, M. *Carbohydr. Res.* **1982**, *101*, 263-270.
62. Paulsen, H.; Reck, F.; Brockhausen, I. *Carbohydr. Res.* **1992**, *236*, 39-71.
63. Fernandez-Mayoralas, A.; Martin-Lomas, M. *Carbohydr. Res.* **1986**, *154*, 93-101.
64. Munavu, R. M.; Szmant, H. H. *J. Org. Chem.* **1976**, *41*, 1832-1835.
65. Nakano, T.; Ito, Y.; Ogawa, T. *Carbohydr. Res.* **1993**, *243*, 43-69.
66. Baer, H. H.; Wu, X. *Carbohydr. Res.* **1993**, *238*, 215-230.
67. Takeo, K.; Shibata, K. *Carbohydr. Res.* **1984**, *133*, 147-151.

68. Nashed, M. A.; Chowdhary, M. S.; Anderson, L. *Carbohydr. Res.* **1982**, *102*, 99-110.
69. Helm, R. F.; Ralph, J.; Anderson, L. *J. Org. Chem.* **1991**, *56*, 7015-7021.
70. Bredenkamp, M. W.; Holzapfel, C. W.; Swanepoel, A. D. *Tetrahedron Lett.* **1990**, *31*, 2759-2762.
71. Wu, X.; Kong, F. *Carbohydr. Res.* **1987**, *162*, 166-169.
72. Nashed, M. A.; Anderson, L. *Carbohydr. Res.* **1977**, *56*, 419-422.
73. Youssef, R. H.; Silwanis, B. A.; El-Sokkary, R. I.; Nematalla, A. S.; Nashed, M. A. *Carbohydr. Res.* **1993**, *240*, 287-293.
74. David, S.; de Sennyey, G. *Carbohydr. Res.* **1979**, *77*, 79-97.
75. Thiem, J.; Wessel, H.-P. *Liebigs Ann. Chem.* **1983**, 2173-2184.
76. Kim, M. H.; Yang, J. W.; Lee, B. J.; Kim, H. O.; Chun, M. W. *Soul Taehakkyo Yakhak Nonmunjip* **1986**, *11*, 71-73.
77. Tsuda, Y.; Nishimura, M.; Kobayashi, T.; Sato, Y.; Kanemitsu, K. *Chem. Pharm. Bull.* **1991**, *39*, 2883-2887.
78. Tsuda, Y.; Haque, M. E.; Yoshimoto, K. *Chem. Pharm. Bull.* **1983**, *31*, 1612-1624.
79. Monneret, C.; Gagnet, R.; Florent, J.-C. *J. Carbohydr. Chem.* **1987**, *6*, 221-229.
80. Getman, D. P.; DeCrescenzo, G. A.; Heintz, R. M. *Tetrahedron Lett.* **1991**, *32*, 5691-5692.
81. Mereyala, H. B.; Kulkarni, V. R. *Carbohydr. Res.* **1989**, *187*, 154-158.
82. Halcomb, R. L.; Wittman, M. D.; Olson, S. H.; Danishefsky, S. J.; Golik, J.; Wong, H.; Vyas, D. *J. Am. Chem. Soc.* **1991**, *113*, 5080-5082.
83. Kong, X.; Grindley, T.B. *Organometallics*, in press.
84. Augé, C.; David, S.; Veyrières, A. *J. Chem. Soc. , Chem. Commun.* **1976**, 375-376.
85. Nagashima, N.; Ohno, M. *Chem. Pharm. Bull.* **1991**, *39*, 1972-1982.
86. Smid, P.; de Reuter, G. A.; van der Marel, G. A.; Rombouts, F. M.; van Boom, J. H. *J. Carbohydr. Chem.* **1991**, *10*, 833-849.
87. Nashed, M. A. *Carbohydr. Res.* **1978**, *60*, 200-205.
88. Srivastava, V. K.; Schuerch, C. *Tetrahedron Lett.* **1979**, *35*, 3269-3272.
89. Rana, S. S.; Barlow, J. J.; Matta, K. L. *Carbohydr. Res.* **1980**, *85*, 313-317.
90. Hong, N.; Funabashi, M.; Yoshimura, J. *Carbohydr. Res.* **1981**, *96*, 21-28.
91. Varma, A. J.; Schuerch, C. *J. Org. Chem.* **1981**, *46*, 799-803.
92. Eis, M. J.; Ganem, B. *Carbohydr. Res.* **1988**, *176*, 316-323.
93. Dasgupta, F.; Garegg, P. J. *Synthesis* **1989**, 626-628.
94. Halcomb, R. L.; Boyer, S. H.; Danishefsky, S. J. *Angew. Chem. Int. Ed. Engl.* **1992**, *31*, 338-340.
95. Gurjar, M. K.; Mainkar, A. S. *Tetrahedron* **1992**, *48*, 6729-6738.
96. Gurjar, M. K.; Reddy, K. R. *J. Chem. Soc., Perkin Trans. I* **1993**, 1269-1272.
97. Severn, W. B.; Richards, J. C. *J. Am. Chem. Soc.* **1993**, *115*, 1114-1120.
98. Kosemura, S.; Yamamura, S.; Kakuta, H.; Mizutani, J.; Hasegawa, K. *Tetrahedron Lett.* **1993**, *34*, 2653-2656.

99. Garegg, P. J.; Olsson, L.; Oscarson, S. *J. Carbohydr. Chem.* **1993**, *12*, 955-967.

100. Borbás, A.; Lipták, A. *Carbohydr. Res.* **1993**, *241*, 99-116.

101. Kováč, P. *Carbohydr. Res.* **1993**, *245*, 219-231.

102. Nashed, M. A.; Anderson, L. *Carbohydr. Res.* **1977**, *56*, 325-336.

103. Slife, C. W.; Nashed, M. A.; Anderson, L. *Carbohydr. Res.* **1981**, *93*, 219-230.

104. Wetzel, P.; Bulian, H.-P.; Maulshagen, A.; Muller, D.; Snatzke, G. *Tetrahedron* **1984**, *40*, 3657-3666.

105. Kováč, P.; Glaudemans, C. P. J.; Taylor, R. B. *Carbohydr. Res.* **1985**, *142*, 158-164.

106. Takeo, K.; Aspinall, G. O.; Brennan, P. J.; Chatterjee, D. *Carbohydr. Res.* **1986**, *150*, 133-150.

107. Jacqinet, J. C.; Sinaÿ, P. *Carbohydr. Res.* **1987**, *159*, 229-253.

108. Pozsgay, V.; Jennings, H. J. *Carbohydr. Res.* **1988**, *179*, 61-75.

109. Baer, H. H.; Mateo, F. H.; Siemsen, L. *Carbohydr. Res.* **1989**, *187*, 67-92.

110. Nishimura, S.-I.; Murayama, S.; Kurita, K.; Kuzuhara, H. *Chem. Lett.* **1992**, 1413-1416.

111. Zegelaar-Jaarsveld, K.; van der Marel, G. A.; van Boom, J. H. *Tetrahedron* **1992**, *48*, 10133-10148.

112. Zuurmond, H. M.; Veeneman, G. H.; van der Marel, G. A.; van Boom, J. H. *Carbohydr. Res.* **1993**, *241*, 153-164.

113. Dupradeau, F.-Y.; Allaire, S.; Prandi, J.; Beau, J.-M. *Tetrahedron Lett.* **1993**, *34*, 4513-4516.

114. Ferro, V.; Skelton, B. W.; Stick, R. V.; White, A. H. *Austral. J. Chem.* **1993**, *46*, 787-803.

115. Martin, A.; Pais, M.; Monneret, C. *Carbohydr. Res.* **1983**, *113*, 21-29.

116. Haque, M. E.; Kikuchi, T.; Yoshimoto, K.; Tsuda, Y. *Chem. Pharm. Bull.* **1985**, *33*, 2243-2256.

117. Yoshimura, J.; Aqeel, A.; Hong, N.; Sato, K.-I.; Hashimoto, H. *Carbohydr. Res.* **1986**, *155*, 236-246.

118. Auzanneau, F.-I.; Charon, D.; Szabo, L.; Merienne, C. *Carbohydr. Res.* **1988**, *179*, 125-136.

119. Izquierdo Cubero, I.; Plaza Lopez-Espinosa, M. T.; Tornel Osorio, P. L. *An. Quim.* **1988**, *84*, 340-343.

120. Kiyoshima, K.; Sakamoto, M.; Ishikura, T.; Fukagawa, Y.; Yoshioka, T.; Naganawa, H.; Sawa, T.; Takeuchi, T. *Chem. Pharm. Bull.* **1989**, *37*, 861-865.

121. Yoshikawa, M.; Murakami, N.; Inoue, Y.; Hatakeyama, S.; Kitagawa, I. *Chem. Pharm. Bull.* **1993**, *41*, 636-638.

122. Boons, G. J. P. H.; van Delft, F. L.; van der Klein, P. A. M.; van der Marel, G. A.; van Boom, J. H. *Tetrahedron* **1992**, *48*, 885-904.

123. Chen, S.-H.; Danishefsky, S. J. *Tetrahedron Lett.* **1990**, *31*, 2229-2232.

124. Auzanneau, F.-I.; Charon, D.; Szilágyi, L.; Szabó, L. *J. Chem. Soc., Perkin Trans. 1* **1991**, 803-809.

125. Gurjar, M. K.; Saha, U. K. *Tetrahedron Lett.* **1992**, *33*, 4979-4982.

126. Boons, G. J. P. H.; van Delft, F. L.; van der Klein, P. A. M.; van der Marel, G. A.; van Boom, J. H. *Tetrahedron* **1992**, *48*, 885-904.
127. van der Klein, P. A. M.; Filemon, W. ; Boons, G. J. P. H.; Veeneman, G. H.; van der Marel, G. A.; van Boom, J. H. *Tetrahedron* **1992**, *48*, 4649-4658.
128. Gigg, J.; Gigg, R.; Payne, S.; Conant, R. *J. Chem. Soc., Perkin Trans. I* **1987**, 1757-1762.
129. Garegg, P. J.; Lindberg, B.; Kvarnström, I.; Svensson, S. C. T. *Carbohydr. Res.* **1988**, *173*, 205-216.
130. Elie, C. J. J.; Verduyn, R.; Dreef, C. E.; Brounts, D. M.; van der Marel, G. A.; van Boom, J. H. *Tetrahedron* **1990**, *46*, 8243-8254.
131. Sawyer, D. A.; Potter, B. V. L. *J. Chem. Soc., Perkin Trans. I* **1992**, 923-932.
132. Lampe, D.; Mills, S. J.; Potter, B. V. L. *J. Chem. Soc., Perkin Trans. I* **1992**, 2899-2906.
133. Marecek, J. F.; Prestwich, G. D. *Tetrahedron Lett.* **1991**, *32*, 1863-1866.
134. Dreef, C. E.; Jansze, J.-P.; Elie, C. J. J.; van der Marel, G. A.; van Boom, J. H. *Carbohydr. Res.* **1992**, *234*, 37-50.
135. Gou, D.-M.; Liu, Y.-C.; Chen, C.-S. *Carbohydr. Res.* **1992**, *234*, 51-64.
136. Danishefsky, S. J.; Lee, J. Y. *J. Am. Chem. Soc.* **1989**, *111*, 4829-4837.
137. Vogel, C.; Steffan, W.; Ott, A. Y.; Betaneli, V. I. *Carbohydr. Res.* **1992**, *237*, 115-129.
138. Alais, J.; Maranduba, A.; Veyrières, A. *Tetrahedron Lett.* **1983**, *24*, 2383-2386.
139. Anisuzzaman, A. K. M.; Anderson, L.; Navia, J. L. *Carbohydr. Res.* **1988**, *174*, 265-278.
140. Ekberg, T.; Magnusson, G. *Carbohydr. Res.* **1993**, *246*, 119-136.
141. Qin, H.; Grindley, T. B. *J. Carbohydr. Chem.* **1993**, in press.
142. Ogawa, T.; Kaburagi, T. *Carbohydr. Res.* **1982**, *103*, 53-64.
143. Fernandez-Mayoralas, A.; Marra, A.; Trumtel, M.; Veyrières, A.; Sinaÿ, P. *Carbohydr. Res.* **1989**, *188*, 81-95.
144. Takeo, K.; Nakaji, T.; Shinmitsu, K. *Carbohydr. Res.* **1984**, *133*, 275-287.
145. Pedretti, V.; Veyrières, A.; Sinaÿ, P. *Tetrahedron* **1990**, *46*, 77-88.
146. Zuurmond, H. M.; van der Klein, P. A. M.; van der Marel, G. A.; van Boom, J. H. *Tetrahedron Lett.* **1992**, *33*, 2063-2066.
147. Verduyn, R.; Douwes, M.; van der Klein, P. A. M.; Mösinger, E. M.; van der Marel, G. A.; van Boom, J. H. *Tetrahedron* **1993**, *49*, 7301-7316.
148. Zuurmond, H. M.; van der Klein, P. A. M.; van der Marel, G. A.; van Boom, J. H. *Tetrahedron* **1993**, *49*, 6501-6514.
149. Ichikawa, Y.; Monden, R.; Kuzuhara, H. *Tetrahedron Lett.* **1986**, *27*, 611-614.
150. Ichikawa, Y.; Monden, R.; Kuzuhara, H. *Carbohydr. Res.* **1988**, *172*, 37-64.
151. Takeo, K.; Tei, S. *Carbohydr. Res.* **1985**, *141*, 159-164.
152. Jütten, P.; Dornhagen, J.; Scharf, H.-D. *Tetrahedron* **1987**, *43*, 4133-4140.
153. Coleman, R. S.; Fraser, J. R. *J. Org. Chem.* **1993**, *58*, 385-392.

154. Monneret, C.; Gagnet, R.; Florent, J.-C. *Carbohydr. Res.* **1993**, *240*, 313-322.
155. Chahoua, L.; Baltas, M.; Gorrichon, L.; Tisnès, P.; Zedde, C. *J. Org. Chem.* **1992**, *57*, 5798-5801.
156. Hashimoto, H.; Araki, K.; Saito, Y.; Kawa, M.; Yoshimura, J. *Bull. Chem. Soc. Jpn.* **1986**, *59*, 3131-3136.
157. Yoshikawa, M.; Murakami, N.; Inoue, Y.; Hatakeyama, S.; Kitagawa, I. *Chem. Pharm. Bull.* **1993**, *41*, 636-638.
158. Prandi, J.; Beau, J.-M. *Tetrahedron Lett.* **1989**, *30*, 4517-4520.
159. Chen, S.-H.; Horvath, R. F.; Joglar, J.; Fisher, M.; Danishefsky, S. J. *J. Org. Chem.* **1991**, *56*, 5834-5845.
160. Bredenkamp, M. W.; Holzapfel, C. W.; Toerien, F. *Synth. Comm.* **1992**, *22*, 2459-2477.
161. Valverde, S.; Garcia-Ochoa, S.; Martin-Lomas, M. *J. Chem. Soc., Chem. Commun.* **1987**, 1714-1715.
162. Pedretti, V.; Mallet, J.-M.; Sinaÿ, P. *Carbohydr. Res.* **1993**, *244*, 247-257.
163. Ogawa, T.; Horisaki, T. *Carbohydr. Res.* **1983**, *123*, C1-C4.
164. Marzi, M.; Misiti, D. *Tetrahedron Lett.* **1989**, *30*, 6075-6076.
165. Auzanneau, F.-I.; Charon, D.; Szabó, L. *J. Chem. Soc., Perkin Trans. 1* **1991**, 509-517.
166. Bauer, F.; Ruess, K.-P.; Liefländer, M. *Liebigs Ann. Chem.* **1991**, 765-768.
167. Duréault, A.; Portal, M.; Depezay, J. C. *Synlett* **1991**, 225-226.
168. Ogawa, T.; Beppu, K.; Nakabayashi, W. *Carbohydr. Res.* **1981**, *93*, C6-C9.
169. Ogawa, T.; Nakabayashi, S.; Sasajima, K. *Carbohydr. Res.* **1981**, *95*, 308-312.
170. Ogawa, T.; Matsui, M. *Carbohydr. Res.* **1977**, *86*, C17-C21.
171. Gerz, M.; Matter, H.; Kessler, H. *Angew. Chem. Int. Ed. Engl.* **1993**, *32*, 269-271.
172. Sato, T.; Fujita, Y.; Otera, J.; Nozaki, H. *Tetrahedron Lett.* **1992**, *33*, 239-242.
173. Keinan, E.; Sahai, M.; Roth, Z.; Nudelman, A.; Herzig, J. *J. Org. Chem.* **1985**, *50*, 3558-3566.

RECEIVED April 27, 1994

Chapter 5

Flexibility of Biomolecules

Implications for Oligosaccharides

Laura Lerner

Department of Chemistry, University of Wisconsin,
1101 University Avenue, Madison, WI 53706

Recent examples of the application of spectroscopic methods for the detection and characterization of flexibility in oligosaccharides and other biomolecules are reviewed. Nuclear magnetic resonance spectroscopy is the most versatile method for this purpose, although other methods can provide important information. Possibilities for using these methods to study the role of flexibility in binding interactions are discussed. More accurate descriptions of oligosaccharide flexibility are made possible by the inclusion of coupling constant and relaxation rate data.

The original lock-and-key model has undergone considerable revision since first proposed by Emil Fischer a century ago (*1*). It has long been recognized that flexibility could enhance or reduce specificity in the interactions of receptors and ligands. In 1946, Linus Pauling suggested that an enzyme could catalyze a reaction by causing the substrate to adopt its transition state conformation, thereby lowering the activation energy for the reaction (*2*). Twenty years later, Koshland, Némethy, and Filmer (*3*) proposed that a rigid substrate could induce a favorable fit with a flexible protein. As experimental methods for determining structure improved, numerous examples of flexible keys and/or flexible locks have appeared. In this article, I will survey recent progress in dealing with the determination and significance of flexiblity in recognition. The focus is on the application of nuclear magnetic resonance (NMR) methods to carbohydrate ligands, but other methods and molecules will be mentioned when appropriate.

A central paradigm in modern biochemistry is that there is a direct relationship between structure and function (*2*). If the chemical structure of a compound is known, its three-dimensional structure (i.e., conformation) should be predictable. If its conformation is known, it should provide an explanation for how the molecule functions. Many decades have been spent trying to complete the structure = function equation. Although some general rules and algorithms for predicting protein conformations *a priori* have been developed (*4-6*), we are clearly not at the point of designing proteins at will by specifying amino acid sequence. Knowledge about the effects of base sequence on nucleic acid conformation or sugar sequence on oligosaccharide conformation is even more scarce. At this point, there are also no general rules for predicting flexibility from primary structure.

Research on molecular flexibility must address three major questions: (**1**) Is the molecule rigid or flexible? That is, does it have a single conformation or does it exchange between multiple conformations? (**2**) What kind(s) of motion(s) is the

0097–6156/94/0560–0077$08.00/0

molecule undergoing, and on what timescale(s) ? (3) What effect do these motions have on the binding of the molecule to its target host or ligand? Spectroscopy offers many potentially powerful approaches for answering all these questions. This article reviews some recent efforts to realize this potential.

(1) Is the molecule rigid or flexible?

Any method that is sensitive to molecular conformation can also provide information about the existence of multiple conformations. In X-ray crystallography, temperature factors and uncertainty in the allocation of electron density may be interpreted in terms of local flexibility, provided that artifacts are taken into account. For example, Yamada and coworkers (7) compared the cell adhesion activity of RGD-containing peptides (arginine-glycine-aspartate) of various lengths which correspond to human vitronectin and were inserted into human lysozyme. The three-dimensional structures of two of the inserted regions were determined to 1.8 Å resolution, but both contained undefined regions of electron density, suggesting flexibility. The authors proposed that such flexibility was required for induced fit of RGD regions into the binding pocket of the integrin receptor.

The major method for determination of molecular conformation in solution is NMR spectroscopy. The use of NMR spectroscopy to determine conformations of proteins, nucleic acids, carbohydrates and other biomolecules has been thoroughly reviewed elsewhere (8-16). NMR is ideally suited for studying flexibility because samples can be studied in solution, and because many accurate models exist for extracting motional information from NMR relaxation parameters. NMR also offers more specific information about individual atoms than most other spectroscopic methods. Three-dimensional structures of small proteins can be well-defined by a combination of distance contraints (from ^{1}H-^{1}H nuclear Overhauser enhancements (NOE's)), hydrogen bond contraints (from ^{1}H-exchange rates), and torsion angle contraints (from scalar coupling constants). Regions of a protein that have minimal constraints are often assumed to be flexible (17).

For carbohydrates, the predominant sources of information about probable conformations have been ^{1}H-^{1}H NOE's and spin-coupling constants (^{1}H-^{1}H and/or ^{1}H-^{13}C), usually in conjunction with potential energy minimization (11-12,16). In some cases, a single conformation fits the available data adequately; whereas in others, conformational averaging must be invoked (18,19 and references therein). It is not productive to argue whether or not oligosaccharides as a class are rigid or flexible, when there are relatively few examples of either case, and most importantly, when there is no reason to presume that all oligosaccharides must behave the same.

Oligosaccharide conformation can be described from several points of view: the individual rings, the side groups, and the orientation of one ring relative to another. Spin-coupling constants are the richest source of information on conformation of an individual ring and its sidegroups (20,21). Recent work by Serianni and coworkers (22-24) provides useful relations between ^{1}H-^{13}C and ^{13}C-^{13}C spin-coupling constants and ring conformation, glycosidic bond conformation, and hydroxymethyl orientation for furanose and pyranose systems. If spin-coupling constants are intermediate in value between expected values for particular conformations, or if they exhibit dependences on temperature and solvent, such behavior may be interpreted in terms of flexibility, as has been done for idose (see below). However, to extract relative populations of conformers from spin-coupling constants, the conformations must be known so that a Karplus relation between dihedral angle and spin-coupling constant can be applied.   Usually, such information comes from X-ray crystallographic studies of model compounds. For example, the spin-coupling constants between the proton at position 5 of a hexopyranose ring and methylene protons at position 6 are sensitive to the relative populations of rotamers (*25-27*). Lack of averaging of spin-coupling constants can be used to infer restricted mobility. An example of using lack of averaging to infer rigidity is the relatively small linewidth of the hydroxyl proton at position 3 on D-glucuronate derivatives. Our group (*28*) and Heatley and Scott (*29,30*) have reported values of 3-4 Hz for the coupling between this hydroxyl and the ring proton at position 3 (J_{HOCH}) for sodium D-glucuronate as the monosaccharide and in a hyaluronan disaccharide and tetrasaccharide, in various solvent mixtures. If there were unrestricted rotation about the C-O bond of the hydroxyl group, the value of J_{HOCH} would be closer to 6 Hz, as is seen for other hydroxyl protons in these molecules.

Long-range ^{1}H-^{13}C coupling constants across glycosidic linkages can be measured (*31,32*), but correlation between relatively small changes in these values (e.g., 4.5 Hz vs. 4.8 Hz) and ϕ and ψ values is not well-established. Recently, some progress in exploiting one-bond ^{1}H-^{13}C (*33*) and ^{13}C-^{13}C (*34*) coupling constants for conformational information has been reported. Direct correlations between coupling constants and conformation will require better Karplus relations specific for carbohydrates (*23,35-38*).

There are some nearly insurmountable difficulties in defining oligosaccharide conformations with certainty by experiment. In contrast to proteins, oligosaccharides offer few constraints. The interglycosidic angles can be determined by a combination of long-range ^{1}H-^{13}C coupling constants and/or ^{1}H-^{1}H NOE's across the glycosidic linkage.   In some cases, interresidue NOE's between protons distant from the glycosidic linkage may provide important additional information. For example, Several groups (*39-41*) have reported a rotating frame NOE (ROE) or regular NOE between the methyl protons or H-5 of the fucose and H-2 on the galactose of the sialyl Lewis[x] tetrasaccharide or related compounds, which could arise if the oligosaccharide folded such that the two sugar rings, separated by a 2-acetamido-2-deoxyglucopyranose (GlcNAc) residue, were stacked. Scarsdale and coworkers (*42*) observed NOE's between distant residues in gangliosides in dimethyl sulfoxide which could be interpreted in terms of hydrogen bonding between residues. NOE's between the amide proton on GlcNAc residues to adjacent glucuronate residues have proven useful for determining the solution conformation of hyaluronan oligosaccharides (Holmbeck, Petillo, and Lerner, in preparation). More commonly, angles must be inferred from only one or two constraints which may not be sensitive enough to distinguish among a range of ϕ, ψ values.

The heart of the problem in defining solution conformations by ^{1}H-^{1}H NOE's is this: you need either very well-defined constraints, or a large number of loose constraints, as is typical for globular proteins. Unfortunately, only a small number of poorly-defined constraints can be extracted from most oligosaccharides. This makes it difficult to unequivocably determine whether an oligosaccharide is flexible or rigid based solely on ^{1}H-^{1}H NOE's.

Sometimes inconsistencies in different NMR parameters can be interpreted in terms of flexibility; that is, the presence of local motions superimposed on the overall tumbling in solution. For example, Cumming and Carver (43) demonstrated that ^{1}H-^{1}H NOE and ^{1}H longitudinal relaxation time (T$_1$) data for a series of disaccharides could best be fit by assuming a Boltzmann-weighted ensemble of conformations. Poppe and van Halbeek (44) suggested that inconsistent ^{1}H-^{1}H NOE's and ROE's in sucrose could be explained by flexibility. Hricovini and coworkers (19) calculated order parameters, based on ^{1}H NOE's and ^{13}C T$_1$'s, at two different field strengths for specifically deuterated Manα(1-3)Manβ-OMe and Xylβ(1,4)Xylβ-OMe. Field dependence was interpreted in terms of incomplete radial averaging of dipolar interactions, resulting from internal motions on the timescale of overall molecular tumbling. Similarly, Herve du Penhoat and coworkers (45) calculated correlation times for carbons and protons in a series of oligosaccharides, based on ^{13}C and ^{1}H T$_1$'s, and also concluded that internal and overall motions were on the same timescale.

The increasing availability of powerful computers and sophisticated molecular dynamics programs has encouraged researchers to predict molecular flexibility based on empirical force fields (46,47). Brady has written an excellent review of computational aspects of studying flexibility in mono- and disaccharides (47). The typical publication reporting the solution conformation of an oligosaccharide includes a combination of NMR parameters and a demonstration that the conformation(s) consistent with the NMR parameters is also a low-energy conformer. Meyer and coworkers (48) used the GEGOP program (49) and Metropolis Monte Carlo simulations to predict the conformation of xyloglucan, and to assess sidechain flexibility. As discussed later, the most probable conformations for idose residues in heparin fragments have been predicted by fitting experimental ^{1}H-^{1}H spin-coupling constants and minimization of potential energy, and by molecular dynamics calculations (50-52).

Hydrogen exchange rates, readily measured by ^{1}H-NMR, can sometimes be interpreted in terms of molecular flexibility. For example, Guéron and coworkers (53) have interpreted anomalous hydrogen exchange rates in terms of local perturbation of oligonucleotide structures. Englander and coworkers (54,55) and Woodward and coworkers (56,57) have used this approach to define flexible and rigid regions of globular proteins, which have relatively faster and slower hydrogen exchange rates, respectively. Relatively high exchange rates for hydrogens in the interior of a globular protein may be interpreted in terms of greater accessibility to water, and this can be correlated with the presence of water molecules determined by X-ray crystallography or NMR spectroscopy (58). It will be much more difficult to relate hydrogen exchange rates to flexibility in oligosaccharides. Because of their extended structure in solution, it is much less likely that dramatic differences in hydrogen exchange rates will be observed, especially in aqueous solutions where hydrogen bonds with water will be entropically favored. A relatively slow exchange rate can be interpreted as being caused by an intramolecular hydrogen bond (59), but could also arise from electronic effects.

(2) What is the nature and timescale of the motion(s)?

NMR spectroscopy supplies a wealth of detailed information about molecular dynamics, in addition to primary chemical structure and conformation. Relaxation of nuclear spins depends on the motion of the spins relative to the source of relaxation: other spins, either scalar or dipolar coupled; the electric field gradient (for quadrupolar relaxation); the chemical-magnetic environment (for chemical shift anisotropy). The efficiency of a particular relaxation pathway depends in large part on the spectral density function $J(\omega)$ (the Fourier transform of the autocorrelation function; for detailed explanations of these concepts see *60-62*). Motions occurring close to the resonance (Larmor) frequency, ω_0, or twice this frequency, $2\omega_0$, can contribute to enhanced spin-lattice $(1/T_1)$ and spin-spin $(1/T_2)$ relaxation rates. Very slow motions, on the order of $\omega \sim 0$, can also contribute to $1/T_2$, and hence, linewidth. The spin-lattice relaxation rate in the rotating frame, $1/T_{1\rho}$, is also sensitive to motions on the order of the spin-lock field strength, $\omega_1 = \gamma B_1$. The heteronuclear NOE factor for ^{1}H-^{13}C pairs depends on motions of the ^{1}H-^{13}C bond vector at frequencies close to ω_C, $(\omega_C + \omega_H)$, and $(\omega_H - \omega_C)$. Motions at other frequencies could be monitored by direct spectral density mapping of other heteronuclear pairs, such as ^{1}H and ^{15}N (*63*). The motional dependence of various NMR parameters is shown schematically in Figure 1 (*64*).

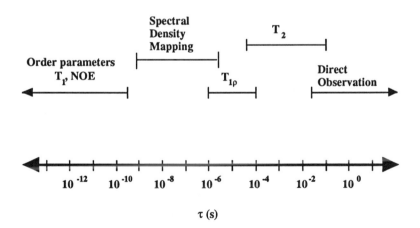

Figure 1. Different NMR parameters are sensitive to different timescales of motion (τ, in seconds).

We are using a battery of NMR parameters to characterize flexibility in idose derivatives. It has been observed that epimerization of the hydroxymethyl group at the C-5 position changes the populations of conformers of idose relative to glucose, which is found overwhelmingly as the $^4C_1$ conformer in solution (*35,65*). The only direct experimental evidence which indicates there is interconversion among chair and boat conformers in idose derivatives is the unusual ^{1}H-^{1}H coupling constants observed for idose and its derivatives in heparin-like oligosaccharides (51,64). In sharp contrast to glucose, these coupling constants in idose are intermediate in value between those expected for ^{1}C$_4$ and ^{4}C$_1$ conformations, and they are temperature- and solvent-sensitive. It has been suggested (66) that this flexibility of idose is somehow related to the ability of 2-O-sulfated idose residues to bind to proteins such as antithrombin III (67) and fibroblast growth factor (68). In our laboratory, we have sought direct quantification of molecular motions in idose by a combination of NMR methods and molecular dynamics (65). Thus far, our results indicate that ring interconversion of idose is faster than milliseconds and slower than picoseconds. For α-D-idose pentaacetate, we estimate a lifetime of roughly 10^{-7} to 10^{-9} seconds over the temperature range 263 - 293 K for interconversion between the chair and skew-boat conformers. This estimate is based on $T_{1\rho}$ (69,70) and line-broadening measurements over a wide temperature range for an equimolar mixture of α-D-idose pentaacetate and α-D-glucose pentaacetate in acetone.

Berglund and coworkers (71) measured a combination of ^{15}N T_1's, $T_{1\rho}$'s and steady-state {^{1}H}-^{15}N NOE's for most of the backbone amide nitrogens in a region of the glucocorticoid receptor DNA-binding domain (GR BDB). These data were used to calculate order parameters and effective correlation times, and it was concluded that the backbone was relatively rigid and ordered. This approach has been used by several groups since it was introduced by Torchia and coworkers (72).

Gorenstein and coworkers (73) used ^{31}P chemical shifts to assess the role of phosphate ester backbone flexibility in protein-DNA recognition. They claim there is a correlation between ^{31}P chemical shifts and $J_{H3'-P}$ coupling constants, which in turn are sensitive to the C4'C3'-O-P and C3'-O3'-P-O5 torsion angles, so that conformer populations can be inferred from small changes in chemical shifts. The use of NMR methods to characterize dynamics in nucleic acids was recently reviewed by Lane (74).

(3)   What role does flexibility play in binding?

A complete NMR analysis of an oligosaccharide, including ^{1}H-^{1}H NOE's, homo- and heteronuclear coupling constants, and relaxation measurements, can provide a detailed picture of the amplitude of the molecular motions, the timescale over which they occur, and the effects of various sidegroups. But such an analysis by itself can say nothing about how this flexibility affects the specific interactions of flexible substrates with their biological receptors. This is by far the hardest question to answer about flexibility, although many exciting results are emerging from recent studies reviewed below. As usual, more studies on proteins and oligonucleotides are published than on oligosaccharides, but obviously methods that apply to one class of biomolecules can usually be adapted to others. The studies generally focus on flexibility in either the host molecule or the ligand.

Most published studies compare binding affinities or other functional properties for rigid vs. flexible ligands. The flexibility of a ligand is usually evaluated from ^{1}H-^{1}H NOE data. For example, Tsikaris and coworkers (75) compared Torpedo acetylcholine receptor fragments from the main immunogenic region (MIR) with one or two specific amino acid substitutions. Relative affinities for an anti-MIR antibody were interpreted in terms of the need to maintain a rigid structure of the fragments for binding.

Gierasch and coworkers (*76*) described a mobile loop region of GroES that loses its mobility and accessibility upon binding to GroEL, a molecular chaperone protein from *E. coli*. They suggested that mobility allowed the GroES seven-mer to bind simultaneously to seven GroEL subunits. The one-dimensional ^{1}H NMR spectrum of GroES is unusual for a such a large protein (molecular weight 70,000) because it has sharp peaks superimposed on an envelope of broader resonances, indicative of some regions of high mobility.

Miller and coworkers (*77*) used isotope-edited NMR spectroscopy to observe the amide resonances of selectively ^{15}N-labelled amino acid residues in wild-type p21ras and an oncogenic mutant, (G12D)p21ras. They proposed that guanosine triphosphate analogs could couple flexible loops of the protein to a rigid loop in the active site. Lim and coworkers (*78*) used a combination of NOE and ROE data to extract interproton distances and correlation times for Ac-(D)Phe-Pro-boroArg-OH, a potent inhibitor of thrombin. They concluded that this peptide analog was rigid in solution, in a conformation very similar to the conformation of H-(D)Phe-Pro-ArgCH$_2$Cl bound to the active site of thrombin (as determined by X-ray crystallography). They proposed that this rigidity contributed significantly to the ten-fold higher affinity observed for the boro-analog, by reducing the entropic cost of binding.

In an earlier study, Roberts (*79*) used the pH-dependence of ^{1}H chemical shifts in a series of complexes of dihydrofolate reductase (native and mutant) with inhibitors and cofactors. In general, inhibitors bound only in non-productive conformations whereas substrates bound in two conformations, one of which could lead to catalysis.

Such conformational selectivity leads to the question, does flexibility enhance binding by facilitating a change of conformation necessary for binding and/or activity ? Glaudemans and coworkers (*80*) used transferred NOE spectroscopy to show that the conformation of a disaccharide ligand bound to a monoclonal immunoglobulin was not the predominant conformation of the ligand when free in solution. Either the antibody altered the conformation upon binding, or it selected a rare conformer out of the solution.

A large number of structure-activity studies address the role of flexibility indirectly, by probing which functional groups at what particular orientations are required for binding. For example, Lemieux and co-workers have mapped the required number and orientation of hydroxyl groups on the tetrasaccharide known as Le^b-OMe (the Lewis b human blood group determinant) for recognition by the plant lectin IV of *Griffonia simplicifolia* (*81,82*). More recently, Hays and coworkers (*83*) used a series of rigid hexahydrofluorenamine analogs of PCP to determine which structural features enhanced binding to the noncompetitive site of the N-methyl-D-aspartate (NMDA) receptor.

Hydrogen exchange rates have been used to examine which regions are perturbed upon binding. Englander and coworkers (*55*) found that the binding of the E3 and C3 antibodies to horse cytochrome c affected the hydrogen-exchange behavior of regions within and beyond the immediate binding site, perhaps by restricting local conformational flexibility. The binding sites of both antibodies were mapped by perturbations in hydrogen exchange rates, on the assumption that binding would slow exchange. Patel and coworkers (84,85) used hydrogen exchange rates to characterize binding in several DNA-drug complexes.

Flexibility can enhance binding if the sites of interaction must be distorted, but it can add to the free energy cost of binding by contributing a negative change in entropy upon binding (86). The relative contributions of enthalpy and entropy to the free energy of binding can be determined from the temperature dependence of the binding constant. The phenomenon of enthalpy-entropy compensation is widespread in biology (87), and Carver (86) and Lemieux (82) have suggested that it plays an important role in carbohydrate-protein interactions.

Other Methods

Spectroscopic methods that rely on the transfer of quantized energy between a sample and the environment are potentially sensitive to the motion of the sample. In NMR, molecular motions which are close to multiples of the Larmor frequency are efficient contributors to spin-lattice and spin-spin relaxation of nuclear spins. Similar mechanisms prevail in electron spin resonance (ESR) and fluorescence spectroscopy, and thus yield information about motions on different timescales, because of the different transitions involved. Furthermore, different spectroscopic methods can offer insights into different modes of motion; i.e., rotation vs. vibrations. For example, time-resolved Fourier transform infrared spectroscopy (FTIR) can be used to detect changes in vibrational modes down to a resolution of a few nanoseconds. Hochstrasser and coworkers (88) used this approach to detect slow conformational changes in bacteriorhodopsin as it undergoes its photoinduced reaction cycle. Haris and coworkers (89) recently reported the utility of selective ^{13}C-labelling of proteins to improve the resolution and specificity of FTIR. Such labelling might be extremely useful for identifying motions in specific regions of oligosaccharides.

Hochstrasser and coworkers (90) have also used time-resolved fluorescence energy transfer between a donor and acceptor dye attached to opposite 5′ termini to show that there was a broad distribution of end-to-end distances in a nine base-pair duplex DNA oligonucleotide. Dorovska-Taran and coworkers (91) attached an anthraniloyl group to the active site of α-chymotrypsin so that the overall rotational-correlation time in aqueous solution and in reversed micelles could be compared. Gallay and coworkers (92) used time-resolved fluorescence to distinguish two separate environments for two tryptophan residues in recombinant human epidermal growth factor, and to demonstrate that the C-terminal region of the protein is rather flexible. Lee and coworkers (93,94) demonstrated the applicability of time-resolved energy transfer to the study of flexibility in glycopeptides.

Morrow and coworkers (95) reported a combined ^{2}H-NMR and nitroxide ESR study of order parameters of glycosphingolipid chains in unsaturated phosphatidylcholine bilayers.

Summary

Undaunted by the emerging complexity of structure-function relationships, researchers continue to add to the data-base of molecular conformations. For many biomolecules, single, rigid conformations cannot be well-defined by current spectroscopic methods, which may reflect their flexibility rather than artifacts of the methods. Partly because of the small number of constraints, it is difficult to prove that oligosaccharides exist in rigid conformations. Depending on the particular molecule and force field used, molecular mechanics and dynamics calculations may yield deep, narrow potential wells and constant trajectories, suggesting a single, rigid conformation, or broad, shallow potential wells and varying trajectories, suggesting many low-energy conformers are possible.

For characterizing the flexibility of oligosaccharides, two lines of evidence should be pursued: inference of multiple conformations from ^{1}H-^{1}H NOE's and coupling constants, and measurement of the timescales and amplitudes of the molecular motions by relaxation measurements. Every bit of conformational information must be extracted to increase the number of constraints defining oligosaccharide conformations. The progress which is being made in measuring and relating coupling constants to conformation will be very valuable. Molecular modelling can provide useful models for motion, but is too dependent on force field parametrization to provide truly independent evidence for flexibility.

In the centennial year of Fischer's lock-and-key model, we have the tools in hand for quantifying flexibility and its role in specificity.

Acknowledgments

The author wishes to thank Phil Hajduk for his very helpful advice during preparation of this manuscript, and for preparing Figure 1.

Literature Cited

1. Fisher, E. *Ber. Dt. Chem. Ges.* **1894**, *27*, 2985-2993.
2. Pauling, L. *Chem. Eng. News* **1946**, *24*, 13 75-1377.
3. Koshland, D. E.; Némethy, G.; Filmer, D. *Biochemistry* **1966**, *5*, 365-
4. Benner, S. A.; Gerloff, D. L. *FEBS-Lett.* **1993**, *325*, 29-33.
5. Rost, B.; Schneider, R.; Sander, C. *Trends Biochem. Sci.* **1993**, *18*, 120-3.
6. Godzik, A.; Skolnick, J. *Proc. Natl. Acad. Sci. U. S. A.* **1992**, *89*, 12098-102.
7. Yamada, R.; Matsushima, M.; Inaka, K.; Ohkubo, T.; Uyeda, A.; Maeda, T.; Titani, K.; Sekigushi, K.; Kikuchi, M. *J. Biol. Chem.* **1993**, *268*, 10588-10592.
8. Gronenborn, A. M.; Clore, G. M. *Prog. NMR Spectrosc.* **1985**, *17*, 1-32.
9. Wemmer, D. E.; Reid, B. R. *Annu. Rev. Phys. Chem.* **1985**, *36*, 105-137.
10. Wüthrich, K. *NMR of Proteins and Nucleic Acids*; John Wiley and Sons: New York, NY, 1986.
11. Serianni, A. S. In *Glycoconjugates: Composition, Structure and Function*; Allen, H. J.; Kisailus, E. C., Eds.; Marcel-Dekker, New York, NY, 1992..
12. van Halbeek, H.; Poppe, L. *Magn. Reson. Chem.* **1992**, *30*, S74-S86.
13. Clore, G. M.; Robien, M. A.; Gronenborn, A. M. *J. Mol. Biol.* **1993**, *231*, 82-102.
14. Bax, A.; Grzesiek, S. *Acc. Chem. Res.* **1993**, *26*, 131-138.

15. Emsley, L.; Dwyer, T. J.; Spielmann, H. P.; Wemmer, D. E. *J. Am. Chem. Soc.* **1993**, *115*, 7765-7771.
16. Homans, S. W. *Biochem. Soc. Trans.* **1993**, *21*, 449-52.
17. Powers, R.; Garrett, D. S.; March, C. J.; Frieden, E. A.; Gronenborn, A. M.; Clore, G. M. *Biochemistry* **1993**, *32*, 6744-6762.
18. Rutherford, T. J.; Partridge, J.; Weller, C. T.; Homans, S. W. *Biochemistry* **1993**, *32*, 12715-12724.
19. Hricovini, M.; Shah, R. N.; Carver, J. P. *Biochemistry* **1992**, *31*, 10018-10023.
20. Altona, C; Haasnoot, C. A. G. *Org. Magn. Reson.* **1980**, *13*, 417-429.
21. Barker, R.; Nunez, H. A.; Rosevear, P.; Serianni, A. S. *Meth. Enzym.* **1982**, *83, part D*, 58-69.
22. Duker, J. M.; Serianni, A. S. *Carbohydr. Res.* **1993**, *249*, 281-303.
23. Bandyopadhyay, T.; Wu, J.; Serianni, A. S. *J. Org. Chem.* **1993**, *58*, 5513-5517.
24. Wu, J.; Bondo, P. B.; Vuorinen, T.; Serianni, A. S. *J. Am. Chem. Soc.* **1992**, *114*, 3499-3505.
25. Brisson, J. R.; Carver, J. P. *Biochemistry* **1983**, *22*, 3680-3686.
26. Poppe, L. *J. Am. Chem. Soc.* **1993**, *115*, 8421-8426.
27. Marchessault, R. H.; Perez, S. *Biopolymers* **1979**, *18*, 2369-2374.
28. Sicinska, W.; Adams, B.; Lerner, L. *Carbohydr. Res.* **1993**, *242*, 29-51.
29. Scott, J. E.; Heatley, F.; Hull, W. E. *Biochem. J.* **1984**, *220*, 197-205.
30. Heatley, F.; Scott, J. E.; Casu, B. Carbohydr. Res. **1979**, *72*, 13-23.
31. Poppe, L.; York, W. S.; van Halbeek, H. *J. Biomol. NMR* **1993**, *2*, 81-9.
32. Adams, B.; Lerner, L. *J. Magn. Reson.* **1993**, *Series A 103*, 97-102.
33. Tvaroska, I.; Taravel, F. R. *Carbohydr. Res.* **1991**, *221*, 83-94.
34. Carmichael, I.; Chipman, D. M.; Podlasek, C. A.; Serianni, A. S. *J. Am. Chem. Soc.* **1993**, *115*, 10863-10870.
35. Cano, F. H.; Foces-Foces, C. *J. Org. Chem.* **1987**, *52*, 3367-3372.
36. Haasnoot, C. A. G. *J. Am. Chem. Soc.* **1993**, *115*, 1460-68.
37. Plavec, J.; Tong, W. M.; Chattopadhyaya, J. *J. Am. Chem. Soc.* **1993**, *115*, 9734-46.
38. Tvaroska, I.; Hricovini, M.; Petrakova, E. *Carbohydr. Res.* **1989**, *189*, 359-362.
39. Miller, K. E.; Mukhopadhyay, C.; Cagas, P.; Bush, C. A. *Biochemistry* **1992**, *31*, 6703-6709.
40. Ball, G.; O'Neill, R. A.; Schultz, J. E.; Lowe, J. B.; Weston, B. W.; Nagy, J. O.; Brown, E. G.; Hobbs, C. J.; Bednarski, M. D. *J. Am. Chem. Soc.* **1992**, *114*, 5449-5451.
41. Lin, Y.-C.; Hummerl, C. W.; Huang, D.-H.; Ichikawa, Y.; Nicolaou, K. C.; Wong, C.-H. *J. Am. Chem. Soc.* **1992**, *114*, 5452-5454.
42. Scarsdale, J. N.; Prestegard, J. H.; Yu, R. K. *Biochemistry* **1990**, *29*, 9843-9855.
43. Cumming, D. A.; Carver, J. P. *Biochemistry* **1987**, *26*, 2664-6676.
44. Poppe, L.; van Halbeek, H. *J. Am. Chem. Soc.* **1992**, *114*, 1092-1094.
45. Braccini, I.; Michon, V.; Herve du Penhoat C.; Imberty, A.; Perez, S. *Int. J. Biol. Macromol.* **1993**, *15*, 52-55.
46. Homans, S. W.; Forster, M. *Glycobiology* **1992**, *2*, 143-51.
47. Brady, J. W. *Adv. Biophys. Chem.* **1990**, *1*, 155-202.
48. Levy, S.; York, W. S.; Stuike-Prill, R.; Meyer, B.; Staehelin, L. A. *Plant J.* **1991**, *1*, 195-215.
49. Stuike-Prill, R.; Meyer, B. *Eur. J. Biochem.* **1990**, *194*, 903-19.
50. Ragazzi, M.; Ferro, D. R.; Provasoli, A. *J. Comp. Chem.* **1986**, *7*, 105-112.

51. Ferro, D. R.; Provasoli, A.; Ragazzi, M.; Casu, B.; Torri, G.; Bossennec, V.; Perly, B.; Sinay, P. *Carbohydr. Res.* **1990**, *195*, 157-167.
52. Forster, M. J.; Mulloy, B. *Biopolymers* **1993**, *33*, 575-588.
53. Leroy, J.-L.; Gehring, K.; Kettani, A.; Guéron, M. *Biochemistry* **1993**, *32*, 6019-6031.
54. Englander, S. W.; Mayne, L. *Annu. Rev. Biophys. Biomol. Struct.* **1992**, *21*, 243-265.
55. Mayne, L.; Paterson, Y.; Cerasoli, D., Englander, S. W. *Biochemistry* **1992**, *31*, 10678-10685.
56. Kim, K. S.; Woodward, C.K. *Biochemistry* **1993**, *32*, 9609-13.
57. Kim, K. S.; Fuchs, J. A.; Woodward, C. K. *Biochemistry* **1993**, *32*, 9600-08.
58. Qian, Y. Q.; Otting, G.; Wüthrich, K. *J. Am. Chem. Soc.* **1993**, *115*, 1189-1190.
59. Poppe, L.; van Halbeek, H. *J. Am. Chem. Soc.* **1991**, *113*, 363-365.
60. Ernst, R. R. *Angew. Chemie* **1992**, *31*, 805-930.
61. Abragam, A. *The Principles of Nuclear Magnetism*; Clarendon Press, Oxford, England, 1961.
62. Becker, E. D. *High Resolution NMR*; Academic Press, New York, NY, 1980, 2nd ed.; pp. 184-198.
63. Peng, J. W.; Wagner, G. *J. Magn. Reson.* **1992**, *98*, 308-332.
64. Hajduk, P. Ph.D. thesis, University of Wisconsin, November 1993.
65. Hajduk, P.; Horita, D. A.; Lerner, L. *J. Am. Chem. Soc.* **1993**, *115*, 9196-9201.
66. Casu, B.; Petitou, M.; Provasoli, M.; Sinay, P. *Trends in Biol. Sci.* **1988**, *13*, 221-225.
67. Lindahl, U.; Thunberg, L.; Backstrom, G.; Riesenfeld, J.; Nordling, K.; Bjork, I. *J. Biol. Chem.* **1984**, *259*, 12368-12376.
68. Maccarana, M.; Casu, B.; Lindahl, U. *J. Biol. Chem.* **1993**, *268*, 23898-23905.
69. Wang, Y.-S. *Concepts Magn. Reson.* **1992**, *4*, 327-337.
70. Wang, Y.-S. *Concepts Magn. Reson.* **1993**, *5*, 1-18.
71. Berglund, H.; Kovacs, H.; Dahlman-Wright, K.; Gustafsson, J.-A.; Hard, T. *Biochemistry* **1992**, *31*, 12001-12011.
72. Kay, L. E.; Bax, A.; Torchia, D. A. *Biochemistry* **1989**, *28*, 8972-8979.
73. Botuyan, M. V.; Keire, D. A.; Kroen, C.; Gorenstein, D. G. *Biochemistry* **1993**, *32*, 6863-6874.
74. Lane, A. N. *Progr. NMR Spectroscopy* **1993**, *25*, 481-505.
75. Tsikaris, V.; Detsikas, E.; Sakarellos-Daitsiotis, M.; Sakarellos, C.; Vatzaki, E.; Tzartos, S. J.; Marruad, M.; Cung, M. T. *Biopolymers* **1993**, *33*, 1123-34.
76. Landry, S. J.; Zellstra-Ryalls, J.; Fayet, O., Georgopoulos, C.; Gierasch, L. M. *Nature* **1993**, *364*, 255-258.
77. Miller, A. F.; Halkides, C. J.; Redfield, A. G. *Biochemistry* **1993**, *32*, 7367-76.
78. Lim, M. S.; Johnston, E. R.; Kettner, C. A. *J. Med. Chem.* **1993**, *36*, 1831-8.
79. Roberts, G. C. K. In *Host-Guest Molecular Interactions: from Chemistry to Biology*, Chadwick, D. J.; Widdows, K. Eds.; Ciba Foundation Symposium 158 John Wiley and Sons Ltd., West Sussex, UK, 1991; pp. 169-181.
80. Glaudemans, C. P. J.; Lerner, L.; Daves, Jr., G. D.; Kovác, P.; Venable, R.; Bax, A. *Biochemistry* **1990**, 29, 10906-10911
81. Lemieux, R. U.; Delbaere, L. T. J.; Beierbeck, H.; Spohr, U. In *Host-Guest Molecular Interactions: from Chemistry to Biology*, Chadwick, D. J.; Widdows, K. Eds.; Ciba Foundation Symposium 158, John Wiley and Sons Ltd., West Sussex, UK, 1991; pp. 231-244
82. Lemieux, R. U. *ACS Symposium Series* 519; American Chemical Society, Washington, D. C., 1993; pp. 5-18.

83. Hays, S. J.; Novak, P. M.; Ortwine, D. F.; Bigge, C. F.; Colbry, N. L.; Johnson, G.; Lescosky, L. J.. Malone, T. C.; Michael, A.; Reily, M. D.; Coughenour, L. L.; Brahce, L. J.; Shilis, J. L.; Probert, Jr., A.*J. Med. Chem.* **1993**, *36*, 654-670.
84. Leroy, J. L.; Gao, X. L.; Misra, V.; Guéron, M.; Patel, D. J. *Biochemistry* **1992**, *31*, 1407-15.
85. Leroy, J. L.; Gao, X. L.; Guéron, M.; Patel, D. J. *Biochemistry* **1991**, *30*, 5653-61.
86. Carver, J. P.; Michnick, S. W.; Imberty, A.; Cumming, D. A. In *Symposium on Carbohydrate Recognition in Cellular Function*, Bock, G.; Harnett, S. Eds. Ciba Foundation Symposium 145, John Wiley & Sons, Chichester, UK, 1989; pp. 6-17.
87. Lumry, R.; Rajender, S. *Biopolymers* **1970**, *9*, 1125-1227.
88. Diller, R.; Iannone, M.; Cowen, B. R.; Maiti, S.; Bogomolni, R. A.; Hochstrasser, R. M. *Biochemistry* **1992**, *31*, 5567-5572.
89. Haris, P. I.; Robillard, G. T.; van Dijk, A. A.; Chapman, D. *Biochemistry* **1992**, *31*, 6279-6284.
90. Hochstrasser, R. A.; Chen, S. M.; Millar, D. P. *Biophys. Chem.* **1992**, *45*, 133-41.
91. Dorovska-Taran, V. N.; Veeger, C.; Visser, A. J. *Eur. J. Biochem.* **1993**, *211*, 47-55.
92. Gallay, J.; Vincent, M.; de la Sierra, I. M.; Alvarez, J.; Ubieta, R.; Madrazo, J.; Padron, G. *Eur. J. Biochem.* **1993**, *211*, 213-219.
93. Rice, K. G.; Wu, P.; Brand, L.; Lee, Y. C. *Biochemistry* **1993**, *32*, 7264-7270.
94. Wu, P. G.; Rice, K. G.; Brand, L.; Lee, Y. C. *Proc. Natl. Acad. Sci U. S. A.* **1991**, *88*, 9355-9359.
95. Morrow, M. R.; Singh, D.; Lu, D.; Grant, C. W. *Biophys. J.* **1993**, *64*, 654-64.

RECEIVED March 23, 1994

Synthetic Oligosaccharides
as Tools in the Life Sciences

Chapter 6

Synthetic Oligosaccharides as Probes for Investigating the Binding of Heparin to Antithrombin III

M. Petitou

**Department of Carbohydrate Chemistry, Sanofi Recherche,
9, rue du Président Salvador Allende, F–94256 Gentilly, France**

Antithrombin III (AT III) is a member of the serine protease inhibitors superfamily (serpins) which is present in plasma in an almost inactive form. AT III can be activated by the glycosaminoglycan heparin. A unique pentasaccharide sequence in this polysaccharide is responsible for the binding and activation of AT III. We have identified this sequence and synthesized the corresponding pentasaccharide and several analogs, in order to precisely assess the role of the different structural features of this molecule in the interaction. The results indicate a highly specific interaction. This demonstrates for the first time that the interaction of a complex glycosaminoglycan and a protein may be mediated, at the polysaccharide side, by a short and unique oligosaccharide sequence.

Heparin belongs to the family of glycosaminoglycans. These polyanionic polysaccharides have a highly complex chemical structure which is not yet fully elucidated, and their biological functions are almost unexplored. The poor knowledge that we have about these compounds may be explained by the lack of the sophisticated biophysical tools which are required to explore such complex molecules and also by the lack of interest, from biologists, for compounds that were once believed to only play a structural function in living organisms. Owing to its use as an anticoagulant and antithrombotic drug for more than fifty years, heparin has been the most studied of these glycosaminoglycans.

Heparin: Structure, Biological Activity.

Heparin is biosynthesised as a proteoglycan of very high molecular weight (1,000,000 daltons) which is then enzymically processed to yield polysaccharide chains in the range 3,000 - 30,000 daltons (1). Its true biological function, so far unknown, is probably linked to the anticoagulant activity which was "responsible" for its discovery, in 1916, by McLean, who was actually seeking after pro-coagulant substances (2)!

0097–6156/94/0560–0090$08.00/0

The chemical structure of heparin (Figure 1) has progressively emerged through the work of several groups of investigators, but major structural features have only recently been revealed and others may, thus far, have been overlooked. Standard heparin is a mixture of polysaccharide chains consisting in the repetition of a basic disaccharide sequence made up of a uronic acid (UA) and a glucosamine (GN) which are 1-4 linked. Depending on the length of the chain, between 10 and 30 disaccharides are found. 2-*O*-Sulfate-α-L-iduronic acid (G,I) and 6-*O*-sulfate-*N*-sulfate-α-D-glucosamine (B,H,J) are the predominant monosaccharides. The repetition of trisulfated disaccharides like GH and IJ constitute the so-called "regular regions", which account for a major part of the structure. Other monosaccharide residues like α-L-iduronic acid (C) 6-*O*-sulfate-*N*-acetyl-α-D-glucosamine (D) β-D-glucuronic acid (E) and 3,6-di-*O*-sulfate-*N*-sulfate-α-D-glucosamine (F) occur less often in the "irregular regions". The latter are apparently related to the biological activity as indicated by the critical role played by units E and F in the binding of heparin to its "receptor" antithrombin III (see hereunder). 2-*O*-Sulfated-β-D-glucuronic acid is present in some heparin preparations and a certain number of glucosamine units (less than 20% in general) are not sulfated at position 6. A total of 10 different monosaccharides (4 uronic acids and 6 glucosamines) appear in heparin, making the overall structure a very complex one (*3*).

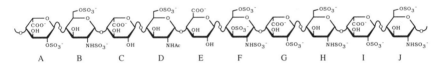

Figure 1. Structure of heparin.

A Plasma Heparin Cofactor: Antithrombin III.   Soon after the discovery of heparin's anticoagulant effect, it was proposed that some plasma cofactor was required for this activity. However, the precise nature of this cofactor, antithrombin III (AT III) was only revealed later (*4, 5*). AT III, a member of the serine proteinase inhibitor superfamily (*6*) (serpins), is a 432 amino acid glycoprotein present in plasma at 2-3 μM concentration. AT III, which by itself poorly inhibits some of the procoagulant enzymes of the blood coagulation system, becomes a very potent inhibitor in the presence of heparin. The complex between the polysaccharide and the protein is the true inhibitor. When the protein is cleaved by thrombin, heparin is released and can enter another catalytic cycle (*5*).

A Specific Sequence in Heparin Binds to Antithrombin III.   Following the identification of AT III, different groups of biochemists (*7-9*) carried out affinity chromatography experiments which allowed standard heparin to be separated into one fraction with affinity for AT III and a second one devoid of affinity. The former had kept the anticoagulant properties while the latter was practically inactive on blood clotting. Moreover, the two fractions had very similar overall chemical compositions, suggesting that a precise arrangement, possibly an AT III binding site, could be present on some heparin molecules only (*10*).

Experiments to identify this site were immediately undertaken by different groups, adopting similar approaches consisting of partial depolymerisation of the polysaccharide, followed by affinity chromatography and gel filtration.

Heparin chains were split either chemically or enzymically into fragments of various sizes (Figure 2). Nitrous acid cleavage converts *N*-sulfate-α-D-glucosamine residues into 2,5-anhydro-D-mannose that constitute the reducing end of the newly formed oligosaccharides (compound **3**, Figure 2). In the case of heparinase cleavage, α,β-unsaturated-uronic acid residues constitute the non reducing end of the new oligosaccharides (compounds **2** and **4**, Figure 2). Partial degradation with nitrous acid of a heparin chain containing the AT III binding sequence gives fragments that contain this sequence, as well as fragments that do not. The former were selected by affinity chromatography using agarose bound AT III. Further fractionation by gel filtration yielded octa-, deca-, dodecasaccharides etc...

The structure of the shortest fragments (octasaccharides, *e.g.* compound **3**, Figure 2) were determined by chemical methods (*11-13*). In the same way, heparinase-derived oligosaccharides were isolated and characterized. When submitted to a "high dose" of enzyme, a hexasaccharide (compound **4**, Figure 2) was obtained. Considering that the conformation of the reducing end unit in this hexasaccharide was altered by loss of the asymmetry at C-5, it was hypothesized that the pentasaccharide sequence D-H could represent the true binding site of heparin to antithrombin III (*14*). Similar conclusions were reached independently by another group of investigators (*13*).

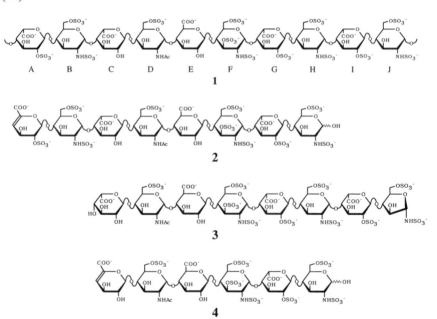

Figure 2. Heparin fragments obtained after chemical (**3**)
or enzymatic (**2** and **4**) cleavage of a heparin chain (**1**)

Probing Heparin-Antithrombin III Interaction with Synthetic Oligosaccharides.

However these conclusions relied on the analysis of the binding properties of oligosaccharides larger than or shorter than the pentasaccharide sequence D-H itself. An unambiguous demonstration necessitated the preparation and analysis of this pentasaccharide *per se* or its *N*-sulfated (at unit D) equivalent, initially detected in beef lung heparin but also present in porcine heparin (*15*). We therefore undertook the synthesis of several oligosaccharides which we analysed for their ability to interact with AT III.

Strategy for the Synthesis of Heparin Fragments. The initial synthesis of the biologically active pentasaccharide **5** was carried out according to the strategy depicted in Figure 3 in which a fully protected pentasaccharide **6** is first prepared and then converted into **5**. The selection of the protective groups of **6** was dictated by: i) the need to introduce sulfate (*O*- as well as *N*- linked), the presence of carboxylate groups and hydroxyl groups in the proper positions on the target molecule, and ii) by the state of the art in oligosaccharide synthesis, particularly the use of 2-azido glucose derivatives for introducing α-linked glucosamine units (this sort of glycosylation requires a non participant group at position 2 of the glycosyl donor), and the use of glycosyl donors with a participating groups at position 2 (whenever it is possible) to control the introduction of 1,2-trans interglycosidic bonds (as requested for uronic acids).

Thus the disaccharide building blocks **9** and **10**, the respective precursors of the EF and GH parts of the molecule, were obtained by condensing properly protected glucuronic acid and iduronic acid monosaccharide derivatives (**11** and **13**) with glucosamine precursors (**12** and **14**). All monosaccharide synthons were obtained from glucose or from glucosamine (*16, 17*).

The protected pentasaccharide **6** was converted into **5** using the following sequence of reactions: i) saponification of ester functions of the molecule, ii) sulphation of the free hydroxyl groups thus obtained, iii) hydrogenolysis, to generate amino groups of glucosamine units and to remove the benzyl groups, and iv) selective *N*-sulphation of amino groups.

Other approaches towards **1** or similar compounds have been described (*18*), some using cellobiose or maltose as a starting material, but with the same protective group strategy.

As previously mentioned, conversion of the pentasaccharide **6** into **5**, requires a hydrogenation step (step iii). However under these conditions amino groups are generated very rapidly and may react with the reducing end (aldehydo function), obtained after cleavage of the anomeric benzyl group. This side-reaction yields stable products (particularly after reduction of the Schiff base) which correspond to a mixture of various dimers and trimers of the initial pentasaccharide (*19*). In order to circumvent this nasty side-reaction, a methyl glycoside was introduced to protect the reducing end, mimicking the α- bond present at this position in heparin (*19*). The resulting pentasaccharide (**15**) displays the same biological properties as **5** (Table I). Recently, a new synthesis of this product was developed using the reaction between a trisaccharide imidate and an acceptor disaccharide (see synthesis of **24**, Figure 7).

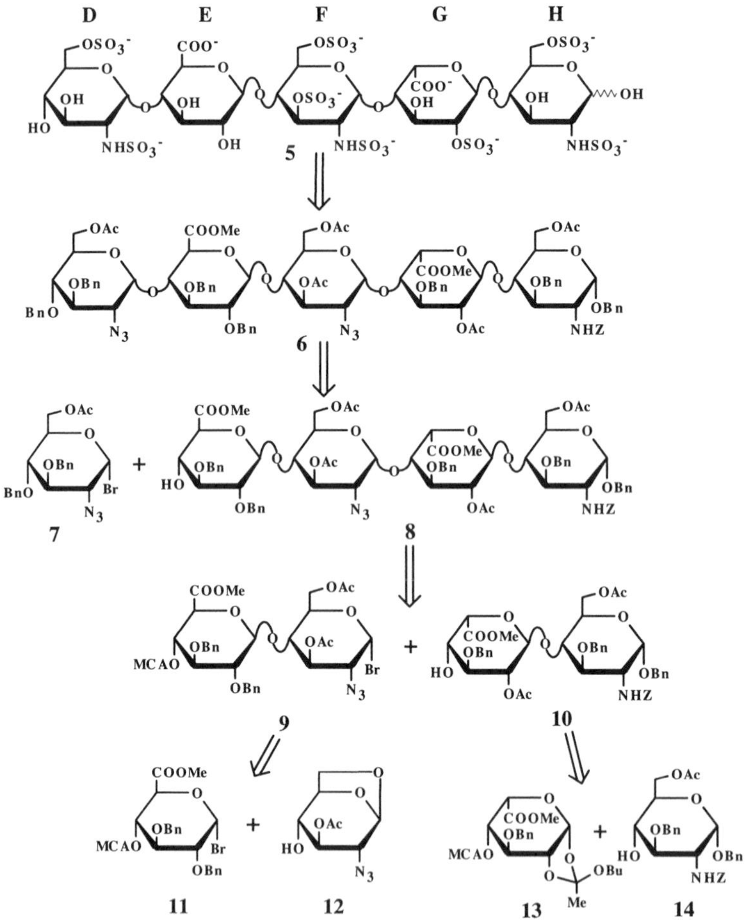

Figure 3. Strategy for the synthesis of heparin fragments.

Synthetic Oligosaccharides to Probe Heparin Interaction with Antithrombin.
The affinity of heparin for AT III and various synthetic heparin fragments was
assessed by determination of the binding constant by spectrofluorimetry. The method
is based on the change in fluorescence emission of tryptophan residues in AT III
when the protein undergoes a conformational change.

The first objective was to prove that a pentasaccharide sequence is the minimum
structure required in heparin for binding to AT III, the pentasaccharide **5** was
prepared first (20). This product binds to AT III with an affinity equal to heparin
(Table I). It also selectively potentiates AT III inhibitory activity against coagulation
factor Xa while no activity can be detected against factor IIa (thrombin). As
mentioned above, the α-methylated pentasaccharide **15** displays the same biological
properties as **5**.

We then synthesized the two tetrasaccharides **16** and **17** (Figure 4) lacking either

Table I. Affinity for AT III, and anti-factor Xa activity of different oligosaccharides. (nd: not determined).

	Dissociation Constant (K_D; µM)	Anti-Xa Activity (unit/mg)
High affinity heparin	0.010	170
Low affinity heparin	>2 000	# 0
5	0.050	700
15	0.058	700
16	24	4
17	8	6
18 (3'' OH)	>2 000	# 0
19 (2', 6 OH)	0.450	127
20 (2' OH)	0.380	124
21 (6 OH)	nd	175

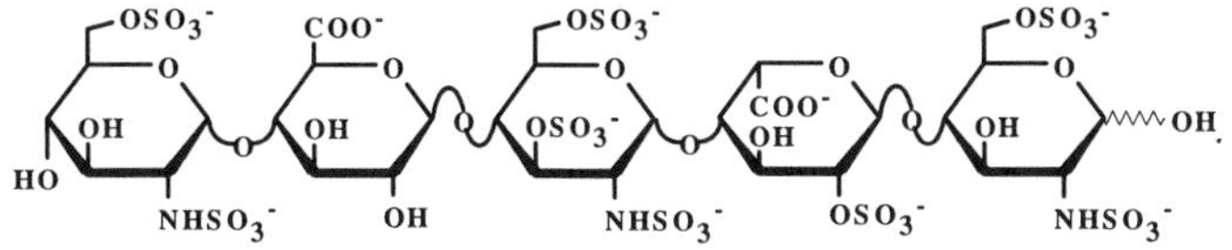

5

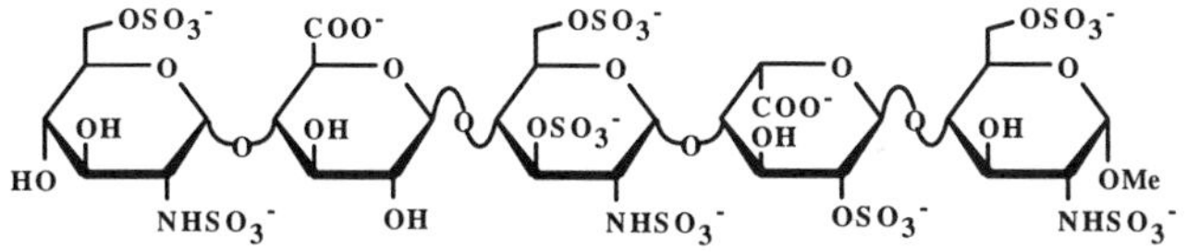

15

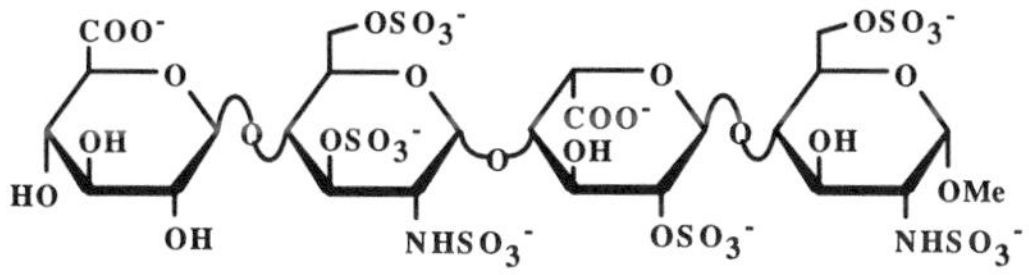

16

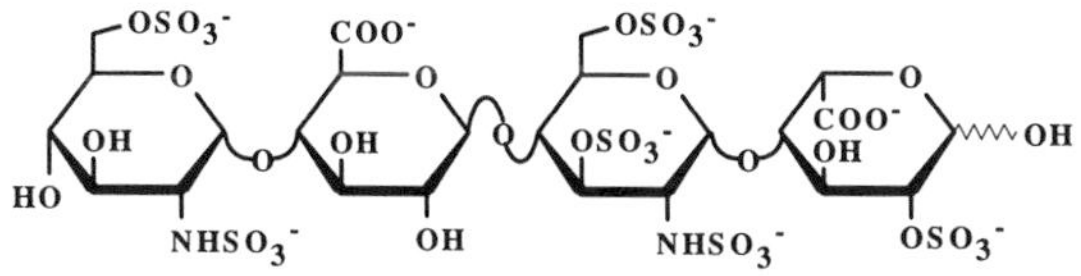

17

D E F G H

Figure 4. Structures of synthetic oligosaccharides used to determine minimum sequence for heparin interaction with antithrombin.

the non-reducing end glucosamine unit (D) or the reducing end glucosamine unit (H). These products only bound weakly to AT III compared to pentasaccharide **5** (Table I). These results definitely settled the minimum length of the oligosaccharide sequence required for optimal binding.

Another major question regarding the heparin binding site on AT III was whether the unique 3-*O*-sulfate group (*11*) present on the F unit in all the oligosaccharides having affinity for AT III (*12*) is indeed required for binding to the protein. To answer this question we synthesized the pentasaccharide **18** (Figure 5) which does not contain this unique sulfate ester. This compound has low affinity for AT III (Table I), which definitely establishes the critical role of this sulfate ester (*21*).

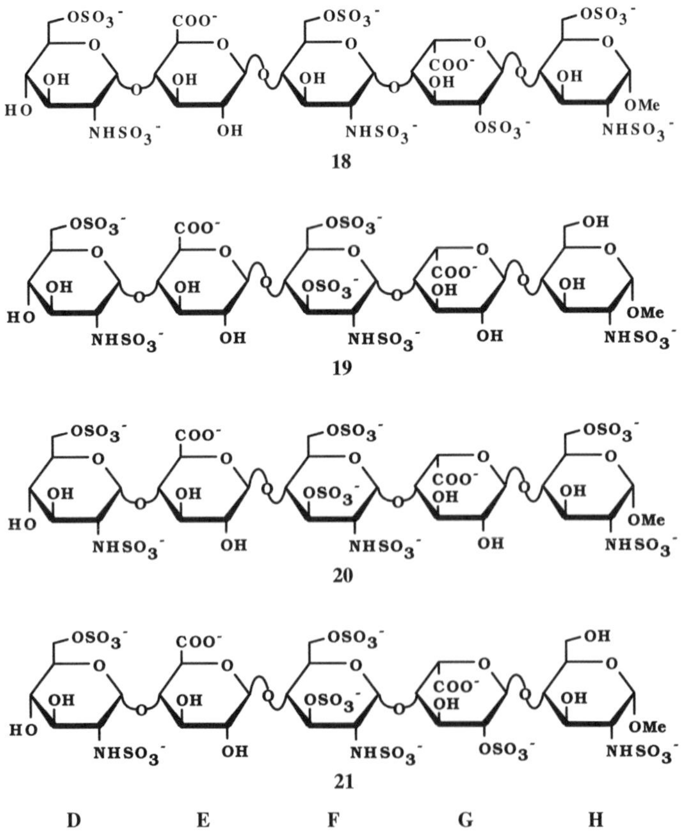

Figure 5. Structures of synthetic oligosaccharides used to determine the role of the sulfate ester moiety in heparin interaction with antithrombin.

Other synthetic derivatives were also required to determine the role played by the other structural elements in the DEFGH pentasaccharide sequence. Chemical synthesis allows structural variations at will, and all the desired products could be obtained following the strategy described above. Using compounds **19**, **20** (Petitou M. *et al.*, unpublished) and **21** (*22*), we found that the 2-*O*-sulfate group on the L-iduronic acid residue G and the 6-*O*-sulfate group on the D-glucosamine unit H are both required. They are present in every oligosaccharide with high affinity for AT III (*13*) and act synergistically to fully express the anti-factor Xa activity of pentasaccharide **5** (Table I). Removal of one of these groups results in the same loss in affinity and activity as removing both of them (*23,24*).

Regarding the role of *N*-sulfate groups at glucosamine units D, F and H (see Figure 6), it was known that removal of the *N*-sulfate either at F or at H leads to inactive compounds (*25*). The results concerning H were confirmed using the hydroxylated pentasaccharides **22** (*18*, van Boeckel *et al.*, unpublished results). Surprisingly we found that replacing the *N*-sulfate group at D by a *N*-acetyl (the usual substituent at this position in heparin of porcine origin, see above), resulted in a two fold decrease in affinity (Table I, compound **24**). Similar results were obtained with **23** (*18*, van Boeckel *et al.*, unpublished results).

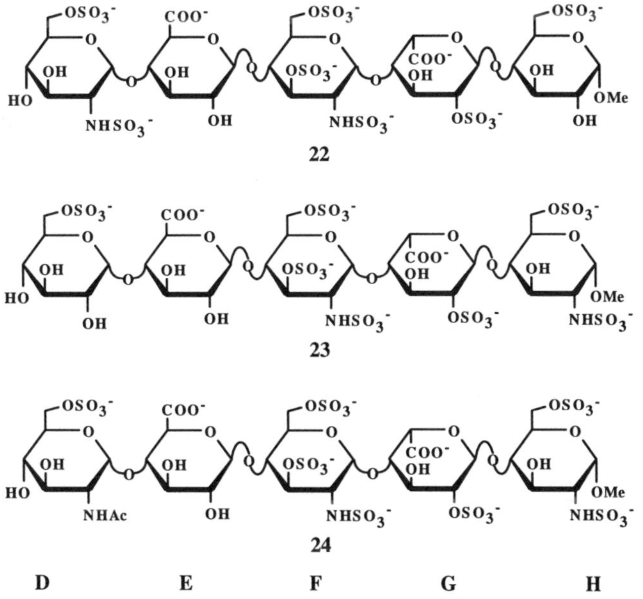

Figure 6. Structures of synthetic oligosaccharides used to determine the role of *N*-sulfate moiety in heparin interaction with antithrombin.

The synthesis of the *N*-acetylated pentasaccharide **24** requires chemical differentiation between the *N*-acetyl glucosamine unit D and the two N-sulphate glucosamine, F and H. This could be performed (Figure 7) using a trisaccharide building block in which the *N*-acetyl group was introduced before coupling to the GH part of the molecule (*26*). Thus, the azido group of **28** was selectively converted into an amine which was then acetylated. Azide opening of the epoxide followed by acetylation, acetolysis, and conversion into the imidate **27** were performed as ususal

(26). The trisaccharide imidate was then condensed with the disaccharide **26** in 40 % yield. The protected pentasaccharide was then converted into **24**, using standard techniques.

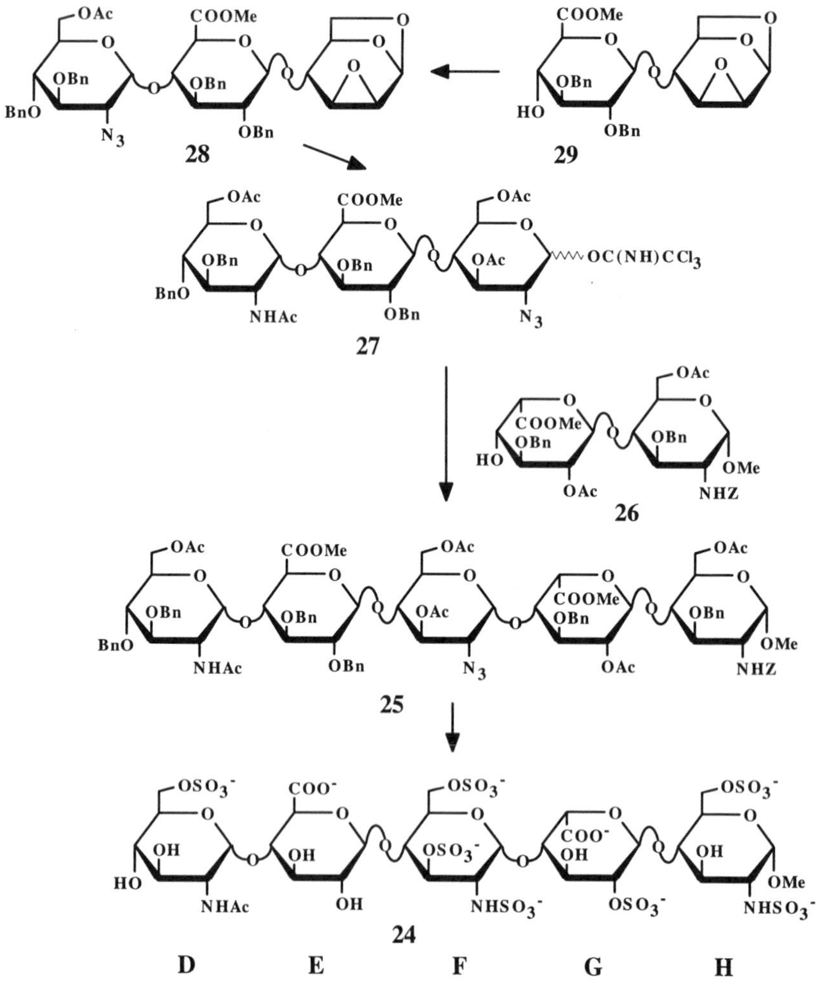

Figure 7. Synthesis of **24**.

Finally, the role of the uronic acid moieties and particularly of the carboxylate groups could be assessed using the synthetic oligosaccharides of Figure 8. Compounds **30** (27) and **31** were obtained as by-products during the preparation of **15** (Petitou *et al.*, unpublished). These derivatives showed very weak affinity for AT III, demonstrating the important role of uronic acid moieties in the interaction of heparin with AT III.

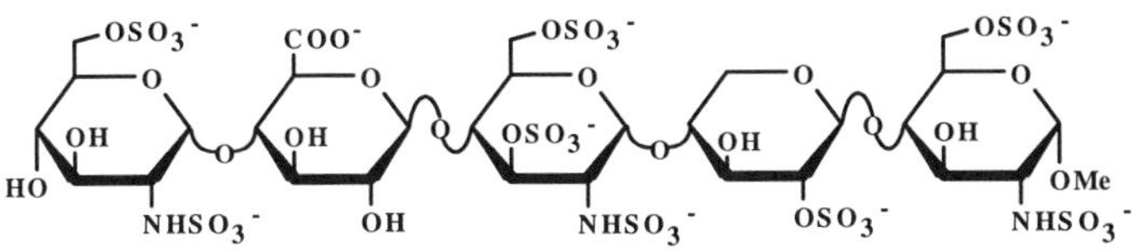

Figure 8. Structures of synthetic oligosaccharides used to determine the role of the uronic acid moiety in heparin interaction with antithrombin.

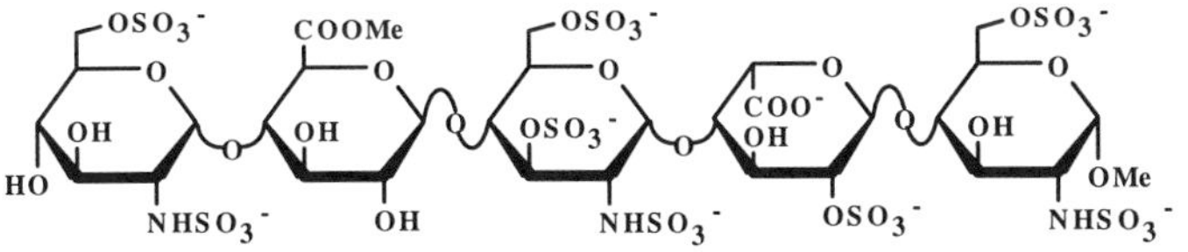

In combining the above data one can assess the role of the different structural features in the pentasaccharide sequence D-H, and consequently in heparin, for optimal binding and activation of AT III (Figure 9). In addition to the carboxylate groups, and the presence of iduronic and glucuronic acids, the 2- and 6-sulfate on D, the 3- and 2-sulfate on F, the 2-sulfate on G, and the 2- and 6-sulfate on H are all required for optimal interaction with AT III. Among these groups, the 6-sulfate of D, 3-sulfate of F, and 2-sulfate of F and H have prominent roles since their removal results in a complete loss of affinity for AT III and complete abolition of anti-factor Xa activity.

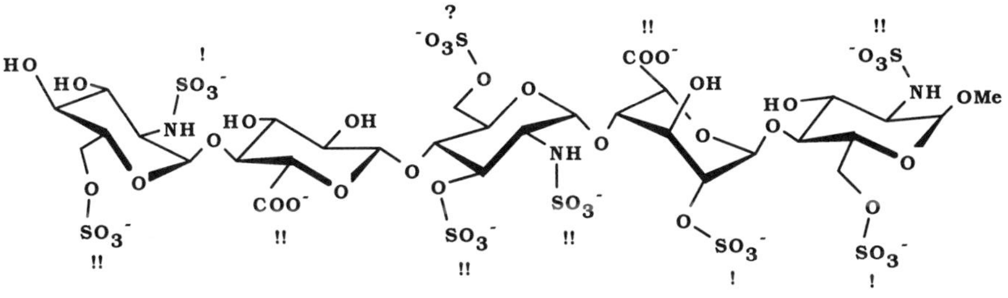

Figure 9. Role of the different sulfate and carboxylate groups in the interaction with AT III.

Conformational properties of synthetic heparin oligosaccharides in relation with the interaction with antithrombin III.

In the course of ^{1}H-n.m.r. analysis of synthetic heparin fragments, previously unobserved interproton coupling constants were revealed for some L-iduronic acid residues. The presence of these peculiar conformational properties of L-iduronic acid could be uncovered by the use of well defined synthetic fragments since the differences in values of coupling constants could hardly have been detected through analysis of full length polysaccharide chains.

The interpretation of the data led to the conclusion that another conformer ($^2S_0$, Figure 10) should be considered in addition to the conformers already recognized ($^1C_4$ and $^4C_1$) to describe the conformational equilibrium of L-iduronic acid in heparin and other glycosaminoglycans (28-31). The existence of these three conformers, and the conformational flexibility of iduronic acid-containing glycosaminoglycans, may explain the remarkable biological properties of this class of polysaccharides (32).

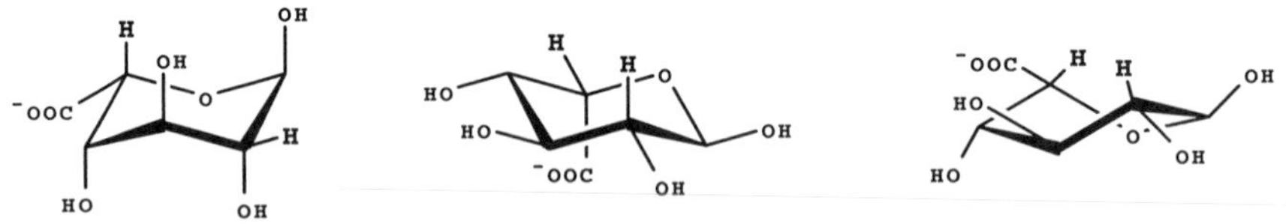

Figure 10. The three conformers of L-iduronic acid present in heparin.

The interaction of heparin oligosaccharides and antithrombin requires, at the oligosaccharide level, a precise positioning of the sulfates and carboxylates involved. The spatial orientation of these groups is mainly governed by the conformation of the carbohydrate backbone of the molecule, which in turn depends on the conformation adopted by the individual monosaccharide residues and the conformational relationships between adjacent residues.

In all the heparin oligosaccharides studied so far, the glucosamine and glucuronic acid residues have invariably been found to adopt the $^4C_1$ conformation and typical interproton coupling constants were observed by ^{1}H-NMR spectroscopy. On the contrary, the conformation of iduronic acid residues is highly dependent both on their substitution pattern (e.g. presence of sulfate esters at position 2) and on the precise nature of the adjacent residues. Thus comparison of ^{1}H-NMR data, particularly coupling constants, obtained on pentasaccharide **15** and its counterpart **18** (21), in which a N-sulfated 6-O-sulfated glucosamine replaces the N-sulfated 3,6-di-O-sulfated glucosamine F, reveals differences in the conformational state of 2-O-sulfated-α-L-iduronic acid residues (Table II, 33). It was also observed that replacement of the terminal reducing end unit N-sulfated 6-O-sulfated glucosamine by a N-sulfated 3,6-di-O-sulfated glucosamine (compound **32**, 34) again induces a new different conformational state for 2-O-sulfated-α-L-iduronic acid residue (33). As shown in Table II the affinity of the different pentasaccharides for AT III seems to be correlated to the occurence of the $^2S_0$ conformer. The precise mechanism behind this observation is not yet known.

Table II. Conformation of iduronic acid and affinity for AT III.

	Conformational State $^1C_4$ / $^2S_0$ / $^4C_1$	Affinity for AT III (K_D; μM)
18	57 / 43 / 0	>2 000
15	36 / 64 / 0	0.050
32	10 / 78 / 12	0.005

The conformation of the synthetic pentasaccharide **15** has been studied (*35*) by combining computational (using a force field derived from the MM2 program) and ^{1}H-NMR experiments. The model obtained shows an asymmetric distribution of the charges that are relevant for the interaction with AT III (Figure 9). One could of course speculate on the role of a conformational change at the iduronic acid level during binding and activation of AT III. But no experimental evidence has been obtained so far on this point.

Conclusion

Synthetic heparin fragments that represent unique sequences of the polymer, and are involved in a specific interaction with the human plasma protein AT III, have been prepared in a homogeneous state. They have proved to be powerful tools in the study of the interaction of heparin and AT III. It has thus been possible to definitively assess the size of the AT III binding sequence in heparin, to definitively prove the essential role in this sequence of the unique 3-sulfated glucosamine unit, and to identify in the so-called AT III binding site the structural features required for interaction with the protein. The results of this research demonstrate for the first time that the interaction of a complex glycosaminoglycan and a protein may be mediated, at the polysaccharide side by a short and unique sequence.

Since related glycosaminoglycans are currently found to have unexpected important biological functions (*36-39*), one may hope that the synthesis of glycosaminoglycan fragments will help the biologists understand these new functions.

Acknowledgments. Part of this research was carried out in collaboration with Organon Scientific Development Group, Oss (The Netherlands) and was sponsored by the EEC Eurêka Programme (EU 237).

Literature cited.

1. Lindahl, U.; Feingold, D. S.; Roden, L. *Trends Biochem. Sci.* **1986,** *11,* 221-225.
2. MacLean, J. *Circulation* **1959,** *19,* 75-78.
3. Casu, B. *Adv. Carbohydr. Chem. Biochem.* **1985,** *43*, 131-134.
4. Abildgaard, U. *Scand. J. Clin. Lab. Invest.* **1968,** *21,* 89-91.
5. Rosenberg, R. D.; Damus, P. S. *J. Biol. Chem.* **1973,** *248,* 6490-6505.
6. Carrell, R. W.; Boswell, D. R. In *Proteinase inhibotors;* Barrett, A. J.; Salvesen, G. Eds.; Elsevier Biochemical Press, Amsterdam, 1986; pp 405-419.

7. Andersson, L. O.; Barrowcliffe, T. W.; Holmer, E.; Johnson, E. A.; Sims, G. E. C. *Thromb. Res.* **1976,** *9,* 575-583.
8. Lam, L.; Silbert, J. E.; Rosenberg, R. D. *Biochem. Biophys. Res. Commun.* **1976,** *69,* 570-577.
9. Höök, M.; Björk, I.; Hopwood, J.; Lindahl, U. *FEBS Lett.* **1976,** *66,* 90-93.
10. Rosenberg, R. D.; Lam, L. *Proc. Natl. Acad. Sci. USA* **1979,** *76,* 1218-1222.
11. Lindahl, U.; Bäckström, G.; Thunberg, L.; Leder, I. G. *Proc. Natl. Acad. Sci. USA* **1980,** *77,* 6551-6555.
12. Casu, B.; Oreste, P.; Torri, G.; Zoppetti, G.; Choay, J.; Lormeau, J. -C.; Petitou, M.; Sinaÿ, P. *Biochem. J.* **1981,** *197,* 599-609.
13. Thunberg, L.; Bäckström, G.; Lindahl, U. *Carbohydr. Res.* **1982,** *100,* 393-410.
14. Choay, J.; Lormeau, J. -C.; Petitou, M.; Sinaÿ, P.; Fareed, J. *Ann. N.Y. Acad. Sci.* **1981,** *370,* 644-649.
15. Lindahl, U.; Thunberg, L.; Bäckström, G.; Riesenfeld, J.; Nordling, D.; Björk, I. *J. Biol. Chem.* **1984,** *259,* 12368-12376.
16. Sinaÿ, P.; Jacquinet, J.-C.; Petitou, M.; Duchaussoy, P.; Lederman, I.; Choay, J.; Torri, G. *Carbohydr. Res.* **1984,** *132,* C5-C9.
17. Petitou, M.; Duchaussoy, P.; Lederman, I.; Choay, J.; Sinaÿ, P.; Jacquinet, J.-C.; Torri, G. *Carbohydr. Res.* **1986,** *147,* 221-236.
18. Petitou, M.; van Boeckel, C. A. A. In *Chemical Synthesis of Heparin Fragments and Analogues;* Herz, W.; Kirby, G. W.; Moore, R. E.; Steglich, W.; Tamm, Ch., Eds.; Progr. Chem. Org. Nat. Prod.; Springer-Verlag: New-York, 1992, Vol. 60; pp 143-210.
19. Petitou, M.; Duchaussoy, P.; Lederman, I.; Choay, J.; Sinaÿ, P.; Jacquinet, J.-C.; Torri, G. *Carbohydr. Res.* **1987,** *167,* 67-75.
20. Choay, J.; Petitou, M.; Lormeau, J.-C.; Sinaÿ, P.; Casu, B.; Gatti, G. *Biochem. Biophys. Res. Commun.* **1983,** *116,* 492-499.
21. Petitou, M.; Duchaussoy, P.; Lederman, I.; Choay, J.; Sinaÿ, P. *Carbohydr. Res.* **1988,** *179,* 163-172.
22. Beetz, T.; van Boeckel, C. A. A. *Tetrahedron Lett.* **1986,** *27,* 5889-5892.
23. Petitou, M.; Lormeau, J.-C.; Choay, J. *Eur. J. Biochem* **1988,** *176,* 637-640.
24. Barzû, T.; Petitou, M.; Jaurand, G.; Lormeau, J.-C.; Choay, J. *Thromb. Haemost.* **1991,** *65,* 934.
25. Riesenfeld, J.; Thunberg, L.; Höök, M.; Lindahl, U. *J. Biol. Chem.* **1981,** *256,* 2389-2394.
26. Duchaussoy, P.; Lei, P. S.; Petitou, M.; Sinaÿ, P. G.; Lormeau, J.-C.; Choay, J. *BioMed. Chem. Lett.* **1991,** *1,* 99-102.
27. van Boeckel, C. A. A.; Lucas, H.; van Aelst, S. F.; van den Nieuwenhof, M. W. P.; Wagenaars, G. N.; Mellema, J.-R. *Recl. Trav. Chim. Pays-Bas* **1987,** *106,* 581-591.
28. Casu, B.; Choay, J.; Ferro, D. R.; Gatti, G.; Jacquinet, J.-C.; Petitou, M.; Provasoli, A.; Ragazzi, M.; Sinaÿ, P.; Torri, G. *Nature* **1986,** *6076,* 215-216.
29. Ferro, D. R.; Provasoli, A.; Ragazzi, M.; Torri, G.; Casu, B.; Gatti, G.; Jacquinet, J.-C.; Sinaÿ, P.; Petitou, M.; Choay, J. *J. Am. Chem. Soc.* **1986,** *108,* 6773-6778.

30. Torri, G.; Casu, B.; Gatti, G.; Petitou, M.; Choay, J.; Jacquinet, J.-C.; Sinaÿ, P. *Biochem. Biophys. Res. Commun.* **1985,** *128,* 134-140.
31. van Boeckel, C. A. A.; van Aelst, S. F.; Wagenaars, G. N.; Mellema, J.-R.; Paulsen, H.; Peters, T; Pollex, A.; Sinnwell, V. *Recl. Trav. Chim. Pays-Bas* **1987,** *106,* 19-29.
32. Casu, B.; Petitou, M.; Provasoli, M.; Sinaÿ, P. *Trends Biochem. Sci* **1988,** *13,* 221-225.
33. Ferro, D. R.; Provasoli, A.; Ragazzi, M.; Casu, B.; Torri, G.; Bossennec, V.; Perly, B.; Sinaÿ, P.; Petitou, M.; Choay, J. *Carbohydr. Res.* **1990,** *195,* 157-167.
34. van Boeckel, C. A. A.; van Aelst, S. F.; Beetz, T. *Tetrahedron Lett.* **1988,** *29,* 803-806.
35. Ragazzi, M.; Ferro, D. R.; Perly, B.; Sinaÿ, P.; Petitou, M.; Choay, J. *Carbohydr. Res.* **1990,** *195,* 169-185.
36. Maimone, M. M.; Tollefsen, D. M. *J. Biol. Chem.* **1990,** *265,* 18263-18271.
37. Turnbull, J. E.; Fernig, D. G.; Ke, Y.; Wilkinson, M. C.; Gallagher, J. T. *J. Biol. Chem.* **1992,** *267,*10337-10341.
38. Yanagishita, M.; Hascall, V. C. *J. Biol. Chem.* **1992,** *267,* 9451-9454.
39. Clowes, A. W.; Karnovsky, M. J. *Nature* **1977,** *265,* 625-626.

RECEIVED December 15, 1993

Chapter 7

Synthesis and Antigenic Properties of Sialic Acid Based Dendrimers

R. Roy, D. Zanini, S. J. Meunier, and A. Romanowska

Department of Chemistry, University of Ottawa, Ontario K1N 6N5, Canada

Solid phase synthesis on Wang resin was used to construct dendritic α-thiosialosides which can be used as inhibitors of influenza virus hemagglutinins. The design of these new hyperbranched clusters is based on the rational scaffolding of L-lysine core structures using well established Fmoc-chemistry and benzotriazolyl esters as coupling procedures. One step chain extension of all the lysyl-amino groups with chloroacetylglycylglycine active ester allowed the introduction of the required functionality necessary for the coupling to α-thiosialoside derivative prepared under improved phase transfer catalyzed conditions. Well defined di-, tetra-, octa- and hexadeca-valent dendritic α-thiosialosides were thus prepared by a straight forward approach. The antigenicity of the dendrimers was compared to a known sialylated polymer used as reference. Regioselective 9-O-acetylation of the octavalent dendrimers was also achieved to provide access to inhibitor of other strains of influenza virus hemagglutinins.

N-Acetylneuraminic acid (NeuAc) represents the most ubiquitous member of the sialic acid family of derivatives present on cell surface glycolipids and glycoproteins (*1*). The wide variety of sialic acids is responsible for the very diversified biological and immunochemical roles ascribed to sialyloligosaccharides. For instance, infections of mammalian host tissues, including human tissues, by influenza viruses are mediated by the binding of the viral membrane glycoprotein hemagglutinin (HA) to the host-cell sialyloligosaccharides present on glycolipids and glycoproteins (*2*). The viral membrane envelopes are also composed of a glycohydrolase having sialidase activities which allows the virus particles to circulate freely within the mucous linings rich in sialylated glycoproteins.

Although different virus isolates have shown slight variations in their receptor specificities including O-acetyl substitutions (*3*), sugar residues and linkage positions (*4*), the epitope recognized as being essential for binding has been associated with sialic acid residues. Inhibition experiments with synthetic α-sialosides (*5*), X-ray data (*6*) and proton NMR spectroscopic studies (*7*) have clearly demonstrated the critical role played by α-sialosides. The binding constant of the viral HAs is, however, very weak (mM range). To circumvent the poor affinities of single α-sialosides toward the

0097–6156/94/0560–0104$08.00/0

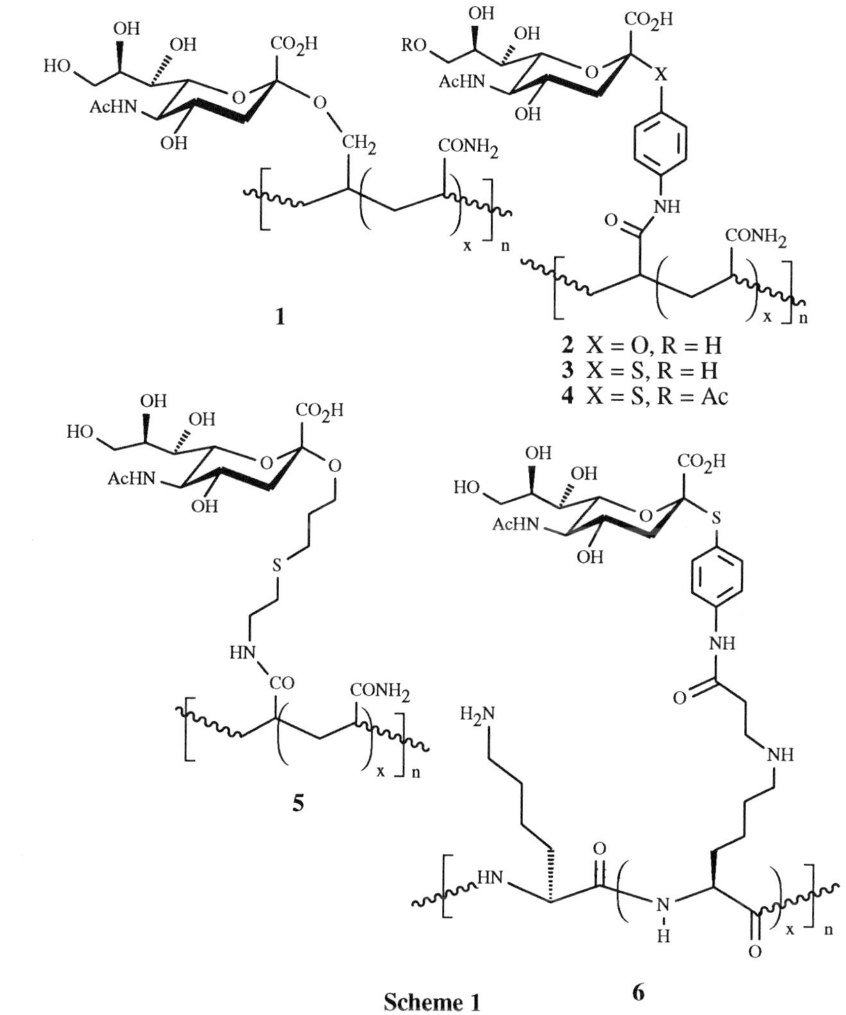

Scheme 1

design of potential inhibitors of viral HAs a strategy based on the cluster effect was successfully proposed. Sialylated neoglycoproteins and polymers, by virtue of their multivalent epitopes, possess the required feature necessary to compensate for the low affinity of individual α-sialosides. The structures of few such glycopolymers are depicted in Scheme 1 (*8-11*). The improved binding (avidity) incorporated by the glycopolymer multivalencies has increased the inhibitory activities by a factor of one thousand (μM range).

These sialylated polymers vary by the length and nature of the spacers incorporated between the polymer backbone and the N-acetylneuraminic acid residues. The polymer **1** is essentially deprived of such spacer and accordingly did not show inhibition activity (*8*). The more rigid polymer **4**, which contains 9-O-acetyl sialic acid residues, was specific against influenza C virus hemagglutinins (*10*). However it did not inhibit influenza A virus HAs. Interestingly, the longer and more accessible sialic acid containing polymer **5** was an efficient inhibitor for the influenza A virus HAs (*9,12*).

The polymer **6** was obtained by a conjugate-1,4 Michael type addition of poly-L-lysine to N-acryloylated phenylthio α-sialoside (*11*). It was designed to confer some elements of bioavailability for better therapeutic purposes. Moreover, the most recent sialylated glycopolymers were also synthesized as thiosialosides in order to provide them with resistance against the viral neuraminidases (*10,11*). In spite of the fact that linear poly-L-lysine used as carriers are available in low molecular weight, it was deemed of interest to generate starburst poly-L-lysine cores (dendrimers) with better defined molecular weight.

Dendritic molecules (*13-17*) may represent ideal clusters because they can be synthesized with the desired carbohydrate density thus mimicking multi-antennnary glycoproteins. The existing starburst or convergent (*16*) synthetic approaches to dendritic macromolecules suffer some technical drawbacks. A solid phase synthesis of sialylated dendrimers is herein reported to circumvent these drawbacks and to get access to potential influenza virus hemagglutinin inhibitors.

Solid Phase Synthesis of Dendritic α-Thiosialosides

Phase Transfer Catalyzed Synthesis of α-Thiosialosides. As mentioned previously, it was deemed useful to prepare the sialylated dendrimers in the form of thioglycoside to provide viral hemagglutinin inhibitors which would also be resistant to the viral neuraminidase. Following improved phase transfer catalyzed conditions (PTC) developed in this laboratory (*17-18*), the synthesis of the known (*19*) α-thio sialosyl derivative **8** was achieved at room temperature in a two phase system using ethyl acetate, 1 M sodium carbonate and the lipophilic tetrabutylammonium hydrogen sulfate as phase transfer catalyst. The reaction was entirely stereospecific and provided crystalline **8** in 66% yield. The reaction was very useful and was applied to a wide range of other nucleophiles (*17-18*). The PTC method was also used for the preparation of the reference phenylthio α-sialoside **10** (*20*).

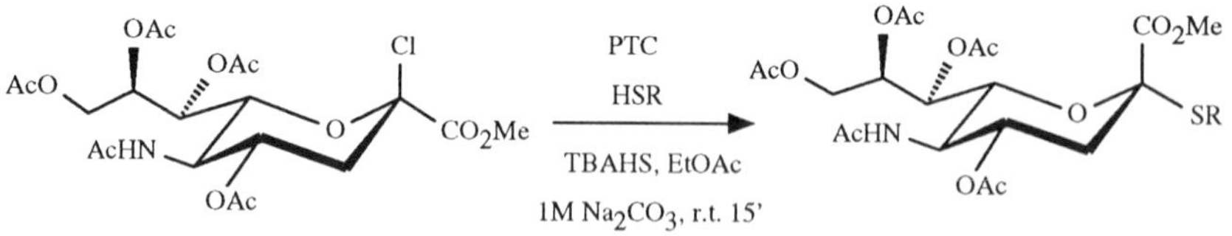

Scheme 2

7

8 R = Ac
9 R = H
10 R = Ph

Chemoselective hydrolysis of the thioacetate group in **8** was achieved following a slight modification of published procedure (*21*) using mild trans-esterification conditions (0.95 equiv. NaOMe, MeOH, -40°C,15 min). Compound **9** was obtained in 88% yield after purification by silica gel column chromatography. The product was relatively stable to both anomerization and to oxidation into disulfide. The thioglycoside **9** is however usually freshly prepared for the coupling to the N-chloroacetylated polymer-bound dendrimers.

Solid Phase Synthesis of the N-Chloroacetylated Dendrimers. The striking advantages of the present approach in comparison to previous dendrimers syntheses rely on the fact that L-lysine was used as core unit upon which the scaffolding of the subsequent generations was constructed (*22*). The synthesis of each dendritic structure was based on well established and high yielding solid phase peptide chemistry using 9-fluorenylmethoxycarbonyl (Fmoc) amino-protecting groups and benzotriazolyl esters as coupling reagents.

Thus, dendritic L-lysine cores were elaborated on *p*-benzyloxybenzyl alcohol (Wang) resin (0.58 mmol/g) to which was anchored a β-alanyl spacer using the above Fmoc/benzotriazolyl ester strategy (Fmoc-β-Ala-OBt, 2 equiv.,0.5 equiv. DMAP, DMF, 2.5 hr). N^α,N^ϵ-Di-Fmoc- L-lysine was synthesized in 68% yield using a well established procedure with 9-fluorenylmethyl chloroformate in 10% sodium bicarbonate (*24*). The corresponding benzotriazolyl ester derivative **11** was freshly prepared in N,N-dimethylformamide (DMF) with one equivalent each of N-hydroxybenzotriazole (HOBt) and diisopropylcarbodiimide (DIC, 0°C, then 25°C for 1 hr). In each cycle, the Fmoc-protecting groups were removed by the usual β-elimination process using 20% piperidine in DMF. The extent of coupling can be established by the spectrophotometric quantitation of the released dibenzofulvene chromophore at 300 nm following the piperidine treatment.

The products resulting from each sequential generation were directly treated with pre-formed chloroacetylglycylglycine benzotriazolyl ester **14** prepared by the above procedure. Interestingly, the required chloroacetylglycylglycine is commercially available and did not necessitate individual couplings of glycine residues and capping with chloroacetic anhydride as commonly done. The completion of full derivatization was determined by the usual ninhydrin test. Using this solid phase approach, di-(**15**), tetra-(**16**), octa-(**17**) and hexadeca-valent (**18**) chloroacetylated dendrimers were obtained in the first, second, third and fourth generations respectively. For quality and structural determination purposes, the corresponding unbound chloroacetylated acid derivatives **19-22** were released from the polymer support by treatment with aqueous trifluoroacetic acid (95% TFA, 1.5 hr). The synthetic sequence is illustrated in Scheme 3. The dendrimers **19-22** were generally obtained in >90% yields with 90-95% purity.

Coupling of Sialic Acid to the Dendrimers. While still attached to the resin, each dendrimer generation was treated with an excess of 2-thiosialic acid derivative **9** (1% triethylamine/DMF, 16h, 25°C). Before the bulk of the dendrimers were released from the polymeric support, aliquots were withdrawn and hydrolyzed as above. The completeness of the couplings was estimated from the ^{1}H-NMR spectrum of the sialylated dendrimers which showed characteristic signals for any residual chloroacetyl-methylene groups at 4.12 ppm (DMSO-d_6). Where required, the couplings were repeated (Fig. 1).

The polymer bound peracetylated sialyl dendrimers **23**, **27**, **31** and **36** were released from the polymer support as above for **15-18** and obtained in 66-99% yields after removal of the solvent under reduced pressure. The ^{1}H- and ^{13}C-NMR spectra (DMSO-d_6, 500 MHz) of the dendrimers revealed the integrity of the α-sialoside linkages as well as the ratio of the β-alanyl residues relative to those of both L-lysyl and sialyl signals. Each of the protected dendrimers (**24**, **28**, **32**, **37**) were de-esterified with NaOMe/MeOH (25°C, 1h) followed by 0.05 M NaOH (25°C, 2hr) with H$^+$ resin treatment after each step to afford dendrimers **26**, **30**, **34** and **39** in essentially quantitative yields. The larger octa- (**34**) and hexadeca-meric (**39**) dendrimers can be dialyzed using benzoylated dialysis tubing (M.W. cutoff 2 kDa).

Octavalent Dendrimer with 9-O-Acetyl Sialic Acid Residues. In order to also gain access to influenza C virus HA inhibitors, the octavalent dendrimer **34**, released from the polymer support, was treated with an excess of trimethylorthoacetate in dimethyl sulfoxide containing a catalytic amount of trifluoroacetic acid. After stirring overnight at room temperature, the reaction mixture was exhaustively dialyzed against distilled water. Following lyophilization, the 9-O-acetylated dendrimer **35** was obtained in quantitative yield. The high field proton NMR spectra (500 MHz) of **35** revealed only a single positional isomer as judged by the unique O-acetyl signal

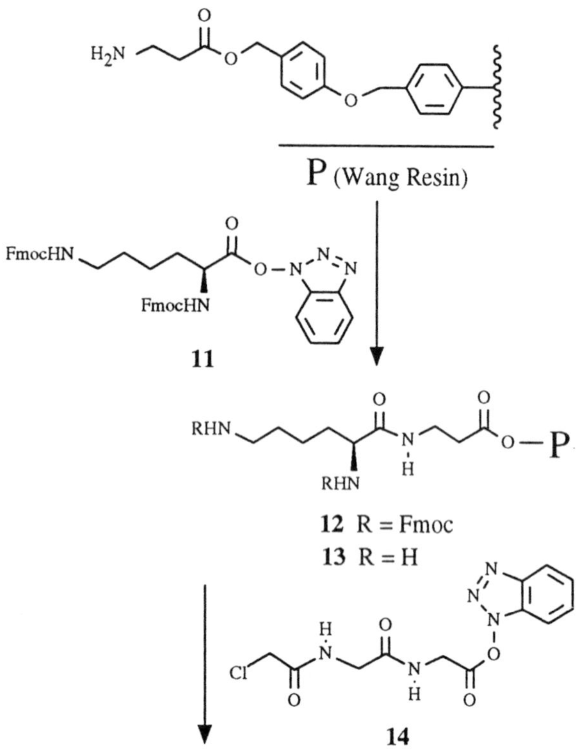

N^{α},N^{ϵ}-(Cl-CH_2-CO-Gly-Gly-$)_2$-Lys-β-Ala-O-R

15 R = **P**

19 R = H

N^{α},N^{ϵ}-(Cl-CH_2-CO-Gly-Gly-$)_4$-Lys_2-Lys-β-Ala-O-R

16 R = **P**

20 R = H

N^{α},N^{ϵ}-(Cl-CH_2-CO-Gly-Gly-$)_8$-Lys_4-Lys_2-Lys-β-Ala-O-R

17 R = **P**

21 R = H

N^{α},N^{ϵ}-(Cl-CH_2-CO-Gly-Gly-$)_{16}$-Lys_8-Lys_4-Lys_2-Lys-β-Ala-O-R

18 R = **P**

22 R = H

Scheme 3

appearing at 2.24 ppm and integrating for three protons (8x). The intensity of the above O-acetyl signal was found identical to that of the N-acetyl signal at 2.13 ppm thus demonstrating the regioselectivity of this process. The chemical shift of one of the corresponding H-9 proton (4.18 ppm) in **34** was accordingly shifted downfield to 4.38 ppm in **35** (Fig. 2).

23 R = Ac, R' = Me, R" = **P**
24 R = Ac, R' = Me, R" = H
25 R = R" = H, R' = Me
26 R = R' = R" = H

27 R = Ac, R' = Me, R" = **P**
28 R = Ac, R' = Me, R" = H
29 R = R" = H, R' = Me
30 R = R' = R" = H

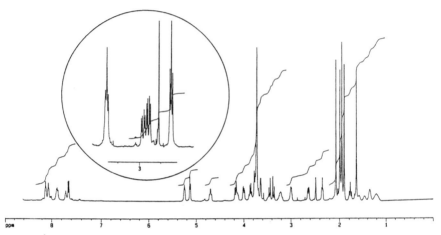

36 R = Ac, R' = Me, R" = **P**
37 R = Ac, R' = Me, R" = H
38 R = R" = H, R' = Me
39 R = R' = R" = H

Fig. 1 ^{1}H-NMR spectrum (500 MHz, DMSO-d$_6$) of the per-O-acetylated divalent sialyl-dendrimer **24**.   Inset: deprotected divalent dendrimer **30** showing the H$_{3e}$-region.

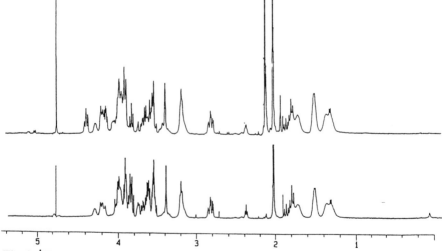

Fig. 2 ^{1}H-NMR spectra (500 MHz, D$_2$O) of the octavalent sialylated-dendrimers **34** (bottom) and 9-O-acetylated dendrimer **35** (up).

31 R = Ac, R' = Me, R" = P
32 R = Ac, R' = Me, R" = H
33 R = R" = H, R' = Me
34 R = R' = R" = H
35 9-O-Acetylated **34**

Antigenicity of Dendritic α-Thiosialosides

The antigenic properties of the dendritic α-thiosialosides were first established in a model study using the plant lectin wheat germ agglutinin (WGA) in a solid phase immunoassay format. To this end, it was of interest to evaluate the dendrimers capacity to be used as coating haptens in microtiter plates in order to determine their antigenicity by a direct enzyme linked lectin assay (ELLA). Thus, di-, tetra-, octa- and hexadeca-valent sialylated dendrimers together with poly(acrylamide-co-p-N-acrylamidophenylthio α-sialoside) (3) taken as a positive reference control were used to coat the wells of microtiter plates. After the usual incubation time, blockings and washings, the presence of adsorbed sialoside residues was then confirmed by treating the plates with horseradish peroxidase labelled wheat germ agglutinin (HRP-WGA) using 2,2'-azinobis(3-ethylbenzothiazoline -6-sulfonic acid) (ABTS) and hydrogen peroxide as enzyme substrates. The results of these first control experiments are illustrated in Figure 3. From these results, it appeared obvious that the octa- **34** and hexadeca-valent **39** dendrimers were almost as efficient coating antigens as the phenylthio α-sialoside copolymer **3**. The poor coating properties of the di- and tetra-valent dendrimers **26** and **30** was mainly attributed to their lack of lipophilic components. Although less efficient as coating antigens, even the di- and tetra-valent dendrimers showed promises as multivalent or clustered haptenic carbohydrate structures.

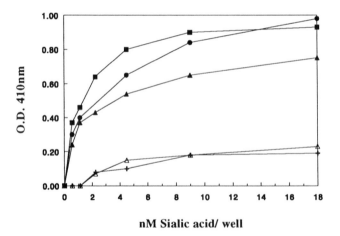

Fig. 3 Enzyme linked lectin assays (ELLA) using di- (Δ), tetra- (+), octa- (▲), hexadeca- (●) valent dendrimers **26, 30, 34, 39** and polymer **3** (■) as coating antigens in microtiter plates with horseradish peroxidase labelled wheat germ agglutinin (HRP-WGA) and ABTS-H_2O_2 as peroxidase substrates.

In a second set of experiments, the dendrimers were evaluated for their relative usefulness to inhibit the strong binding interactions between wheat germ agglutinin and the glycopolymer **3**. Again, enzyme linked lectin assay was used in competitive inhibition experiments. Thus the glycopolymer **3** was first taken as coating antigen by adsorption in the wells of the microtiter plates. Pre-incubated mixtures of phenylthio α-sialoside or dendrimers at various concentrations with horseradish peroxidase-labeled WGA were then added to each well for the competitive inhibition experiments. The residual enzymatic activity was measured with the ABTS-hydrogen peroxide couple and the results were plotted as a function of percentage inhibitions. The results are illustrated in Figure 4. These results clearly demonstrate that the dendrimers were far superior to the monosialoside in the inhibition experiments, thus confirming the value of dendritic sialosides. The above results were also plotted as a function of dendrimer concentrations required for 50% inhibition relative to the dendrimer valency (Figure 5). As previously demonstrated with other clustered glycosides, the inhibition potential of clusters follows an exponential growth for a linear increase in the cluster size.

Preliminary experiments with influenza A virus (strain X-31) showed that the dendrimers were potent inhibitors of hemagglutination of human erythrocytes. For instance, dendrimers **26**, **30**, **34** and **39** showed inhibition at 625, 312.5, 156 and 91 μM respectively. The inhibition experiments with the divalent dendrimer **26** (625 μm) illustrate that even at this minimum level of clustering the divalent dendritic structure is at least five times more efficient than monosialosides (~3 mM). The hexadecavalent dendrimer **39** is as potent as previously described sialopolymer.

Experimental

General methods. Melting points were determined on a Gallenkamp apparatus and are uncorrected. The ^{1}H- and ^{13}C-NMR spectra were recorded on Varian XL-300 or on Bruker AMX-500 spectrometers. The proton chemical shifts (δ) are given relative to internal chloroform (7.24 ppm) for CDCl$_3$ solutions, to HOD (4.75 ppm) for D$_2$O solutions or to DMSO (2.49 ppm) for DMSO-d$_6$ solutions. The carbon chemical shifts are given relative to deuterochloroform (77 ppm) and to DMSO-d$_6$ (39.5 ppm). The analyses were done as a first-order approximation and assignments were based on COSY, HETCOR and/or HMQC experiments. Optical rotations were measured on a Perkin-Elmer 241 polarimeter and were run at 23°C for 1% solutions. Mass spectra were recorded on a VG 7070-E spectrometer (CI, ether) and Kratos Concept IIH (FAB-MS, thioglycerol). Thin-layer chromatography (TLC) was performed using silica gel 60 F-254 and column chromatography on silica gel 60 (230-400 mesh, E. Merck No. 9385). HPLC profiles were run on a Waters Model 991/625LC system equipped with a diode-array detector. A Waters μBondapak C18 reverse-phase (10μm) column (300 x 3.9 mm) was used at a flow rate of 1mL/min with solvent A (0.1% aq. TFA) and B (0.1% TFA in CH$_3$CN). An isocratic elution with 0% solvent B for 20 min was followed by a linear gradient of 0-40% B from 20 to 30 min.

Methyl   5-acetamido-4,7,8,9-tetra-O-acetyl-2-S-acetyl-3,5-   dideoxy-2-thio-*D*-glycero-α-*D*-galacto-2-nonulopyranosonate (8). To a solution of freshly prepared acetochloroneuraminic acid **7** (3.39 g, 7.73 mmol) (*25*) in ethyl acetate (20 mL) was added a solution of sodium thioacetate (12.8 mmol) and tetrabutylammonium hydrogen sulfate (2.89 g, 7.73 mmol) in 1 M sodium carbonate (20 mL). The mixture was stirred at room temperature for 30 min, the reaction being monitored by TLC (ethyl acetate). The reaction mixture was then diluted with 75 mL each of ethyl acetate and saturated aqueous sodium bicarbonate. The organic layer was separated and washed with saturated aqueous NaHCO$_3$ (2x 75 mL) followed by saturated

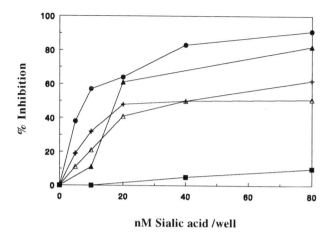

Fig. 4 Inhibition of binding of sialylated polymer **3** used as coating antigen to HRP-WGA in microtiter plate enzyme linked lectin assays. Inhibitors: Phenylthio-α-sialoside (deprotected **10**) (■), di- (Δ), tetra- (+), octa- (▲), and hexadeca- (●) valent dendrimers **26, 30, 34,** and **39.**

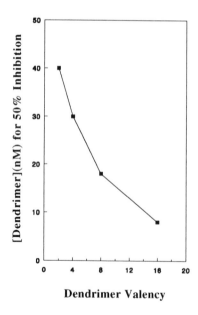

Fig. 5 Effect of dendrimer valencies on inhibition of binding of polymer **3** to HRP-WGA. Data taken from Fig. 2 at 50% inhibition.

sodium chloride (50 mL). The dried organic phase (Na_2SO_4) was concentrated near dryness under vaccum, and purified by crystallization from ethyl acetate/hexanes (2.8 g, 66%). The white needles had m.p. 89-92°C, $[\alpha]_D$ = +52.0° (c 1.0, $CHCl_3$), (lit. (*19*), mp.75-80°C, $[\alpha]_D$ +46.7°); R_f(EtOAc) = 0.38; MS(CI) for $C_{22}H_{31}NO_{13}S$: 550.0 (M + H, 55.7%), 474.0 (M-aglycone, 31.6%); ^1H-NMR ($CDCl_3$) δ: 1.88 (s, 3H, NAc), 2.02 (dd, 1H, $J_{3a,3e}$ = 13.0, $J_{3a,4}$ = 11.7 Hz, H - 3a), 2.01, 2.03, 2.12, 2.13 (4s, 12H, 4 OAc), 2.27 (s, 3H, SAc), 2.62 (dd, 1H, $J_{3e,3a}$ = 13.0, $J_{3e,4}$ = 4.6 Hz, H - 3e), 3.79 (s, 3H, MeO), 4.02 (dd, 1H, $J_{8,9}$ = 6.2, $J_{9,9'}$ = 12.4 Hz, H-9), 4.09 (m, 1H, H-5), 4.40 (dd, 1H, $J_{8,9'}$ = 2.5, $J_{9,9'}$ = 12.4 Hz, H-9'), 4.65 (dd, 1H, $J_{5,6}$ = 10.8, $J_{6,7}$ = 2.3 Hz, H-6), 4.90 (ddd, 1H, $J_{3a,4}$ = 11.7, $J_{3e,4}$ = 4.6, $J_{4,5}$ = 10.4 Hz, H-4) 5.22 (ddd, 1H, $J_{8,9'}$ = 2.5, $J_{8,9} \cong J_{7,8} \cong$ 6.2 Hz, H-8), 5.35 (dd, 1H, $J_{6,7}$ = 2.3, $J_{7,8}$ = 6.2 Hz, H-7), 5.40 (d, 1H, $J_{5,NH}$ = 10.1 Hz, NH); ^{13}C-NMR ($CDCl_3$) δ: 20.7, 20.8, 20.9 (4 OAc), 23.1 (NAc), 30.1 (SAc), 37.5 (C-3), 49.0 (C-5), 53.4 (MeO), 62.4 (C-9), 67.8 (C-7), 68.9 (C-4), 70.6 (C-8), 75.2 (C-6), 84.5 (C-2), 169,4, 170.1, 170.3, 170.4, 170.7, 170.8 (C=O), 191.5 (SAc).

Anal. Calcd. for $C_{22}H_{31}NO_{13}S$: C, 48.08; H, 5.68; N, 2.55; S, 5.83. Found: C, 48.04; H, 5.79; N, 2.49; S, 6.02.

Methyl   5-acetamido-4,7,8,9-tetra-O-acetyl-3,5-dideoxy-   2-thio-*D*-glycero-α-*D*-galacto-2-nonulopyranosonate (9). To a solution of **8** (143.5 mg, 0.261 mmol) in dry methanol (5 mL), cooled to -40°C, was added a 1 M solution of sodium methoxide (250 µl, 0.249 mmol). The reaction mixture was stirred for 10 min at -40°C after which it was treated with H^+ resin (Amberlite IR-120) at -40°C for 15 min. The solution was filtered and evaporated at room temperature under vaccum to afford compound **9**. The oily crude product was purified by column chromatography with a gradient of 3:1 to 9:2 of ethyl acetate/hexanes to afford 117 mg (88%) of **9** as a white foam. M.p. (amorphous solid) 82-86°C, $[\alpha]_D$ = +15.6° (c 0.99, $CHCl_3$), MS(CI) for $C_{20}H_{29}NO_{12}S$: 507.9 (M + H, 100%), 476 (M +H)-32, 85%); ^1H-NMR ($CDCl_3$) δ: 1.88 (s, 3H, NAc), 2.04 (dd, 1H, $J_{3a,3e}$ = 12.8, $J_{3a,4}$ = 11.7 Hz, H-3a), 2.03, 2.05, 2.13, 2.14 (4s, 12H, 4 OAc), 2.80 (dd, 1H, $J_{3e,3a}$ = 12.8, $J_{3e,4}$= 4.8 Hz, H-3e), 3.17 (s, 1H, SH), 3.73 (dd, 1H, $J_{6,7}$= 1.9, $J_{5,6}$ = 10.8 Hz, H-6), 3.83 (s, 3H, OMe), 4.06 (m, 1H, H-5), 4.10 (dd, 1H, $J_{9,9'}$ = 12.4, $J_{8,9}$ = 5.5 Hz, H-9), 4.49 (dd, 1H, $J_{8,9'}$ = 2.1, $J_{9,9'}$ = 12.4 Hz, H-9'), 4.91 (ddd, 1H, $J_{4,5}$ = 10.3, $J_{3a,4}$ = 11.7, $J_{3e,4}$ = 4.8 Hz, H-4), 5.17 (d, 1H, $J_{NH,5}$ = 10.5 Hz, NH), 5.28-5.33 (m, 2H, H-7 and H-8); ^{13}C-NMR ($CDCl_3$) δ: 20.8, 20.8, 21.1 (4 OAc), 23.1 (NAc), 38.9 (C-3), 49.1 (C-5), 53.4 (MeO), 62.2 (C-9), 67.7 (C-7), 69.4 (C-4), 70.1 (C-8), 75.1 (C-6), 81.6 (C-2), 170.0, 170.1, 170.4, 170.4, 170.6, 170.9 (C=O).

Synthesis of N^α,N^ϵ-di-Fmoc-L-Lysine. To an ice cold solution of L-Lysine hydrochloride (0.692 g, 3.8 mmol) in 10% aqueous $NaHCO_3$ (10 mL) was added 9-fluorenylmethyl chloroformate (1.96 g, 7.6 mmol) dissolved in the minimum volume of dioxane. The solution was stirred at 0° C for 2.5 hr during which time a precipitate had formed. Water (200 mL) was then added to the heterogenous mixture and the resulting solid compound was washed with a limited volume of ether to remove small quantities of 9-fluorenylmethanol and the high melting polymer of dibenzofulvene (*24*). The white precipitate was collected and successively washed with 1N HCl and water until the filtrate was found neutral. The title compound was dried in a dessicator over P_2O_5. It was obtained in 68% yield (1.52 g) and had m.p. 120-124 °C; $[\alpha]_D$ = -6.5° (c 1, DMF); ^1H-NMR (DMSO-d_6) δ: 1.12-1.80 (m, 6H, β,γ,δ-CH_2), 2.97 (m, 2H, ϵ-CH_2), 3.91 (m, 1H, α-CH), 4.12-4.38 (m, 6H, Fmoc-CHCH$_2$), 7.21-7.95 (m, aryl and NH).

Solid Phase Synthesis of Poly-L-Lysine Dendrimers. The solid phase synthesis was accomplished manually using a glass funnel equipped with a coarse sintered glass filter and a bottom stopcock T-valve. Wang resin (2.51 g) having 0.58 mmol/g of hydroxyl group substitution (100-200 mesh, Advanced Chemtech, LO, KY) was suspended in DMF. The resin was allowed to swell for 1 hr at room temperature by bubbling nitrogen through the funnel. This process was repeated three times. After draining the DMF, pre-formed Fmoc-β-Ala-OBt (2.92 mmol, see below) and DMAP (0.09 g, 0.73 mmol) in DMF were added to the resin which was agitated by bubbling N_2 for 2.5 hr. The excess reagents and DMF were evacuated and the resin was washed with fresh DMF (5x 1 min). The Fmoc-β-Ala-OBt coupling was repeated until aliquots treated with 20% piperidine showed greater than 85% substitution as judged by the quantitative spectrophotometric analysis of the dibenzofulvene chromophore. The capping of unsubstituted hydroxyl groups was achieved by converting the remaining groups with acetic anhydride (276 μL, 2.92 mmol) in DMF (6mL) containing DMAP (0.18 g, 1.46 mmol). After acetylation, the resin was washed with DMF (5x 1 min). The Fmoc- protecting groups were removed by treatment with 20% piperidine in DMF (1x 3 min, 2x 5 min). The addition of pre-formed N^α,N^ϵ-di-Fmoc-L-Lys-OBt **11** (see below) was achieved as described above for the β-alanyl spacer. The reaction was again allowed to proceed at room temperature for 2.5 hr. The resin was washed with DMF after each cycle (5x 1 min). A Kaiser-ninhydrin test was performed after each Fmoc-deprotection sequence to ensure complete susbtitution (26). Fmoc-deprotection and subsequent scaffolding with activated L-lysyl derivative **11** were repeated until di-, tetra-, octa- and hexadeca-valent dendrimers were synthesized. The last sequence of each deprotected lysyl cores were then derivatized with chloroacetylglycylglycine residues using pre-formed benzotriazolyl ester derivative **14** as described above. The cleavage of each chloroacetylated dendrimer generation was accomplished by treating the resin with 95% TFA (see below). Compounds **19-22** were obtained in >90% yields. Due to the repetitive structures of the dendrimers, selected data only are reported. Compound **19**: ^{1}H NMR (DMSO-d_6) δ 1.16 (m, 2H, lysyl γ-CH$_2$), 1.33 (m, 2H, lysyl δ-CH$_2$), 1.46 and 1.60 (2m, 2H, lysyl β-CH$_2$, unequiv.), 2.36 (t, 2H, J=7 Hz, β-alanyl α-CH$_2$), 2.99 (m, 2H, lysyl ε-CH$_2$), 3.20 (m, 2H, β-alanyl β-CH$_2$), 3.65 and 3.74 (2d, 4H, J=5.8 Hz, glycyl CH$_2$'s), 3.77 (d, 4H, J=5.8 Hz, glycyl CH$_2$'s), 4.12 (d, 4H, J=2.6 Hz, ClCH$_2$), 4.13 (m, 1H, lysyl α-CH), 7.72 (t, 1H, J=5.6 Hz, lysyl ε-NH), 7.92 (m, 2H, β-alanyl NH and lysyl α-NH), 8.18 (m, 2H, glycyl NH), 8.45 (m, 2H, glycyl NH), 10.80 (bs, 1H, CO$_2$H); ^{13}C NMR (by HMQC) δ 22.7 (lysyl γ-C), 28.7 (lysyl δ-C), 31.7 (lysyl β-C), 33.8 (β-alanyl α-C), 34.3 (β-alanyl β-C), 38.7 (lysyl ε-C), 41.7 (4 x glycyl CH$_2$), 42.4 (ClCH$_2$), 52.5 (lysyl α-C); FAB-MS (pos.) calcd. for C$_{21}$H$_{33}$Cl$_2$N$_7$O$_9$ 598.4, found 598.3 (7.9% base peak). Compound **20**: FAB-MS (pos.) calcd. for C$_{45}$H$_{71}$Cl$_4$N$_{15}$O$_{17}$ 1236.0, found 1236.9 (M+1).

Synthesis of Fmoc-Amino Acid-OBt Active Esters. The appropriate Fmoc-amino acid (Novabiochem, Ca), for example Fmoc-β-Ala (0.99 g, 3.19 mmol) was dissolved in the minimum amount of DMF (3mL). N-Hydroxybenzotriazole (HOBt, 0.43 g, 3.19 mmol) also dissolved in DMF (8mL) was added to the Fmoc-amino acid solution that was cooled to 4°C (ice bath). Diisopropylcarbodiimide (497 μL, 3.19 mmol) was added dropwise to the cooled solution and the mixture was stirred at 4°C for 15 min. The solution was then left at room temperature with stirring for 1 hr and added directly to the Wang resin as required. The benzotriazolyl ester formation was monitored by TLC (CHCl$_3$:MeOH:HOAc / 85:10:5).

Coupling of α-Thiosialoside and Peptidic Dendrimer. The resin to which the chloroacetylated dendrimer cores were anchored was washed with DMF (5 x 1 min). The α-thiosialoside (**9**) (1.2 equiv./chloroacetyl group) dissolved in 1% Et$_3$N/DMF was added to the resin which was agitated by N$_2$ bubbling overnight (approx. 16 hr). The resin was drained and washed with DMF (5x 1 min).

Hydrolysis of Dendritic α-Thiosialosides from the Resin. The resin bound dendritic α-thiosialosides were treated with 5% trifluoroacetic acid (approximately 1 mL/10 mg) and the mixture was stirred at room temperature for 1.5 hr. The filtrate was collected and the resin beads rinsed 3 times with neat TFA. The combined filtrates were evaporated under vacuum and the last traces of TFA were co-evaporated with ether (2 times). The isolated dendrimers were obtained as white solids in yields ranging between 66 and 90%. All the products gave consistent NMR and mass spectral data. Compound **24**: peptide backbone, identical to **19** except for -SCH$_2$ at 3.37 and 3.47 (2d, 2 x 2H, J=7.1 Hz), NeuAc residues: 1.78 (dd, 2H, 2 x J≅12.1 Hz, H-3a), 1.65 (s, 6H, NAc), 1.91, 1.96, 1.99, 2.06 (4s, 24H, OAc), 2.64 (dd, 2H, J=4.6 and 12.5 Hz, H-3e), 3.72 (s, 6H, CO$_2$Me), 3.77 (m, 2H, H-6), 3.86 (m, 2H, H-5), 4.02 (dd, 2H, J-5.8 and 12.2 Hz, H-9), 4.16 (dd, 2H, J=3.1 and 12.2 Hz, H-9'), 4.70 (ddd, 2H, J=10.8 and 4.6 Hz, H-4), 5.12 (dd, 2H, J=2.0 and 8.3 Hz, H-7), 5.25 (m, 2H, H-8), 7.62-8.16 (m, 9H, NH's); ^{13}C NMR δ: 20.6 (OAc's), 22.6 (NAc), 22.8 (lysyl γ-C), 28.8 (lysyl δ-C), 31.7 (lysyl β-C), 32.1 (2 x SCH$_2$), 33.8 (β-alanyl α-C), 34.9 (β-alanyl β-C), 37.4 (C-3), 38.5 (lysyl ε-C), 42.0, 42.5, and 42.6 (glycyl CH$_2$'s), 47.8 (C-5), 52.5 (lysyl α-C), 53.1 (MeO), 61.9 (C-9), 67.1 (C-7), 67.9 (C-4), 69.5 (C-8), 73.7 (C-6), 82.4 (C-2), 167.7 to 172.8 (C=O). Compound **32**: consistent spectral data with compound **24** except in the ^{13}C-NMR (DMSO-d$_6$, 600 MHz), the β-alanyl residues became part of the background, ^{13}C-NMR : 20.7, 20.7, 20.9,21.0 (OAc), 22.7 (NAc), 23.0 (lysyl γ-C), 28.9 (lysyl δ-C), 31.9 (lysyl β-C), 32.2 (S-CH$_2$), 37.5 (C-3), 38.6 (lysyl ε-C), 42.1 and 42.6 (4x glycyl CH$_2$'s), 47.9 (C-5), 52.7 (lysyl α-C), 53.2 (MeO), 62.0 (C-9), 67.1 (C-7), 68.0 (C-4), 69.6 (C-8), 73.8 (C-6), 82.5 (C-2), 167.8 to 170.3 (C=O). HPLC analysis of compounds **24, 28, 32** and **37** showed purities >90% (% area) with optimized retention times being 5.10, 4.90, 5.00 and 4.90 min respectively.

De-O-Acetylation of Dendritic α-Thiosialosides. The dendritic α-thiosialoside (10 mg) was dissolved in MeOH (2 mL) and 10% 1 N NaOMe/MeOH was added. The solution was stirred at room temperature for 1 hr. The solution was then treated with H$^+$ resin (Amberlite IR-120) and the filtrate taken to dryness under vacuum. The residue was then freeze dried. The resulting white powder usually obtained in quantitative yields, gave consistent NMR and mass spectral data.
    Compound **25**: ^{1}H NMR (D$_2$O) δ peptide backbone 1.42 (m, 2H, lysyl γ-CH$_2$), 1.64 (m, 2H, lysyl δ-CH$_2$), 1.78 and 1.84 (2m, 2H, lysyl β-CH$_2$, unequiv.), 2.69 (m, 2H, β- alanyl α-CH$_2$), 3.34 (m, 2H, lysyl ε-CH$_2$), NeuAc-residues: 1.94 and 1.96 (2dd, 2H, J's 2x 12.1 Hz, H-3a), 2.10 (1s, 6H, NAc), 2.91 and 2.96 (2dd, 2H, J=12.6 and J=4.7 Hz, H-3 e), 3.50-4.33 (m, 35, peptide backbone, NeuAc- residues excluding above, 3.83 (1s, 6H, CO$_2$Me) and NH's and CO$_2$H).

Hydrolysis of the α-Thiosialoside Methyl Ester. A solution of the methyl ester of the dendritic α-thiosialoside (10 mg) dissolved in 0.05 M NaOH (2 mL) was stirred at room temperature for 2 hr. The solution was then treated with H$^+$ resin and the filtrate freeze dried. The products were obtained as white, lyophilized powder in quantitative yields and gave consistent spectral data.
    Compound **26**: ^{1}H-NMR (D$_2$O) δ: identical to compound **25** except as listed NeuAc-residues: 1.90 and 1.96 (2dd, 2H, J's 12.1 Hz, H-3a), 2.12 and 2.13 (2s, 6H,

NAc), 3.50-4.33 (m, 29H, peptide backbone, NeuAc-residues excluding above and NH's and CO_2H's); ^{13}C NMR (D_2O, by HMQC) δ: 18.8 (NAc), 19.2 (lysyl γ-C), 24.6 (lysyl δ-C), 27.4 (lysyl β-C), 29.5 (2 x SCH_2), 30.8 (β-alanyl α-C), 32.4 (β-alanyl β-C), 35.8 (lysyl ε-C), 36.7 and 37.1 (C-3 and C-3'), 39.4 (4 x glycyl CH_2's), 48.2 (C-5), 50.7 (lysyl α-C), 59.9 (C-9), 65.0 (C-7), 69.1 (C-4), 72.1 (C-8), 72.9 (C-6); FAB-MS (pos.) calc for $C_{43}H_{69}N_9O_{25}S_2$ 1176.2, found 1176.4 (% base peak 0.8).

Compound **34**: Consistent spectral data with compound **26** except NeuAc-residues at 2.09 (s, 24H, NAc).

9-O-Acetylation of Octameric α-Thiosialoside. The octavalent α-thiosialoside (5.3 mg) was dissolved in DMSO (0.5 mL) and trimethylorthoacetate (22.4 μL, 20 equiv.) was added. Three drops of TFA were added and the solution was stirred overnight at room temperature. The solution was dialyzed (dialysis tubing MW cut off 2000) against H_2O. The ester **35** was obtained as a lyophilized white powder (6.2 mg, ~ quant.); ^{1}H NMR (D_2O): identical to compound **34** except for the 9-OAc residues which appeared at 2.24 ppm (s, 24H) and H-9 multiplet at 4.38 ppm.

Enzyme Linked Lectin Assay (ELLA). Linbro (Titertek) microtitration plates were coated with 10 μg/well of poly(acrylamide-co-*p*-N-acrylamidophenylthio sialoside) polymer **3** at room temperature overnight. The wells were then blocked with 200 μL/well of 1% bovine serum albumin/ phosphate buffer saline (BSA/PBS) for 60 minutes. After washing 5 times (4 times with PBS-Tween and once with PBS), the wells were then filled with 100 μL/well of serial dilutions of wheat germ agglutinin/horseradish peroxidase (WGA/HRP) from 10^{-2} to 10^{-5} mg/mL in PBS and incubated at room temperature for three hours. The plates were washed with PBS-Tween (5x) and with PBS (2x). Then 50 μL/well of 2,2'-azinobis (3-ethylbenzothiazoline- 6-sulfonic acid) diammonium salt (ABTS) (1 mg/4 mL) in citrate-phosphate buffer (0.2 M, pH 4.0 with 0.015% H_2O_2) was added. After 15 minutes, the optical density (O.D.) was measured at 410 nm relative to 570 nm. Blank wells did not contain conjugates. The results are reported in Figure 1.

Inhibition experiments. The microtiter plates were coated overnight at room temperature with the copolymer **3** (10 μg/well). The plates were then washed and blocked with BSA as described above. Each of the following inhibitors were used as stock solutions of 1 mg/mL PBS: phenylthio α-sialoside (de-O-acetylated **10**) as reference monovalent compound, di-(**26**), tetra-(**30**), octa-(**34**) and hexadeca-(**39**) valent dendrimers. Each inhibitor were added in serial two fold dilutions (50 μL/well) in PBS with 50 μL of WGA/HRP (2.5 μg/mL) at 25°C for 60 min. The above solutions were then transferred to the copolymer-coated plates which were incubated for 2 hr at room temperature. The plates were washed as described above and the ABTS substrate was then added (50 μL/well). The O.D. was measured at 410 nm relative to that at 570 nm after 15 min. The pourcent inhibitions were calculated as follows: % Inhibition = $A_{(without\ inhibitor)}$ - $A_{(with\ inhibitor)}$ X 100/ $A_{(without\ inhibitor)}$.

Acknowledgments. We are thankful to Prof. R. Schauer and Dr. S. Kelm (Kiel, Germany) for generously providing us with the preliminary data on the inhibitions of hemagglutination of human erythrocytes by influenza virus. Support from the Natural Sciences and Engineering Research Council of Canada (NSERC) is also gratefully acknowledged.

Literature cited.

1. Corfielf, A.P.; Schauer, R. In *Sialic Acids,Chemistry, Metabolism and Funcion;* Schauer, R., Ed.; Cell Biology Monograph, Vol. 10; Springer-Verlag, New York, NY, **1982**, pp.5-50.
2. Wiley, D.C.; Skehel, J.J. *Annu. Rev. Biochem.* **1987**, *56*, 365-394.
3. Herrler, G.; Rott, R.; Klenk, H.-D.; Muller, H.-P.; Shukla, A.K.; Schauer, R. *EMBO J.* **1985**, *4*, 1503-1506.
4. Paulson, J.C.; Rogers, G.N.; Caroll, S.M.; Higa, H.H.; Pritchett, T.; Milks, G.; Sabesan, S. *Pure Appl. Chem.* **1984**, *56*, 797-805.
5. Pritchett, T.J.; Brossmer, R.; Rose, U.; Paulson, J.C. *Virology*, **1987**, *160*, 502-506.
6. Weiss, W.; Brown, J.H.; Cusack, S.; Paulson, J.C.; Skehel, J.J.; Wiley, D.C. *Nature*, **1988**, *333*, 426-421.
7. Sauter, N.K.; Bednarski, M.D.; Wurzburg, B.A.; Hanson, J.E.; Whitesides, G.M.; Skehel, J.J.; Wiley, D.C. *Biochemistry*, **1989**, *28*, 8388-8396.
8. Roy, R.; Laferrière, C.A.; Gamian, A.; Jennings, H.J. *J. Carbohydr. Chem.* **1987**, *6*, 161-165.
9. Roy, R.; Laferrière, C.A. *Carbohydr. Res.* **1988**, *177*, C1-C4.
10. Roy, R.; Andersson, F.O.; Harms, G.; Kelm, S.; Schauer, R. *Angew. Chem. Int. Ed. Engl.* **1992**, *31*, 1478-1481.
11. Roy, R.; Pon, R.A.; Tropper, F.D.; Andersson, F.O. *J.Chem. Soc., Chem. Commun.* **1993**, 264-265.
12. Gamian, A.; Chomik, M.; Laferrière, C.A.; Roy, R. *Can. J. Microbiol.* **1991**, *37*, 233-237.
13. Denkewalter, R.G.; Kolc, J.F.; Lukasavage, W.J. *US Patent* 4,410,688, **1983**; *Chem. Abstr.* **1984**, *100*, 103,907p.
14. Aharoni, S.M.; Crosby, C.R.,III; Walsh, E.K. *Macromolecules* **1982**, *15*, 1093-1098.
15. a) Tomalia, D.A.; Naylor, A. M.; Goddard III, W.A. *Angew. Chem. Int. Ed. Engl.* **1990**, *29*, 138-175; b) Newkome, G.R.; Moorefield, C.N.; Baker, G.R. *Aldrichimica Acta* **1992**, *25*, 31-38; c) Mekelburger, H.-B.; Jaworek, W.; Vögtle, F. *Angew. Chem. Int. Ed. Engl.* **1992**, *31*, 1571-1576.
16. Hawker, C.J.; Fréchet, J.M.J. *J. Am. Chem. Soc.* 1990, *112*, 7638-7647.
17. Tropper, F.D.; Andersson, F.O.; Cao, S.; Roy, R. *J. Carbohydr. Chem.* **1992**, *11*, 741-750.
18. Tropper, F.D.; Andersson, F.O.; Grand-Maître, C.; Roy, R. *Carbohydr. Res.* **1992**, *229*, 149-154.
19. Rothermel, J.; Faillard, H. *Biol. Chem. Hoppe-Seyler* **1989**, *370*, 1077-1084.
20. Roy, R.; Andersson, F.O.; Letellier, M. *Tetrahedron Lett.* **1992**, *33*, 6053-6056.
21. Hasegawa, A.; Nakamura, J.; Kiso, M. *J. Carbohydr. Chem.* **1986**, *5*, 11-19.
22. Tam. J.P. *Proc. Natl. Acad. Sci. USA* **1988**, *85*, 5409-5413.
23. Lu. G.; Mojsov, S.; Tam, J.P.; Merrifield, R.B. *J. Org. Chem.* **1981**, *46*, 3433-*3436*.
24. Carpino, L.A.; Han, G.Y. *J. Org. Chem.* **1972**, *37*, 3404-3409.
25. a) Kuhn, R.; Lutz, P.; McDonald, D.L. *Chem. Ber.* **1966**, *99*, 611-617; b) Roy, R.; Laferrière, C.A. *Can. J. Chem.* **1990**, *68*, 2045-2054.
26. Solid Phase Peptide Synthesis, A Practical Approach; Atherton, E.; Sheppard, R.C., eds.; IRL Press, Oxford University Press, England.

RECEIVED December 20, 1993

Chapter 8

Synthetic Acceptors
for α-L-Fucosyltransferases

**Khushi L. Matta, Conrad F. Piskorz, Gurijala V. Reddy,
E. V. Chandrasekaran, and Rakesh K. Jain**

**Department of Gynecologic Oncology, Roswell Park Cancer Institute,
Buffalo, NY 14263**

An elegant synthesis of Galß1→3(α-L-Fuc1→4)GlcNAcß-0-allyl and 3-0-SO$_3^-$Na$^+$.Galß1→3(α-L-Fuc1→4)GlcNAcß-0-allyl have been accomplished through use of a key glycosyl donor, namely, methyl 2,3,4-tri-0-(4-methoxybenzyl)-1-thio-ß-L-fucopyranoside. These compounds can be employed as unique oligosaccharide ligands for human selectin molecules.

The synthesis of 3-0-SO$_3^-$Na$^+$.Galß1→3/4GlcNAcß-0R as an acceptor for α-L-fucosyltransferase is also described. The expression of L-fucose containing carbohydrate structures in a majority of human cancers (*1-5*) provoked an enormous interest in the characterization of α-L-fucosyltransferase activities present in various tissues and cell lines. Such investigations have even led to the discovery of new types of α-L-fucosyltransferases based on their requirement of specific structures in the acceptor molecules for optimal activity (*1,6-17*). Likewise, the biosynthetic pathways of these fucosyl linked glycoconjugates have become a target of investigations in many laboratories. An aberration in α-L-fucosylation as well as α(2,3)-sialylation appears to be responsible for the synthesis of sialyl Le^x and sialyl Le^a type structures. Thus, there has been a surge of interest in the study of glycosyltransferases in normal and tumor tissues and various cell lines. Compared to the sialylated and fucosylated glycoconjugates reports on cancer-associated sulfoglycoconjugates are scanty (*18-24*). Nevertheless, these anionic glycoconjugates, especially the sulfated mucin type structures (*25-29*)

0097–6156/94/0560–0120$08.00/0

have attracted much recent attention. Interest in the sulfoglycoconjugates has further stemmed from the fact that these compounds containing sulfate, fucose and even sialic acid are being reported to be likely ligands for selectins, the family of cell adhesion molecules (*29-35*).

An examination of the structure of sulfated and sialylated oligosaccharides, especially those occurring in glycolipids (*36*), reveals that sulfate groups in most of these glycoconjugates are generally located where sialic acid occurs in the corresponding sialylated glycolipids and glycoproteins. We have observed that the specificity of α-L-fucosyltransferase (*17*) is similar for both sulfated and corresponding sialylated acceptors. Based upon these observations, we have synthesized allyl 0-(3-0-sulfo-β-D-galactopyranosyl)-(1→3)-2-acetamido-2-deoxy-β-D-glucopyranoside sodium salt (**13**), allyl 0-(β-D-galactopyranosyl)-(1→3)-[0-(α-L-fucopyranosyl)-(1→4)]-2-acetamido-2-deoxy-β-D-glucopyranoside (**17**), allyl 0-(3-0-sulfo-β-D-galactopyranosyl)-(1→3)-[0-(α-L-fucopyranosyl)-(1→4)]-2-acetamido-2-deoxy-β-D-glucopyranoside   sodium salt (**24**), benzyl   0-(3-0-sulfo-β-D-galactopyranosyl)-(1→3)-2-acetamido-2-deoxy-β-D-glucopyranoside sodium salt (**25**),   and   benzyl   0-(3-0-sulfo-β-D-galactopyranosyl)-(1→4)-2-acetamido-2-deoxy-β-D-glucopyranoside sodium salt (**26**).

Compounds **13**, **25** and **26** can be used as substrates for α-L-fucosyltransferase and **17** and **24** as ligands for selectin molecules.

The key intermediate **12** was prepared in six steps from   the   known   disaccharide   allyl   0-(β-D-galactopyranosyl)-(1→3)-2-acetamido-2-deoxy-β-D-glucopyranoside (*37*) **8**, which upon treatment with tert-butylchlorodiphenylsilane (*38*), followed by acetalation with 2,2-dimethoxypropane-acetone in the presence of 4-toluenesulfonic acid, then the removal of the tert-butyldiphenylsilyl group with IM tetrabutylammonium fluoride in oxolane (*38*), provided 3',4'-0-isopropylidene compound. Acetylation with pyridine-acetic anhydride, followed by cleavage of the 3',4'-0-isopropylidene group with chloroform-trifluoroacetic acid-water furnished the diol **11** in 71% yield. The diol **11** was converted into its 3,4-(ethyl orthoacetate), which was hydrolysed to give a key   intermediate,   allyl   0-(2,4,6-tri-0-acetyl-β-D-galactopyranosyl)-(1→3)-2-acetamido-4,6-di-0-acetyl-2-deoxy-β-D-glucopyranoside **12** in 75% yield. The ^{1}H-N.M.R. spectrum of **12** exhibited a low field chemical shift at $\delta 5.28$ (d, J = 3.7 Hz), confirming that diol **11** had been acetylated at O'-4 to give **12**.   Sulfation of **12** with sulfur   trioxide-pyridine   complex   followed   by   de-0-acetylation   with   methanolic   sodium   methoxide   gave compound **13**. A similar reaction sequence was employed for the synthesis of benzyl O-(2,4,6-tri-0-acetyl-ß-D-galactopyranosyl)-(1→3)-2-acetamido-4,6-di-0-acetyl-2-deoxy-ß-D-glucopyranoside, a key intermediate for the compound **25**, and benzyl O-(2,4,6-tri-0-acetyl-ß-D-galactopyranosyl)-(1→4)-2-acetamido-3,6-di-0-acetyl-2-deoxy-ß-D-glucopyranoside, a key intermediate for the compound **26**, as that described for the preparation of **13** from **8** (See Scheme 1). The structure of **13**, **25** and **26** was confirmed by ^{13}C-N.M.R. spectroscopy (see Table I) and FABMS (see Experimental Section).

Methyl 2,3,4-tri-0-benzyl-1-thio-ß-L-fucopyranoside, 2,3,4-tri-0-benzyl-ß-L-fucopyranosyl fluoride, and 2,3,4-tri-0-benzyl-α,ß-L-fucopyranosyl trichloroacetamidate (*39-43*) cannot be utilized for the synthesis of compound **17** and **24** since they require hydrogenolysis for removal of the benzyl protecting groups. Methyl 2,3,4-tri-0-(4-methoxybenzyl)-1-thio-ß-L-fucopyranoside **2** provides a more efficient route for α-L-fucosylation because removal of the protecting groups does not require hydrogenolysis. Compound **2** was obtained through alkylation of methyl-1-thio-ß-L-fucopyranoside **1** with 4-methoxybenzyl chloride-potassium hydroxide-18-crown-6 (*44*) in 68% yield. Allyl O-(2,3,4,6-tetra-0-acetyl-ß-D-galactopyranosyl)-(1→3)-2-acetamido-6-0-benzyl-2-deoxy-ß-D-glucopyranoside **7** was obtained by the reducing ring opening (*45*) of the 4,6-0-benzylidene acetal group of **6** in acidic medium, in the presence of sodium cyanoborohydride in 65% yield. Glycosylation under cupric bromide and tetrabutylammonium bromide (*46*) condition of compound **7** with **2** gave **14** in 68% yield after silica gel column chromatography. Removal of 6-0-benzyl group in **14** with 1% (v/v) of trimethylsilyl trifluoromethane sulfonate in acetic anhydride was not successful to give compound **16**. De-0-acetylation of compound **14** with methanolic sodium methoxide followed by the treatment with 1% (v/v) trimethylsilyl trifluoromethane sulfonate in acetic anhydride (*47*) gave fully acetylated trisaccharide **16** in 61% yield. De-0-acetylation with methanolic sodium methoxide provided compound **17** in 89% yield; the ^{13}C-N.M.R. spectrum was in accord with the structure assigned. Treatment of compound **4** with pivaloyl chloride in pyridine provided the 6-0-pivaloyl compound **5** in 90% yield; its [1]H-N.M.R. spectrum was in agreement with the structure proposed. A similar glycosylation of **5** with **2** afforded trisaccharide derivative **18** in 53% yield, the [1]H-N.M.R. spectrum of which showed a doublet at δ5.38 with spacing of 3.1 Hz, confirming the α-D-configuration of the new interglycosidic bond. Removal of the 4-methoxybenzyl group from compound **18** with ceric ammonium nitrate (*48*) in acetonitrile and followed by the treatment with triethylorthoacetate and de-0-acetylation with methanolic sodium methoxide afforded 3",4"-ethylorthoacetate derivative which on hydrolysis with 80% aqueous acetic acid gave the 4"-0-acetyl derivative (**20**) in 46% yield; the ^{1}H-N.M.R. spectrum of which showed a doublet at δ5.15, confirming that **20** had been acetylated at 0"-4. Isopropylidenation of compound **20** under Catelani et al. (*49*) procedure afforded 3',4'-0-isopropylidene derivative **21**. De-0-acetylation of compound **21** followed by acetylation with pyridine-acetic anhydride and hydrolysis with 60% aqueous acetic acid provided diol **22** in 55% yield. It was converted to key intermediate **23** by the orthoester procedure which on sulfation with sulfurtrioxide-pyridine complex and de-0-acetylation with methanolic sodium methoxide gave compound **24** as sodium salt after passing through IR 120 (Na$^+$) resin. The structure of **24** was confirmed by ^{13}C-N.M.R. and FAB mass spectroscopy (See Scheme 2; Table I).

^{13}C-N.M.R. assignments - The assignment of the ^{13}C-N.M.R. resonances for the three sulfated disaccharides **13**, **25** and **26** were made by comparing their spectra with that of compound benzyl 0-(ß-D-galactopyranosyl)-(1→3)-2-acetamido-2-deoxy-ß-D-glucopyranoside reported in Table I. In the ^{13}C-N.M.R. spectra of **13**, **25** and **26** a downfield shift of 7.55-7.78 p.p.m. was observed in the resonance for C-3', along with an upfield shift of 1.62-1.74 p.p.m. for C-4', in comparison to the spectrum of unsulfated disaccharide, confirming 0-3 as the site of sulfation in these compounds. In the spectrum of **24** a downfield shift of 7.9 p.p.m. was observed for the C-3' resonance in comparison to the spectrum of the parent compound **17** evidencing 0-3' as the site of sulfation.

Synthesis of allyl 0-(3-0-sulfo-ß-D-galactopyranosyl)-(1→3)-2-acetamido-2-deoxy-ß-D-glucopyranoside sodium salt (13). Glycosylation of allyl 2-acetamido-2-deoxy-4,6-0-(4-methoxybenzylidene)-ß-D-glucopyranoside with 2,3,4,6-tetra-0-acetyl-α-D-galactopyranosyl bromide, in the presence of mercury cyanide, gave compound **3** which after removal of acetal and acetyl groups afforded known allyl 0-(ß-D-galactopyranosyl)-(1→3)-2-acetamido-2-deoxy-ß-D-glucopyranoside (**8**) (*37*). Compound **8** was treated with tert-butylchlorodiphenylsilane (*38*) in N,N-dimethylformamide to give allyl 0-(6-0-tert-butyldiphenylsilyl-ß-D-galactopyranosyl)-(1→3)-2-acetamido-6-0-tert-butyldiphenylsilyl-2-deoxy-ß-D-glucopyranoside (**9**) in 68% yield; [α]$_D$ -13° (c 0.8, CHCl$_3$): ^{1}H-N.M.R. (CDCl$_3$): δ 7.69-7.25 (m, 20 H, arom.), 6.24 (d, J = 7.6 Hz, 1 H, NH), 5.95-5.85 (m, 1 H, = CH), 5.16 (d, J = 10.5 Hz, 1 H, H-1), 4.78 (d, J = 8.4 Hz, 1 H, H-1'), 2.16 (s, 3 H, NAc), 1.97 and 0.99 (each s, 18 H, 2 x CMe$_3$).

TABLE I

^{13}C-N.M.R. data[a,b] (Proposed Assignment)

Residue	Chemical Shifts							
	C-1	C-2	C-3	C-4	C-5	C-6	NAc	O-Allyl
Gal-β-(1→3)	102.41	70.45	71.44	67.45	74.21	59.96		
GlcNAc-β-OBn	98.65	53.53	81.35	67.68	74.36	59.75	21.15	
Compound 13								
3-0-SO$_3$NaGal-β-(1→3)	102.25	69.45	79.15	65.71	73.84	59.88		
GlcNAc-β-0-Allyl	98.81	53.35	81.86	67.80	74.36	59.75	21.21	132.33, 117.24
Compound 17								
Gal-β-(1→3)	101.82	69.49	71.32	66.77	73.80	60.00		
Fuc-α-(1→4)	97.03	67.33	68.12	70.92	65.81	14.33		
GlcNAc-β-OAllyl	98.93	54.67	75.09	74.44	71.43	58.75	21.24	132.34, 117.18
Compound 24								
3-0-SO$_3$NaGal-β-(1→3)	101.53	69.57	79.22	66.83	73.48	60.46		
Fuc-α-(1→4)	97.00	67.62	68.15	70.94	65.82	14.33		
GlcNAc-β-OAllyl	98.97	54.68	75.18	74.44	71.34	58.77	21.31	132.38, 117.23
Compound 25								
3-0-SO$_3$NaGal-β-(1→3)	102.26	70.82	79.22	65.83	73.93	59.98		
GlcNAc-β-OBn	98.80	53.34	81.77	67.92	74.50	59.88	21.31	
Compound 26								
3-0-SO$_3$NaGal-β-(1→4)	101.44	70.46	78.99	65.80	73.76	59.88		
GlcNAc-β-OBn	98.84	54.04	71.31	77.39	73.88	59.07	21.11	

[a]For solutions in D$_2$O with Me$_4$Si as the external standard; [b]All products described gave satisfactory elemental analysis.

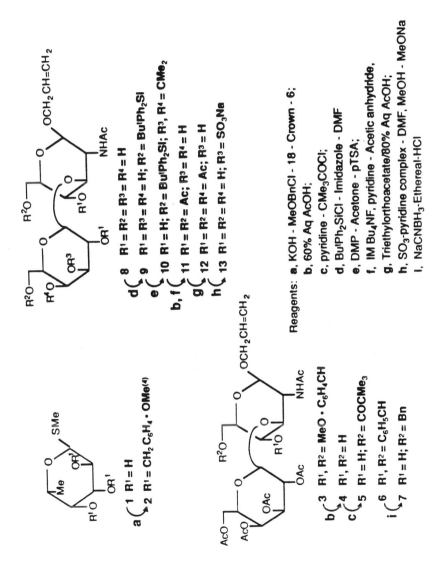

Reagents: **a**, KOH - MeOBnCl - 18 - Crown - 6;
b, 60% Aq AcOH;
c, pyridine - CMe₃COCl;
d, BuᵗPh₂SiCl - Imidazole - DMF
e, DMP - Acetone - pTSA;
f, IM Bu₄NF, pyridine - Acetic anhydride,
g, Triethylorthoacetate/80% Aq AcOH;
h, SO₃-pyridine complex - DMF, MeOH - MeONa
i, NaCNBH₃-Ethereal-HCl

Scheme 1

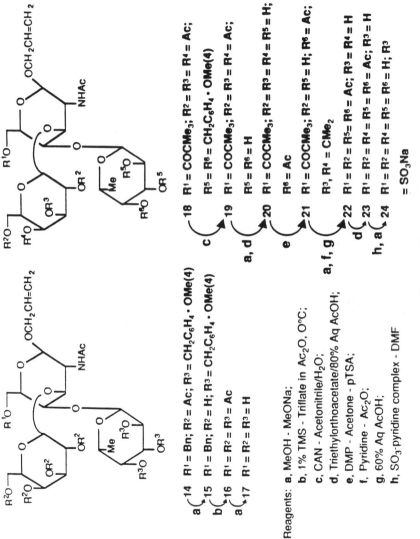

14 R¹ = Bn; R² = Ac; R³ = CH₂C₆H₄ · OMe(4)
15 R¹ = Bn; R² = H; R³ = CH₂C₆H₄ · OMe(4)
16 R¹ = R² = R³ = Ac
17 R¹ = R² = R³ = H

18 R¹ = COCMe₃; R² = R³ = R⁴ = Ac;
    R⁵ = R⁶ = CH₂C₆H₄ · OMe(4)
19 R¹ = COCMe₃; R² = R³ = R⁴ = Ac;
    R⁵ = R⁶ = H
20 R¹ = COCMe₃; R² = R³ = R⁴ = R⁵ = H;
    R⁶ = Ac
21 R¹ = COCMe₃; R² = R⁵ = H; R⁶ = Ac;
    R³, R⁴ = CMe₂
22 R¹ = R² = R⁵ = R⁶ = Ac; R³ = R⁴ = H
23 R¹ = R² = R⁴ = R⁵ = R⁶ = Ac; R³ = H
24 R¹ = R² = R⁴ = R⁵ = R⁶ = H; R³
    = SO₃Na

Reagents: a, MeOH - MeONa;
    b, 1% TMS - Triflate in Ac₂O, 0°C;
    c, CAN - Acetonitrile/H₂O;
    d, Triethylorthoacetate/80% Aq AcOH;
    e, DMP - Acetone - pTSA;
    f, Pyridine - Ac₂O;
    g, 60% Aq AcOH;
    h, SO₃-pyridine complex - DMF

Scheme 2

Isopropylidenation of compound **9** with 2,2-dimethoxypropane-acetone in the presence of 4-toluenesulfonic acid afforded a 3',4'-isopropylidene derivative **10** in 80% yield; $[\alpha]_D$ -0.5° ($\underline{c}$ 0.8, CHCl$_3$): ^{1}H-N.M.R. (CDCl$_3$): δ 7.70-7.25 (m, 20 H, arom.), 6.03 (d, J = 6.9 Hz, 1 H, NH), 5.95-5.89 (m, 1 H, =CH), 4.80 (d, J = 8.4 Hz, 1 H, H-1'), 2.16 (s, 3 H, NAc), 1.48 and 1.28 (each s, 6 H, CMe$_2$), 1.03 and 1.01 (each s, 18 H, 2 x CMe$_3$).

The removal of tert-butyldiphenylsilyl groups with fluoride ion (*38*), provided the 3',4'-isopropylidene compound in 90% yield; $[\alpha]_D$ -2° ($\underline{c}$ 1.3, CHCl$_3$). Acetylation of this compound with pyridine-acetic anhydride, followed by cleavage of the 3',4'-0-isopropylidene group with chloroform-trifluoroacetic acid-water, furnished the diol allyl 0-(2,6-di-0-acetyl-β-D-galactopyranosyl)-(1→3)-2-acetamido-4,6-di-0-acetyl-2-deoxy-β-D-glucopyranoside (**11**), in 71% yield; $[\alpha]_D$ -9° [$\underline{c}$ 0.7, CHCl$_3$:MeOH; 1.0:0.5 (v/v)]; ^{1}H-N.M.R. (CDCl$_3$ + CD$_3$OD): δ 5.87-5.85 (m, 1 H, =CH), 4.77 (d, J = 8.2 Hz, 1 H, H-1'), 2.18-1.99 (cluster of s, 15 H, 4 x OAc and NAc).

Compound **11** was converted into its 3',4'-(ethyl orthoacetate) which was hydrolyzed with 80% aqueous acetic acid to give the key 3'-hydroxy intermediate **12** in 75% yield; $[\alpha]_D$ +1° ($\underline{c}$ 1.0, CHCl$_3$); ^{1}H-N.M.R. (CDCl$_3$): δ 5.90-5.83 (m, 1 H, =CH), 5.80 (d, J = 7.3 Hz, 1 H, NH), 5.28 (d, J = 3.7 Hz, 1 H, H-4'), 4.82 (d, J = 8.1 Hz, 1 H, H-1'), 2.16-2.01 (cluster of s, 18 H, 5 x OAc and NAc).

Sulfation of compound **12** in N,N-dimethylformamide with sulfurtrioxide-pyridine complex at room temperature followed by de-0-acetylation with methanolic sodium methoxide afforded the title compound **13** in 80% yield; $[\alpha]_D$ -16° ($\underline{c}$ 0.7, H$_2$O); C-1 m/z 526.3 (M + 1)$^+$, 548.3 (M + Na)$^+$, 524.4 (M-1)$^-$; for ^{13}C-N.M.R., see Table I.

Allyl 0-(β-D-galactopyranosyl)-(1→3)-0-[0-(α-L-fucopyranosyl)-(1→4)]-2-acetamido-2-deoxy-β-D-glucopyranoside (17). — Condensation of allyl 2-acetamido-4,6-0-benzylidene-2-deoxy-β-D-glucopyranoside with 2,3,4,6-tetra-0-acetyl-α-D-galactopyranosyl bromide, in the presence of mercury cyanide followed by reductive ring opening (*45*) with sodium cyanoborohydride and ethereal-HCl provided allyl 0-(2,3,4,6-tetra-0-acetyl-β-D-galactopyranosyl)-(1→3)-2-acetamido-6-0-benzyl-2-deoxy-β-D-glucopyranoside **7** in 65% yield, $[\alpha]_D$ + 9° ($\underline{c}$ 1.2, CHCl$_3$); ^{1}H-N.M.R. (CDCl$_3$): 7.42-7.30 (m, 5 H, arom.), 5.90-5.69 (m, 1 H, =CH), 2.15-1.98 (cluster of s, 15 H,

4 x OAc and NAc); ^{13}C-N.M.R. (CDCl$_3$): 101.36 (C-1'),
98.09 (C-1), 83.36 (C-3), 66.96 (C-6), 61.54 (C-6').
Methyl 2,3,4-tri-0-(4-methoxybenzyl)-1-thio-β-L-
fucopyranoside (2) was prepared by the alkylation of 1
with 4-methoxybenzyl chloride-KOH-18-Crown-6 (44) in 68%
yield; [α]$_D$ -8° (c 1.0, CHCl$_3$): ^{1}H-N.M.R. (CDCl$_3$): δ 7.30
(d, 6 H, arom.), 6.80 (d, 6 H, arom.), 3.80 (s, 9 H, 3 x
OMe), 2.17 (s, 3 H, SMe), 1.13 (d, 3 H, CMe).

Reaction of 7 with 2 in the presence of cupric
bromide-tetrabutylammonium bromide (46) gave in 68% yield
the fully protected trisaccharide derivative 14; [α]$_D$ -
58° (c 1.4, CHCl$_3$); ^{1}H-N.M.R. (CDCl$_3$): δ 7.33-7.20 (m, 11
H, arom.), 6.89-6.78 (m, 6 H, arom.), 6.50 (d, J = 9.3
Hz, 1 H, NH), 5.87-5.83 (m, 1 H, =CH), 5.37 (d, J = 3.6
Hz, 1 H, H-1"), 4.81 (d, J = 7.9 Hz, 1 H, H-1'), 3.81,
3.79 and 3.76 (each s, 9 H, 3 x OMe), 2.04-1.67 (cluster
of s, 15 H, 4 x OAc and NAc), 1.14 (d, J = 6.5 Hz, 3 H,
CMe).

De-0-acetylation of compound 14 with methanolic
sodium methoxide gave 15 in 65% yield; [α]$_D$ -61° (c 1.2,
CHCl$_3$); ^{1}H-N.M.R. (CDCl$_3$): δ 7.38-7.21 (m, 11 H. arom.),
6.92-6.79 (m, 6 H, arom.), 3.80, 3.79 and 3.76 (each s,
9 H, 3 x OMe), 1.69 (s, 3 H, NAc), 1.09 (d, J = 5.9 Hz,
3 H, CMe).

Treatment of 15 in acetic anhydride-methylene
chloride (2:1 v/v) containing 1% (v/v) of trimethylsilyl
trifluoromethane sulfonate at 0°C gave allyl 0-(2,3,4,6-
tetra-0-acetyl-β-D-galactopyranosyl)-(1→3)-0-[0-(2,3,4-
tri-0-acetyl-α-L-fucopyranosyl)-(1→4)]-2-acetamido-6-0-
acetyl-2-deoxy-β-D-glucopyranoside 16 in 61% yield; [α]$_D$
-92° (c 1.2, CHCl$_3$); ^{1}H-N.M.R. (CDCl$_3$): δ 5.67 (d, J =
8.3 Hz, 1 H, NH), 5.40 (d, J = 3.4 Hz, 1 H, H-1"), 4.87
(d, J = 8.3 Hz, 1 H, H-1'), 2.16-1.96 (cluster of s, 27
H, 8 x OAc and NAc), 1.24 (d, J = 6.5 Hz, 3 H, CMe). De-
0-acetylation with methanolic sodium methoxide provided
title compound 17 in 89% yield; [α]$_D$ -82° (c 1.2, H$_2$O),
m/z: 570.2 (M + 1)$^+$, 592.2 (M + Na)$^+$, 568.4 (M-1)$^-$; For
^{13}C-N.M.R., see Table I.

Allyl 0-(3-0-sulfo-β-D-galactopyranosyl)-(1→3)-[0-
(α-L-fucopyranosyl)-(1→4)]-2-acetamido-2-deoxy-β-D-
glucopyranoside sodium salt (24). Reaction of allyl 0-
(2,3,4,6-tetra-0-acetyl-β-D-galactopyranosyl-(1→3)-2-
acetamido-2-deoxy-β-D-glucopyranoside (4) with
trimethylacetyl chloride (pivaloyl chloride) in pyridine
afforded the 6-0-pivaloyl compound 5 in 90% yield; [α]$_D$
+13° (c 1.0, CHCl$_3$); ^{1}H-N.M.R. (CDCl$_3$): δ 5.83-5.62 (m,
1 H, =CH), 4.53 (d, J = 8.5 Hz, 1 H, H-1'), 2.13-1.63
(cluster of s, 15 H, 4 x OAc and NAc), 1.19 (s, 9 H,
CMe$_3$).

Glycosylation of **5** with **2** under similar conditions as described for the preparation of **14** gave **18** in 53% yield, $[\alpha]_D$ -40° ($\underline{c}$ 1.0, $CHCl_3$); ^{1}H-N.M.R. ($CDCl_3$): δ 7.39-7.17 (m, 6 H, arom.), 6.78-6.69 (m, 6 H, arom.), 3.79 (s, 9 H, 3 x OMe), 2.13-1.67 (cluster of s, 15 H, 4 x OAc and NAc), 1.23 (s, 9 H, CMe_3), 1.07 (d, J = 6 Hz, 3 H, CMe). Removal of the 4-methoxybenzyl group from compound **18** with ceric ammonium nitrate in acetonitrile (*48*) provided compound **19** in 59% yield; $[\alpha]_D$ -71° ($\underline{c}$ 1.0, $CHCl_3$); ^{1}H-N.M.R. ($CDCl_3$): δ 6.60 (d, J = 8.5 Hz, 1 H, NH), 5.86-5.81 (m, 1 H, =CH), 5.38 (d, J = 3.1 Hz, 1 H, H-1"), 4.92 (d, J = 8.1 Hz, 1 H, H-1), 4.70 (d, J = 6.5 Hz, 1 H, H-1'), 2.15, 2.07, 2.05, 2.01 (each s, 8 H, 4 x OAc), 1.97 (s, 3 H, NAc), ^{13}C-N.M.R. ($CDCl_3$): δ 99.99 (C-1'), 98.7 (C-1), 96.85 (C-1").

The triol **19** was converted into its 3",4"-(ethyl orthoacetate) which on de-0-acetylation followed by hydrolysis with 80% aqueous acetic acid gave the intermediate, allyl 0-(ß-D-galactopyranosyl)-(1→3)-[0-(4-0-acetyl-α-L-fucopyranosyl)-(1→4)]-2-acetamido-2-deoxy-6-0-pivaloyl-ß-D-glucopyranoside (**20**) in 46% yield; $[\alpha]_D$ -74° ($\underline{c}$ 1.0, $CHCl_3$); ^{1}H-N.M.R. ($CDCl_3$ +CD_3OD): δ 5.90-5.80 (m, 1 H, =CH), 5.15 (d, 1 H, H-4"), 4.85 (d, J = 8.0 Hz, 1 H, H-1), 2.14 (s, 3 H, OAc), 1.98 (s, 3 H, NAc), 1.22 (s, 9 H, CMe_3), 1.04 (d, J = 6.6 Hz, 3 H, CMe); ^{13}C-N.M.R. ($CDCl_3$ + CD_3OD): δ 102.7 (C-1'), 99.4 (C-1), 97.5 (C-1"), 27.0 (CMe_3).

Isopropylidenation of compound **20** under conditions described by Catelani et al. (*49*) provided the 3',4' acetal compound **21** in 88% yield, $[\alpha]_D$ -44° ($\underline{c}$ 1.0, $CHCl_3$); ^{1}H-N.M.R. ($CDCl_3$): δ 5.90-5.70 (m, 1 H, =CH), 2.10 (s, 3 H, OAc), 1.99 (s, 3 H, NAc), 1.48 and 1.30 (each s, 6 H, CMe_2), 1.21 (s, 9 H, CMe_3), 1.06 (d, 3 H, CMe). De-0-acetylation of compound **21** followed by acetylation with pyridine-acetic anhydride and hydrolysis of isopropylidene with 60% aqueous acetic acid furnished the diol **22** in 55% yield; $[\alpha]_D$ -88° ($\underline{c}$ 0.3, $CHCl_3$). This was transformed into its 3',4' (ethyl orthoacetate) which was hydrolyzed with acetic acid to yield a key intermediate, allyl 0-(2,4,6-tri-0-acetyl-ß-D-galactopyranosyl)-(1→3)-[0-(2,3,4-tri-0-acetyl-α-L-fucopyranosyl)-(1→4)]-2-acetamido-6-0-acetyl-2-deoxy-ß-D-glucopyranose **23**; $[\alpha]_D$ -65° ($\underline{c}$ 0.4, $CHCl_3$). Sulfation of **23** with sulfurtrioxide-pyridine complex in N,N-dimethylformamide, followed by de-0-acetylation with methanolic sodium methoxide provided the title compound **24**; $[\alpha]_D$ -59° ($\underline{c}$ 0.6, H_2O); m/z; 694.1 (M + Na)$^+$, 648.3 (M-Na)$^-$; For ^{13}C-N.M.R., see Table I.

Synthesis of benzyl 0-(3-0-sulfo-ß-D-galactopyranosyl)-(1→3)-2-acetamido-2-deoxy-ß-D-glucopyranoside sodium salt (25). The procedures used for preparing **25** were essentially the same as described for compound **13**, except that benzyl 0-(ß-D-galactopyranosyl)-(1→3)-2-acetamido-2-deoxy-ß-D-glucopyranoside was used as the starting material which was synthesized using the published procedure (*50*). Compound **25** [α]$_D$ -12° (c 0.9, H$_2$O); m/z: 597.9 (M + Na)$^+$, 552.1 (M-Na)$^-$, 574.3 (M-1)$^-$; For ^{13}C-N.M.R., see Table I.

Benzyl 0-(3-0-sulfo-ß-D-galactopyranosyl-3)-(1→4)-2-acetamido-2-deoxy-ß-D-glucopyranoside sodium salt (26). Isopropylidenation of benzyl 0-(ß-D-galactopyranosyl)-(1→4)-2-acetamido-2-deoxy-ß-D-glucopyranoside (*51*) by the procedure of Catelani et al. (*49*) followed by acetylation with pyridine-acetic anhydride afforded benzyl 0-(2,6-di-0-acetyl-3,4-0-isopropylidene-ß-D-galactopyranosyl)-(1→4)-2-acetamido-3,6-di-0-acetyl-2-deoxy-ß-D-glucopyranoside in 50% yield; [α]$_D$ -17° (c 0.8, CHCl$_3$). Cleavage of the 3',4'-0-isopropylidene group with chloroform-trifluoroacetic acid-water, furnished a diol in 92% yield, [α]$_D$ -33° (c 0.7, CHCl$_3$). This diol was converted into a 3,4-(ethyl orthoacetate) derivative, which was hydrolyzed to give the intermediate, benzyl 0-(2,4,6-tri-0-acetyl-ß-D-galactopyranosyl)-(1→4)-2-acetamido-3,6-di-0-acetyl-2-deoxy-ß-D-glucopyranoside in 81% yield; [α]$_D$ -40° (c 0.7, CHCl$_3$). This compound on treatment with sulfur trioxide-pyridine complex in N,N-dimethylformamide followed by 0-deacetylation with methanolic sodium methoxide provided the title compound **26** in 55% yield, [α]$_D$ -15° (c 0.7, H$_2$O), m/z: 598.0 (M + Na)$^+$, 552 (M - Na)$^-$, 574 (M-1)$^-$; For ^{13}C-N.M.R., see Table I.

Acknowledgment: These investigations were supported by grant Nos. CA35329 and CA16056 awarded by the National Cancer Institute.

Literature Cited

1. Hakomori, S. *Advances in Cancer Res.* **1989**, *52*, 257-331.
2. Fukuda, M. *Biochim. Biophys. Acta.* **1985**, *780*, 119-150.
3. Lloyd, K.O.; Old, L.J. *Cancer Res.* **1989**, *49*, 3445-3451.
4. Stults, C.L.M.; Sweeley, C.C; Macher, B.A. *Methods in Enzymology* **1989**, *179[14]*, 167-214.
5. Sell, S. *Human Pathology* **1990**, *21*, 1003-1019.
6. Hansen, G.C.; Zopf, D. *J. Biol. Chem.* **1985**, *260*, 9388-9392.
7. Hanisch, F.G.; Uhlenbruck, G.; Peter-Katalinic, J.; Egge, H. *Carbohydr. Res.* **1988**, *178*, 29-47.
8. Holmes, E.H.; Hakomori, S.; Ostrander, G.K. *J. Biol. Chem.* **1987**, *262*, 15649-15658.

9. Hanisch, F.G.; Mitsakos, A.; Schroten, H.; Uhlenbruck, G. *Carbohydr. Res.* **1988**, *178,* 23-28.

10. Mollicone, R.; Candelier, J.J.; Mennesson, B.; Couillin, P.; Venot, A.P.; Oriol, R. *Carbohydr. Res.* **1992**, *228,* 265-276.

11. Howard, D.R.; Fukuda, M.; Fukuda, M.N.; Stanley, P. *J. Biol. Chem.* **1987**, *262,* 16830-16837.

12. Foster, C.S.; Gillies, D.R.B.; Glick, M.C. *J. Biol. Chem.* **1991**, *266,* 3526-3531.

13. Madiyalakan, R.; Yazawa, S.; Matta, K.L. *Ind. J. of Biochem. Biophys.* **1988**, *25,* 32-35.

14. Weston, B.W.; Nair, R.P.; Larsen, R.D.; Lowe, J.B. *J. Biol. Chem.* **1992**, *267,* 4152-4160.

15. Goetz, S.E.; Hession, C.; Goff, D.; Griffiths, B.; Tizard, R.; Newman, B.; Chi-Rosso, G.; Lobb, R. *Cell* **1990**, *63,* 1349-1356.

16. Basu, M.; Hawes, J.W.; Li, Z.; Ghosh, S.; Khan, F.A.; Zhang, B.; Basu, S. *Glycobiology* **1991**, *1,* 527-535.

17. Chandrasekaran, E.V.; Jain, R.K.; Matta, K.L. *J. Biol. Chem.* **1992**, *267,* 23806-23814.

18. Siddiqui, B.; Whitehead, J.S.; Kim, Y.S. *J. Biol. Chem.* **1978**, *253,* 2168-2175.

19. Gasa, S.; Makita, A.; Hirama, M.; Kawabata, M. *J. Biochem.* **1979**, *86,* 265-267.

20. Yoda, Y.; Gasa, S.; Makita, A.; Fujioka, Y.; Kikuchi, Y.; Hashimoto, M. *J. Natl. Cancer Inst.* **1979**, *63,* 1153-1160.

21. Hattori, H.; Uemura, K.; Taketomi, T. *Biochem. Biophys. Acta.* **1981**, *666,* 361-369.

22. Mutsuyama, T.; Gasa, S.; Taniguchi, N.; Makita, A.; Miyasaka, S.; Matsumura, M.; Tsukada, M.; Ishikura, M. *J. Exp. Clin. Cancer Res.* **1983**, *2,* 25-30.

23. Hiraiwa, N.; Iida, N.; Ishizuka, Itai, S.; Shigeta, K.; Kannagi, R.; Fukuda, Y.; Imura, H. *Cancer Res.* **1988**, *48,* 6769-6774.

24. Hiraiwa, N.; Fukuda, Y.; Imura, H.; Tadano, A.K.; Nagai, K.; Ishizuka, I.; Kannagi, R. *Cancer Res.* **1990**, *50,* 2917-2928.

25. Mawhinney, T.P.; Landrum, D.C.; Gayer, D.A.; Barbero, G.J. *Carbohydr. Res.* **1992**, *235,* 179-197.

26. Slomiany, B.L.; Piotrowski, J.; Nishikawa, H.; Slomiany, A. *Biochem. Biophys. Res. Commun.* **1988**, *157,* 61-67.

27. Capon, C.; Laboisse, C.L.; Wieruszeski, J.M.; Maoret, J.J.; Augeron, C.; Fournet, B. *J. Biol. Chem.* **1992**, *267,* 19248-19257 .

28. Irmura, T.; Matsushita, Y.; Yamori, T.; Hoff, S.D.; McIsaac, A.; Sutton, R.; Cleary, K.R.; Ota, D.M. *XV International Carbohydrate Symposium,* Aug. 12-17, **1990**, Yokohama, Japan.

29. Yuen, C.; Lawson, A.M.; Chai, W.; Larkin, M.; Stoll, M.S.; Stuart, A.C.; Sullivan, F.X.; Ahern, T.J.; Feizi, T. *Biochemistry* **1992**, *31,* 9126-9131.

30. Imai, Y.; Lasky, L.A.; Rosen, S.D. *Nature* **1993**, *361,* 555.
31. Tyrrell, D.; James, P.; Rao, N.; Foxall, C.; Abbas, S.; Dasgupta, F.; Nashed, M.; Hasegawa, A.; Kiso, M.; Asa, D.; Kidd, J.; Brandley, B.K. *Proc. Natl. Acad. Sci. USA* **1991**, *88,* 10372-10376.
32. Tiemeyer, M.; Swiedler, S.J.; Ishihara, M.; Moreland, M.; Schweingruber, H.; Hirtzer, P.; Brandley, B.K. *Proc. Natl. Acad. Sci. USA* **1991**, *88,* 1138-1142.
33. Phillips, M.L.; Nudelman, E.; Gaeta, F.C.A.; Perez, M.; Singhal, A.K.; Hakomori, S.I.; Paulson, J.C. *Science* **1990**, *250,* 1130-1132.
34. Walz, G.; Aruffo, A.; Kolanus, W.; Bevilacqua, M.; Seed, B. *Science* **1990**, *250,* 1132-1135.
35. Suzuki, Y.; Toda, Y.; Tamatani, T.; Watanabe, T.; Suzuki, T.; Nakao, T.; Murase, K.; Kiso, M.; Hasegawa, A.; Tadano-Aritomi, K.; Ishizuka, I.; Miyasaka, M. *Biochem. Biophys. Res. Commun.* **1993**, *190,* 426-434.
36. Nagai, K.; Roberts, D.D.; Toida, T.; Matsumoto, H.; Kushi, Y.; Handa, S., Ishizuka, I. *J. Biol. Chem.* **1989**, *264,* 16220-16239.
37. Diakun, K.R.; Yazawa, S.; Valenzuela, L.; Abbas, S.A.; Matta, K.L. *Immunol. Invest.* **1987**, *16,* 151-163.
38. Hanessian, S.; Lavallee, P. *Can. J. Chem.* **1975**, *53,* 2975-2977.
39. Lemieux, R.U.; Bundle, D.R.; Baker, D.A. *J. Am. Chem. Soc.* **1975**, *97,* 4076-4083.
40. Dejter-Juszynski, M.; Flowers, H.M. *Carbohydr. Res.* **1971**, *18,* 219-226.
41. Sato, S.; Ito, Y.; Nukada, T.; Nakahara, Y.; Ogawa, T. *Carbohydr. Res.* **1987**, *167,* 197-210.
42. Jain, R.K.; Matta, K.L. *Carbohydr. Res.* **1990**, *208,* 51-58.
43. Wegmann, B.; Schmidt, R.R. *Carbohydr. Res.* **1988**, *184,* 254-261.
44. Bessodes, M.; Shamsazar, J.; Antonakis, K. *Synthesis* **1988**, 560-562.
45. Garegg, P.J.; Hultberg, H.; Wallin, S. *Carbohydr. Res.* **1971**, *18,* 219-226.
46. Sato, S.; Mori, M.; Eto, Y.; Ogawa, T. *Carbohydr. Res.* **1986**, *155,* C6-C10.
47. Angibeaud, P.; Utille, Jean-P.S. *Synthesis,* **1991**, 737-738.
48. Johansson, R.; Samuelson, B. *J. Chem. Soc., Chem. Commun.* **1984**, 201-202.
49. Catelani, G.; Colonna, F.; Marra, A. *Carbohydr. Res.* **1988**, *182,* 297-300.
50. Lubineau, A.; Auge, C.; Bouxon, G.; Gautheron, C. *J. Carbohydr. Chem.* **1992**, *11,* 59-70.
51. Rana, S.S.; Matta, K.L. *Carbohydr. Res.* **1983**, *113,* C18-C21.

RECEIVED April 27, 1994

Chapter 9

Oligosaccharide–Polyacrylamide Conjugates of Immunological Interest

Synthesis and Applications

A. Ya. Chernyak[1]

N. D. Zelinsky Institute of Organic Chemistry, Russian Academy
of Sciences, 117913 Moscow, Russia

Well-defined saccharide-polyacrylamide (PAA) conjugates containing the
oligosaccharides (OS) which carry biological information and which are
devoid of any other serologically active components are important
antigenic tools for both immunochemical studies and diagnostic purposes.
A general approach to the preparation of OS-PAA conjugates *via* co-
polymerization of unsaturated ligands (synthetic allyl or acrylamidoalkyl
glycosides) or reducing saccharides isolated from natural sources with
acrylamide has been developed. An alternative approach to OS-PAA
conjugates by nucleophilic displacement of nitrophenyloxy groups in poly-
(4-nitrophenylacrylate) with aminoalkyl glycosides is also reviewed. To
illustrate the first synthetic route, the preparation of OS-PAA conjugates of
synthetic fragments of *Salmonella* and *Proteus* O-antigens and capsular
polysaccharides of *Streptococcus pneumoniae* and *Escherichia coli* is
discussed. The OS-PAA conjugates obtained by both methods are efficient
as coating antigens in EIA assays and were used in immunochemical
studies and diagnosis of infectious diseases (*e.g.* salmonellosis).

The biological roles of the sugar (oligo- and polysaccharide) chains of natural
glycoconjugates range from trivial to crucial for the development, growth, and function
of complex organisms, and their interactions with other organisms in the environment
(*1*). Neoglycoconjugates (NGCs), comprising the oligosaccharides under consideration
here as well as their synthetic analogs, have been used extensively to explore the
biological roles of natural glycoconjugates (*2,3*).

Among neoglycoconjugates, neoglycoproteins, which are semi-synthetic
compounds prepared by covalent attachment of saccharides to a natural protein, are very
popular as research tools (*4*).The protein serves as a carrier which [according to K.
Landsteiner (*5*)] provides the whole molecule with a minimum size required for
displaying antigenic and immunogenic properties. However, proteins themselves can
possess antigenic properties and therefore are not always ideal as carrier molecules for
all the purposes neoglycoconjugates should meet. Non-antigenic and non-immunogenic
synthetic polymeric carriers could be a viable alternative. These compounds could give

[1]Current address: Department of Immunology, Microbiology, Pathology, and Infectious
Diseases, Karolinska Institute, Huddinge Hospital, S–141 86 Huddinge, Sweden

the synthetic chemist greater freedom in designing neoglycoconjugates, the so-called glycopolymers (6), which could have new potentially useful properties.

General Design of Glycopolymers

Two types of glycopolymers derived mainly from accessible mono- and disaccharides have been described: (i) branched molecules composed of an unnatural polymeric backbone with saccharides laterally attached [these comb-like glycopolymers are also called "pseudopolysaccharides" (7)], (ii) linear polymers built up of saccharide units linked in a repetitive manner via unnatural tethers [e.g. (8)] (Figure 1). Glycopolymers of the first type as well as neoglycoproteins resemble natural glycoproteins whereas glycopolymers of the second type are similar in design to natural glycoconjugates of regular repetitive character (bacterial O-antigens, capsular polysaccharides etc.). However, there are very few examples of glycopolymers of the type (ii) described (8) and to the best of our knowledge, they have never been seriously used in immuno- and other assays. So, the comb-like glycopolymers as well as neoglycoproteins mimic both natural glycoproteins and structurally unrelated macromolecular glycoconjugates of non-glycoprotein nature (e.g. bacterial lipo- or capsular polysaccharides).

The differences in design should be taken into account in the latter case. The saccharide chains of bacterial polysaccharides are composed of oligosaccharide repeating units (9) whereas the sugar moiety of glycopolymers is represented by shorter oligosaccharides as branches. Loss of certain structural elements of the natural glycoconjugate may result in attenuation of the immunochemical properties. For example, the internal residues represented by A^1 which are typical for the natural structure (Figure 2) are exposed as immunochemically important (10) terminal residues A enhanced in number in the synthetic glycopolymer as it will be exemplified in the last Section. As a consequence, a problem arises how to choose in a natural glycoconjugate the oligosaccharide sequence best suited for the preparation of the properly functioning glycopolymer.

Synthetic Approaches to Glycopolymers

Synthesis of linear glycopolymers with saccharide units incorporated in the main chain of unnatural polymer is practically uncovered field. From the few examples available, it is worth mentioning the addition polymerisation of unprotected saccharides with ω,ω'-alkylidenediisocyanates (8).

Two general approaches are used for the preparation of comb-like glycopolymers (Figure 3): (i) modification of pre-fabricated polymers with saccharides and (ii) polymerisation or co-polymerisation of double-bond-containing saccharides resulting in de novo formation of a macromolecular carrier. For conjugation of saccharides to polymeric carriers conventional organic reactions (more various in type than those used for the preparation of neoglycoproteins) can be applied e.g. hydrazone formation from polyacrylic hydrazide and reducing oligosaccharides (11) or nucleophilic ring-opening in a protected 6-O-epoxypropyl-sugar with polyvinyl alcohol (7).

Polymerisation seems to be a more general approach to the preparation of glyco-polymers. Unsaturated monoester or monoether derivatives of saccharides are generally used as monomers for polymerisation. If a double-bond-containing substituent occupies any other position than the anomeric one, polymerisation affords a glycopolymer with pendant reducing sugars (12, 13). However, homopolymerisation leads to glycopolymers with a very dense arrangement of saccharide chains located at each monomer unit (14). Co-polymerisation of a carbohydrate monomer with an unsaturated non-carbohydrate monomer seems to be a more flexible approach to glycopolymers with potentials to modify the nature of the polymeric carrier and to control to some extent the number of saccharide chains incorporated.

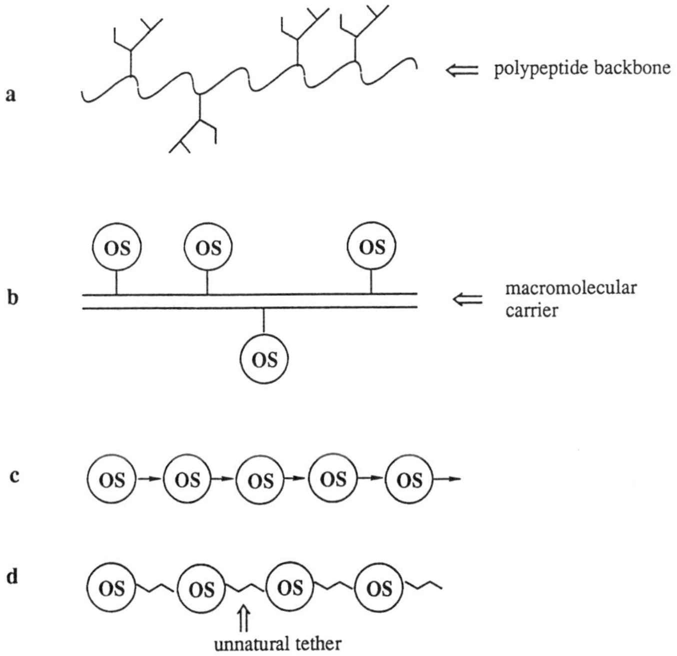

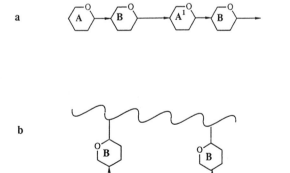

Figure 1. General design of natural glycoconjugates (a - glycoproteins, c - polysaccharides of regular structure) and glycopolymers (b - comb-like glycopolymers, d - glycopolymers with saccharides incorporated into the linear chain).

Figure 2. Immunochemically important difference between bacterial polysaccharides (a) and emulating glycopolymers (b).

Since the first publication of a synthesis of monosaccharide-polyacrylamide polymers manifesting lectin-precipitating properties (*15*), polyacrylamide has almost exclusively been used as a macromolecular carrier for glycopolymers of biological significance. During the last decade several groups in Austria (*16*), Canada (*6*), on Cuba (*17*), in Germany (*18*), Russia (*19, 20*), Sweden (*21*), and the US (*22*) have progressed with the synthesis of oligosaccharide-polyacrylamide (OS-PAA) conjugates which exhibit serological specificity of various bacterial carbohydrate antigens and blood group-specific glycoproteins.

Synthesis of Oligosaccharide-Polyacrylamide Conjugates

Both general approaches to glycopolymers discussed above have been applied to the synthesis of OS-PAA conjugates (Figure 4). Chronologically co-polymerisation (*23, 24*) preceded a recently elaborated procedure of coupling of ω-amino-spacer-containing oligosaccharides to an activated polyacrylic acid (*20*). The first OS-PAA conjugates have been prepared *via* co-polymerisation (*15, 23, 24*). These water-soluble co-polymers, which have advantageous physicochemical and structural properties rapidly gained importance as well-defined, polyvalent glycopolymeric molecules in bio- and immunoassays. It became a stimulus for elaboration of the alternative approach to OS-PAA conjugates (*20*).

Coupling of Saccharides to Polyacrylate Carrier. This approach can be regarded as a traditional one in the sense that it makes use of a ready-made carrier which has been utilized in the synthesis of neoglycoproteins, the technology used for decades (*4*). The polymeric chain of poly(4-nitrophenylacrylate), accessible by AIBN-induced polymerisation of the corresponding monomer is transformed into the polyacrylamide backbone by displacement of the reactive groups with ω-aminoalkyl glycosides, followed by conversion of unreacted ester groups to amides with ammonia (*25*) (Figure 4b). For example, 3-aminopropyl glycosides of synthetic oligosaccharides of blood group specific glycoproteins or glycylaminobenzyl glycosides of sialic acids have been used to prepare artificial blood group substance antigens or polyvalent sialosides (*26*).

Simultaneous or sequential introduction of aminated sugar haptens or non-sugar amino ligands opens new prospects in design of glycopolymers with multiple functionalities. Thus, simultaneous treatment of poly(4-nitrophenylacrylate) with the aminopropyl glycoside of A-disaccharide (Figure 4b, OS = Galβ1-3GalNAcα1-) and (*N*-biotinyl)hexamethylenediamine (Figure 4b, R = -(CH$_2$)$_6$NH-biotin) afforded a biotin-labelled probe for detection and analysis of lectins (*27*). Displacement of reactive *p*-nitrophenyloxy groups in the activated polymeric carrier by the glycoside of the core Le^x trisaccharide (Figure 4b, OS = Galβ1-4[Fucα1-4]GlcNAcβ1-), in combination with phosphatidylethanolamine led to a polymeric polyvalent neoglycolipid resembling components of biological membranes (*25*).

Co-polymerisation of Sugar Monomers with Acrylamide. In the first publication of a synthesis of OS-PAA conjugates, monosaccharidic (D-Man, D-Gal, L-Fuc, D-GalNAc) allyl glycosides have been used in a radical co-polymerisation with acrylamide (*15*). The resulting glycopolymers were found to possess specific lectin-precipitating properties. Application of this technology to allyl or allyloxyalkyl glycosides of complex saccharides that impart specificity of bacterial O-antigens (*28*) or blood group substance (*24*) led to glycopolymers with antigenic properties, which were evident from immunoassays.

Radical co-polymerisation of alkenyl glycosides with acrylamide in aqueous solution can be induced by UV-light or initiated by promoters such as ammonium

reactive groups

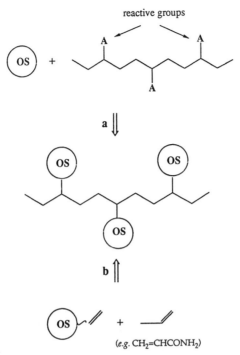

Figure 3. Approaches to the synthesis of glycopolymers (a - modification of a polymeric carrier with saccharides, b - polymerisation or co-polymerisation of saccharide monomers).

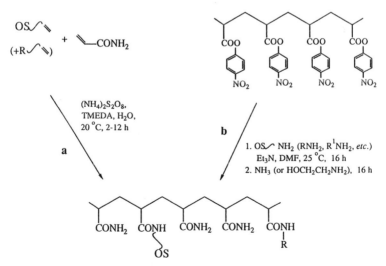

Figure 4. Synthesis of OS-PAA conjugates (a - co-polymerisation of saccharide monomer with acrylamide, b - modification of activated polyacrylic acid).

persulfate (0.1% w/v) and N,N,N',N'- tetramethylethylenediamine (TMEDA, 0.1-0.2% v/v) at 20-40 °C. The OS-PAA conjugates isolated by gel-filtration or precipitation with methanol from aqueous solution have molecular masses of 30-100 kD as estimated by analytical ultracentrifugation, ultrafiltration, or gel-permeation chromatography. The structure of the OS-PAA conjugates with a regular poly-acrylamide backbone and randomly distributed saccharide branches can be easily confirmed from ^{13}C-NMR spectra which match those of parent carbohydrate monomers (signals of the sugar moiety). The ratio of sugar to non-sugar monomer in the final co-polymer can be calculated from the relative intensities of signals attributed to the saccharide units and the methine or methylene groups of the polyacrylamide backbone (29). Similar estimates follow from optical rotation mesurements because it is only the carbohydrate moiety in the polymeric conjugate that possesses optical activity (19). Thus, the OS-PAA conjugates can be considered as structurally well defined glycopolymers for immunochemical studies and other biological applications.

Several sugar monomers can be used at a time in co-polymerisation to prepare multi-specific conjugates (19), or any useful non-sugar ligands can be used. They can be also introduced after post-modification of the PAA carrier. Partial transamidation of amide groups with ethylenediamine in PAA conjugates derived from allyl glycosides afforded primary amino functions which were used for inserting lipophilic residues (30).

Polymerizable Aglycons and Their Precursors. Allyl (or, more generally, alkenyl) glycosides as co-monomers in co-polymerisation with acrylamide have a disadvantage. The reactivity of the double bond in these glycosides is lower than that of acrylamide. As a consequence, an empirically found excess of the sugar monomer should be used to achieve a desired density of saccharides in the OS-PAA conjugate. Therefore in the synthesis of sugar monomers, allyl (alkenyl) glycosides have been in a logical way replaced by glycosides with suitably reactive acrylamide-containing aglycons. The latter are accessible *via* acryloylation from precursors with a temporary protected terminal amino group (31-34) which can also be masked as a nitro (35) or azido group (36) (Scheme 1). For the generation of the amino function in the precursors, conventional reactions like basic hydrolysis for N-trifluoroacetates (31), or hydrogenolysis for N-benzyloxycarbonyl derivatives (34), are used. Hydrogenation smoothly transforms azido and nitro precursors into amino-containing glycosides.

In our syntheses of the OS-PAA conjugates we have used the azidoethyl aglycon as a precursor of the aminoethyl aglycon. The latter can be subjected to N-acryloylation directly or after elongation by reaction with protected glycine or glycine oligomers (37) (Scheme 1). This elongation procedure makes it possible to transform the aminoethyl "prespacer" into a series of non-hydrophobic spacers of increasing length. Azidoethanol, which is readily available from inexpensive 2-chloroethanol was used for the preparation of azidoethyl glycosides (37). Being a liquid, azidoethanol is easy to handle like other low molecular alcohols. All conventional glycosidation procedures can be utilized for the preparation of azidoethyl glycosides. Helferich glycosidation performed in hot azidoethanol as a solvent affords rapidly β-glycosides whereas boron trifluoride assisted glycosidation does not require a preliminary preparation of glycosyl halides (37). Fischer glycosidation catalysed by trifluoromethanesulfonic acid has been used as the simplest procedure for the preparation of azidoethyl α-glycosides of uronic acids (38). 2-Azidoethyl α-glycosides are also accessible from 2-pyridyl 1-thioglycosides (39). The preparation of azidoethyl glycosides of amino sugars *via* oxazolines has been recently reported (40). An alternative approach to azidoethyl glycosides is by displacement of a halogen atom in 2-haloethyl glycosides by azide using tetrabutylammonium azide (37), which is soluble in most organic solvents.

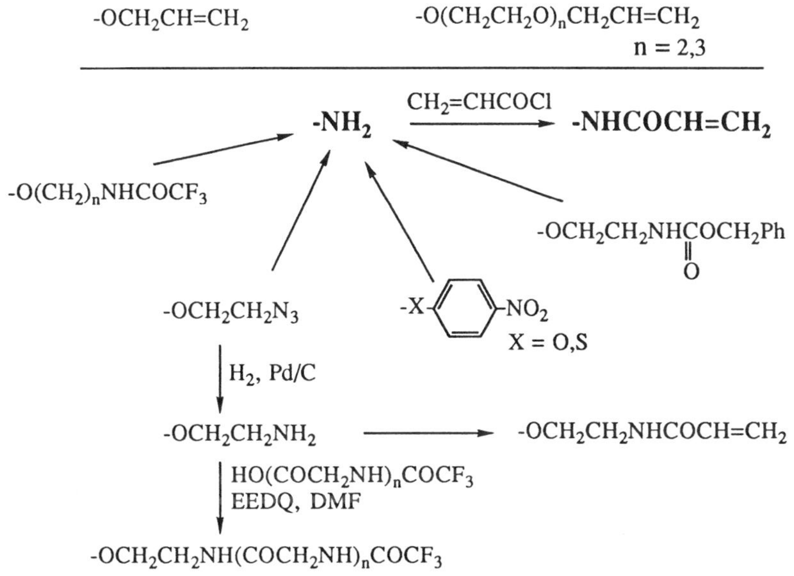

$-OCH_2CH=CH_2$ $-O(CH_2CH_2O)_nCH_2CH=CH_2$
$n = 2,3$

$-NH_2$ $\xrightarrow{\text{CH}_2=\text{CHCOCl}}$ $-NHCOCH=CH_2$

$-O(CH_2)_nNHCOCF_3$

$-OCH_2CH_2NHCOCH_2Ph$
$\overset{\|}{O}$

$-OCH_2CH_2N_3$ $-X-\!\!\!\left\langle\bigcirc\right\rangle\!\!\!-NO_2$
$X = O,S$

$\downarrow$ H_2, Pd/C

$-OCH_2CH_2NH_2$ $\longrightarrow$ $-OCH_2CH_2NHCOCH=CH_2$

$\downarrow$ $HO(COCH_2NH)_nCOCF_3$
EEDQ, DMF

$-OCH_2CH_2NH(COCH_2NH)_nCOCF_3$

SCHEME 1

The allyl aglycon can be transformed into the more reactive acrylamido type (Scheme 2). UV-light induced addition of cysteamine to the allyl group described by Lee & Lee (*41*) increases also the length of the spacer. After *N*-acryloylation this modified spacer is suitable for co-polymerisation with acrylamide and numerous OS-PAA conjugates have been   obtained using this procedure from allyl glycosides of synthetic saccharides containing sialic acids (*42*), 3-deoxy-D-manno-octulosonic acid (KDO) (*16*), or L-glycero-D-manno-heptose in combination with KDO (*18*).

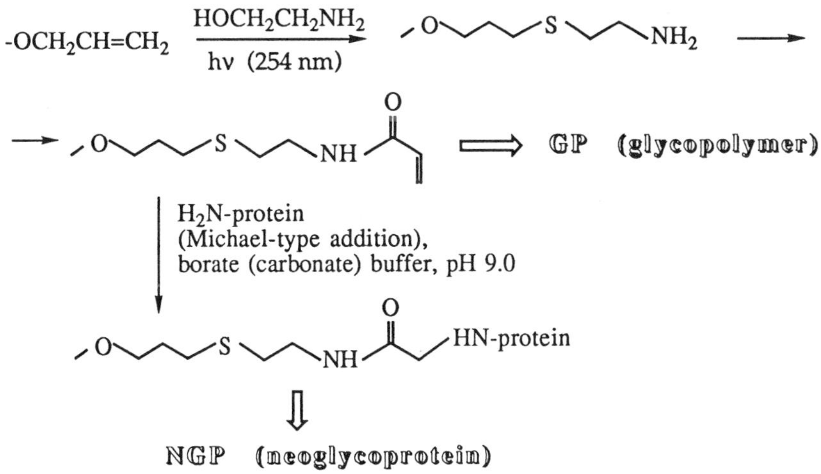

$-OCH_2CH=CH_2$ $\xrightarrow[\text{hv (254 nm)}]{\text{HOCH}_2\text{CH}_2\text{NH}_2}$ $\diagup O\diagdown\!\diagup\!\diagdown S\diagdown\!\diagup\!\diagdown NH_2$ $\longrightarrow$

$\longrightarrow$ $\diagup O\diagdown\!\diagup\!\diagdown S\diagdown\!\diagup NH\diagdown\!\!\overset{\displaystyle O}{\underset{\displaystyle \|}{\text{C}}}$ $\Longrightarrow$ GP (glycopolymer)

$\downarrow$ H_2N-protein
(Michael-type addition),
borate (carbonate) buffer, pH 9.0

$\diagup O\diagdown\!\diagup\!\diagdown S\diagdown\!\diagup NH\diagdown\!\!\overset{\displaystyle O}{\underset{\displaystyle \|}{\text{C}}}\!\!\diagup HN$-protein

$\Downarrow$

NGP (neoglycoprotein)

SCHEME 2

The *N*-acryloyl group is able to act as an acceptor in a Michael-type addition to amino functions in a protein (Scheme 2). This is a very useful reaction and opens up access to both neoglycoproteins and glycopolymers of the PAA type from the same saccharide derivative (*43*).

Conversion of Reducing Saccharides to PAA conjugates. Natural glycoconjugates are a rich source of various oligosaccharides, which are usually isolated in the reducing form. This is why the techniques for preparation of OS-PAA conjugates has been adapted to reducing oligosaccharides. One of elaborated procedures (*44*) seems to be more suitable for longer oligosaccharide sequences because the natural, cyclic form of the monosaccharide residue at the reducing end is lost (Scheme 3).

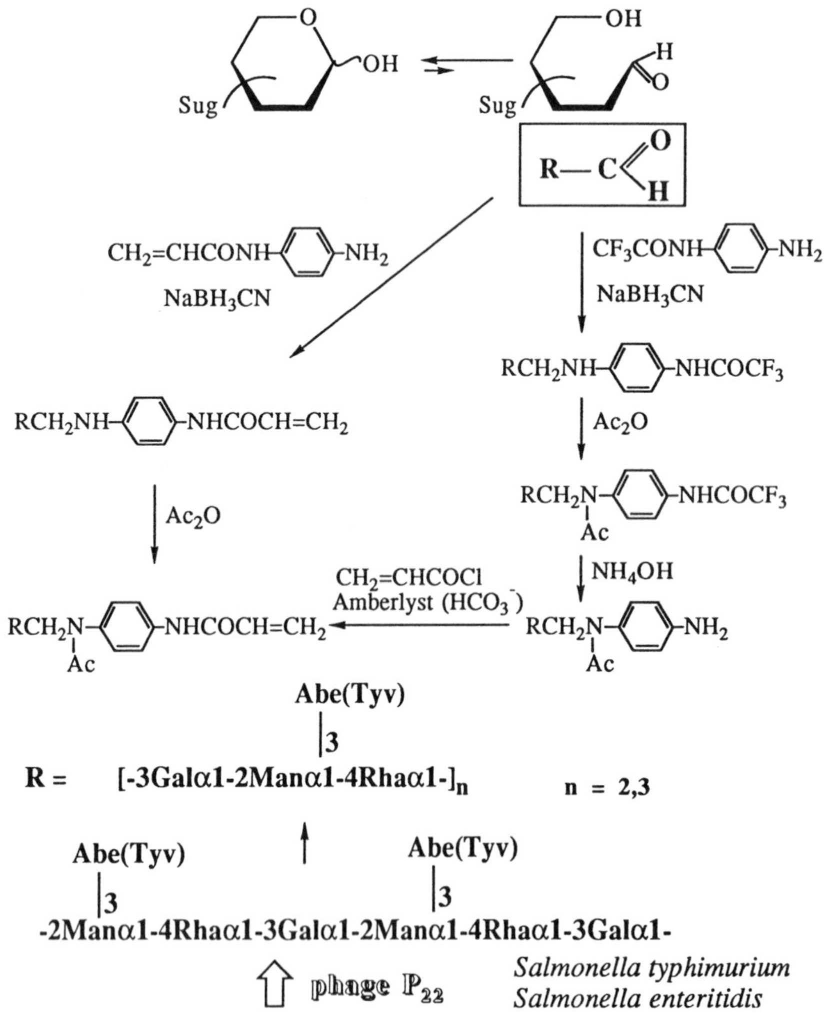
Sug
OH
Sug
OH
H
O
R—C
O
H
CH2=CHCONH
NH2
NaBH3CN
CF3CONH
NH2
NaBH3CN
RCH2NH
NHCOCF3
RCH2NH
NHCOCH=CH2
Ac2O
Ac2O
RCH2N
NHCOCF3
Ac
NH4OH
CH2=CHCOCl
Amberlyst (HCO3-)
RCH2N
NHCOCH=CH2
Ac
RCH2N
NH2
Ac
Abe(Tyv)
3
R = [-3Galα1-2Manα1-4Rhaα1-]n
n = 2,3
Abe(Tyv)
3
Abe(Tyv)
3
-2Manα1-4Rhaα1-3Galα1-2Manα1-4Rhaα1-3Galα1-
phage P22
Salmonella typhimurium
Salmonella enteritidis

SCHEME 3

Reducing oligosaccharides, isolated from *Salmonella* O-antigens by phage degradation, and comprised of two or three repeating units have been converted into derivatives suitable for polymerisation. This was accomplished by the reaction sequence depicted in Scheme 3: reductive amination with *p*-trifluoroacetamidoaniline (TFAN) followed by *N*-acetylation of the secondary amine, generation of the primary amino function using ammonia, and finally *N*-acryloylation. In a more concise route, the same derivative can be obtained by reductive amination with *p*-acrylamidoaniline. The PAA conjugates prepared from these octa- and dodecasaccharide derivatives have been efficiently used as coating antigens in EIA assays (*44*).

Another procedure is based on transformation of reducing oligosaccharides into glycosylamines (*21*) (Scheme 4). A simple reaction (*45*) was utilized for an efficient conversion of the reducing oligosaccharides isolated from human milk into glycosylamines used after *N*-acryloylation in co-polymerisation with acrylamide. The isolated PAA conjugates exhibited strong specific binding to corresponding antibodies (*21*).

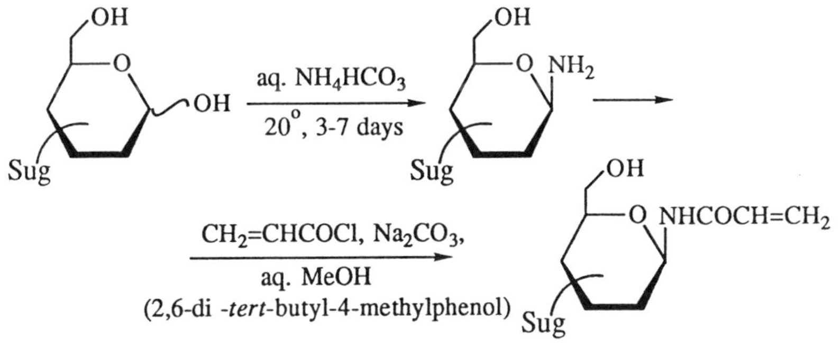

SCHEME 4

In certain cases, the saccharides isolated from natural sources possess an inherent amino function. This was the case with the type-specific hapten from the lipooligosaccharide (LOS) of *Neisseria meningitidis* group B. The amino group of phosphorylethanolamine was subjected to *N*-acryloylation followed by co-polymerisation with acrylamide. The final PAA material exhibited high serological specificity to anti-LOS antibodies, indicating that phosphorylethanolamine is not a part of the dominant epitopes in LOS (*46*)

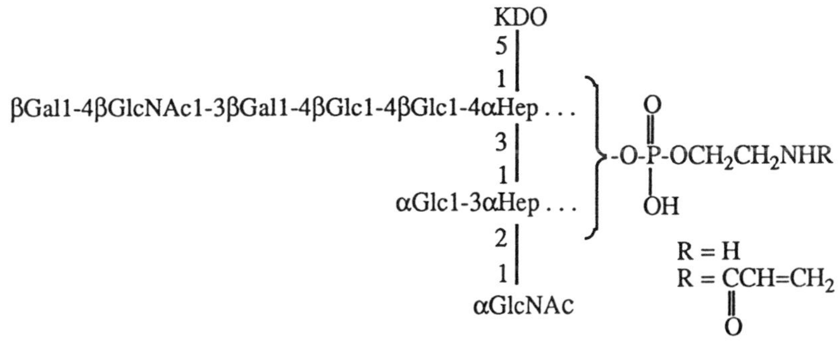

Type-specific oligosaccharide hapten from *Neisseria meningitidis* group B lipooligosaccharide (LOS)

Synthesis of OS-PAA Conjugates Mimicking Bacterial Antigens

The laboratory of carbohydrate chemistry in N.D.Zelinsky Institute of Organic Chemistry (Moscow) is one of the first centres where the technology for the preparation of glycopolymers has been elaborated. A list of glycopolymers (Table I) prepared in this laboratory, partially in cooperative efforts with the Department of Clinical Bacteriology (Huddinge Hospital, Sweden), BioCarb AB (Lund, Sweden), and Indian Institute of Chemical Technology (Hyderabad, India), covers OS-PAA conjugates with the specificity of bacterial antigens from *Salmonella* (*19, 29, 33, 34, 44*), *Streptococcus pneumoniae* (*47-49*), *Proteus* (*38, 50, 51*), *Escherichia coli* (*36, 52*), and *Neisseria meningitidis* (*46*). The complete list of OS-PAA conjugates described in the literature would be much more extensive because the above technology is currently used by several research groups all over the world (*6, 16-22*).

Table I. The OS-PAA conjugates prepared in N. D. Zelinsky Institute of Organic Chemistry (Moscow)

OS-PAA conjugates from:	*Specificity*
1. synthetic di- and trisaccharides as allyl or acrylamidoethyl glycosides	*Salmonella* O-antigens O:2 (A), O:3 (E), O:4 (B), O:8 (C_2, C_3), and O:9 (D_1)
2. reducing octa-, deca-, and dodeca- saccharides isolated from LPS by phage degradation	*Salmonella* O-antigens: *S. typhimurium* (B), *S. enteritidis* (D_1), *S. thompson* (C_1)
3. synthetic di-, tri-, and tetrasaccharides as allyl or acrylamidoethyl glycosides	*Streptococcus pneumoniae* type 3 and 8 capsular poly- saccharides
4. synthetic mono- and disaccharides as acrylamidoethyl glycosides	*Proteus* O-antigens
5. synthetic mono- and disaccharides as acrylamidoethyl glycosides	*Escherichia coli* O6:K54:H10 (K54 capsular polysaccharide)
6. oligosaccharide fraction isolated from lipooligosaccharide	*Neisseria meningitidis* group B

OS-PAA Conjugates with the Specificity of *Salmonella* O-Antigens. The synthesis of disaccharides of the general type 3,6-dideoxyHexα1-3Manα1-, which represent partial O-antigens of *Salmonella* (3,6-dideoxyHex = 3,6-dideoxy-D-*ribo*-hexose, paratose, O-antigen 2; 3,6-dideoxy-D-*xylo*-hexose, abequose, O-antigen 4; 3,6-dideoxy-D-*arabino*-hexose, tyvelose, O-antigen 9) has been performed by glycosidation of either allyl or 2-trifluoroacetamidoethyl glycosides of 3-unprotected mannose derivatives (**1** and **2**). The acylated abequosyl and tyvelosyl bromides were used as glycosyl donors under Helferich conditions (*29, 33*). Paratosyl bromide was generated *in situ* from *O*-benzylated methyl 3,6-dideoxy-α-D-*ribo*-hexopyranoside using trimethylsilyl bromide (Me_3SiBr) and utilized in the glycosidation (of **2**), which was promoted by anhydrous cobalt(II) bromide - tetrabutylammonium bromide (*33*).

The trisaccharide hapten representing *Salmonella* O-antigen 3 (Manβ1-4Rhaα1-3Galβ1) was prepared *via* glycosidation of the allyl galactoside **3** with the glycosyl bromide obtained from per-*O*-acetylated Manβ1-4Rha (*53*).

In order to the study the immunochemistry of the O-antigen 8, both α and β forms of the disaccharide determinant Abeα1-3Rha were synthesized (*54*). Glycosyl acceptors **4** and **5** (the latter prepared by glycosidation of 2-trifluoroacetamidoethanol with the corresponding rhamnosyl bromide using silver carbonate as promoter) were used in the disaccharide syntheses. Trisaccharide sequences (Abeα1-3Rhaα/β1-2Manα1-) comprising the L-rhamnose residue α-1,2 or β-1,2-linked to the α-D-mannose were also obtained (*34*).

After deprotection by conventional methods (and *N*-acryloylation, if necessary) the synthetic *Salmonella* haptens were converted (by co-polymerization with acrylamide) into polymeric NGCs. In formulae **6-15**, the synthetic *Salmonella* glycopolymers are compiled. Those derived from one sugar monomer were monospecific whereas co-polymerisation of two or three sugar monomers led to poly-specific conjugates containing several serogroup-specific haptens in the same macromolecule (*19*).

6   R = Parα1-3Manα1-OCH$_2$CH$_2$NHCO-   (O-antigen 2)
7   R = Abeα1-3Manα1-OCH$_2$-   (O-antigen 4)
8   R = Tyvα1-3Manα1-OCH$_2$-   (O-antigen 9)

9 R = Abeα1-3Rhaα1-OCH$_2$CH$_2$NHCO-

10 R = Abeα1-3Rhaβ 1-OCH$_2$CH$_2$NHCO-   (O-antigen 8)

11 R = Abeα1-3Rhaα1-2Manα1-OCH$_2$CH$_2$NHCO-

12 R = Abeα1-3Rhaβ1-2Manα1-OCH$_2$CH$_2$NHCO-   (O-antigen 8)

13 R = Manβ1-4Rhaα1-3Galβ1-OCH$_2$-   (O-antigen 3)

14 R = Parα1-3Manα1-OCH$_2$CH$_2$NHCO-;

       Abeα1-3Manα1-OCH$_2$CH$_2$NHCO-;

       Tyvα1-3Manα1-OCH$_2$-   (O-antigens 2,4,9)

15 R = Manβ1-4Rhaα1-3Galβ1-OC$_6$H$_4$NHCO-;

       Abeα1-3Manα1-OCH$_2$-;

       Tyvα1-3Manα1-OCH$_2$-   (O-antigens 3,4,9)

The PAA conjugate **7** derived from the hapten O:4 as allyl glycoside was heated with ethylenediamine to cause partial transamidation. Primary amino functions in the modified backbone were N-acylated with palmitoyl chloride and the resulting glycopolymer **16**, containing about 4% of the lipophilic residues, showed an enhanced ability for adsorption on sheep erythrocytes (*30*).

$$CONHR^1 \qquad\qquad CONHR^1$$
$$| \qquad\qquad\qquad |$$
$$[-(CH_2\text{-}CH)_x\text{-}CH_2\text{-}CH\text{-}(CH_2\text{-}CH)_y\text{-}]_n$$
$$|$$
$$CH_2OR$$

    16 R = Abeα1-3Manα1-

          R^1 = H or CH$_3$(CH$_2$)$_{14}$CONHCH$_2$CH$_2$-

OS-PAA Conjugates Comprising Uronic Acids and Amino Acids. The parent glycoconjugates of *S. pneumoniae*, *Proteus*, and *E. coli* listed in Table I have common structural features. They are comprised of uronic acid residues which are usually substituted at position 3. Some of the uronic acid residues are amidated by amino acids, which are present as non-sugar substituents in these O-antigens (*Proteus*) and capsular polysaccharides (*E. coli*). The *Proteus* strains (*55*) are especially rich in glycuronamides of amino acids (alanine, lysine, serine, and threonine) and these unusual, non-sugar substituents can be of potential immunochemical importance.

We have focused our interest on synthesis of fragments of similar bacterial antigens. As examples below are depicted structures of the well known capsular polysaccharide from *S. pneumoniae* type 3, the O-antigen of *Proteus mirabilis* O27 (*56*) which contains even two amidically linked amino acids, and the structure of the capsular polysaccharide from one of uropathogenic strains of *E. coli* (*57*) with variations in substitution with amino acids.

-3GlcAβ1-4Glcβ1-3GlcAβ1-4Glcβ1-

Capsular polysaccharide of *S. pneumoniae* type 3

$$L\text{-}Lys \qquad L\text{-}Ala$$
$$|6 \qquad\quad |6$$
$$-3GlcNAc\beta1\text{-}3GlcA\beta1\text{-}3GalA\alpha1\text{-}$$
$$|6 \qquad\qquad |4$$
$$H_2NCH_2CH_2OP(O)OH \quad \beta GlcNAc$$

O-antigen of *P. mirabilis* O27

$$L\text{-}Thr/L\text{-}Ser \qquad L\text{-}Thr/L\text{-}Ser$$
$$: \qquad\qquad\qquad :$$
$$-3GlcA\beta1\text{-}3Rha\alpha1\text{-}3GlcA\beta1\text{-}3Rha\alpha1\text{-}$$

Capsular polysaccharide K54 from *E. coli* O6:K54:H10
(Approx. 85% of the glucuronic acid residues are substituted with threonine and serine in a ratio of 9:1)

Synthesis of Glycosyl Acceptors by Selective Alcoholysis of Uronolactone Acetates. The bacterial antigens of interest comprise 3-substituted uronic acid residues. This prompted us to develop a simple procedure (*58*) to make accessible 3-unprotected derivatives of uronic acids. Heating of glycosides **17** of glucuronic or galacturonic acid with acetic anhydride, which acts as a dehydrating agent, led to the 6,3-lactone closure and the molecule **18** adopted the conformation $^1C_4$ easily deduced from PMR spectra. In the *manno* series, the situation is more ambiguous - lactonization afforded a mixture of 6,3- and 6,2-lactones in a ratio of about 1:1. Cyclization is accompanied by partial *O*-acetylation, which is complete after addition of pyridine. The lactone ring can be selectively opened by alcoholysis at room temperature and depending on the alcohol used, a basic catalyst like sodium acetate may be required. This simple, two-step procedure permits the preparation of various esters of 3-unprotected uronic acids **19** (*58*)

The utility of this procedure is exemplified by the synthesis of the repeating units of the capsular polysaccharide from *S. pneumoniae* type 3, represented as disaccharide sequences GlcAβ1-4Glcβ1- (**A**) and GlcAβ1-3GlcAβ1- (**B**) overlapped in the polysaccharide chain (*47, 48*). The 6,3-lactone of glucuronic acid in both the furanose and pyranose forms (**20** and **22**) was used in the synthesis of the **B**-sequence. Selective methanolysis of the lactone ring liberated the 3-OH in compounds **21** and **23** for glycosidation with acetobromoglucose. Both paths led to the same allyl glycoside **24** (**B**-sequence), however, the upper way required a furanose-to-pyranose transformation on the disaccharide level.

The **A**-sequence was synthesized from cellobiose by modification of the non-reducing glucose residue (*47*). The dihydroxy derivative **25** prepared by conventional methods, was treated with Jones reagent to give selective oxidation of the primary hydroxyl group in a high yield. Conventional deprotection of the acetylated uronic acid **26** afforded cellobiuronic acid as the allyl glycoside **27** (**A**-sequence). Lactonization of **27** led to the biuronolactone **28**, which after selective methanolysis, has been used in the synthesis of longer sequences of the capsular polysaccharide of *S. pneumoniae* type 3 (*48*). The OS-PAA conjugates derived from these sequences have been used in immunochemical studies (see the last Section).

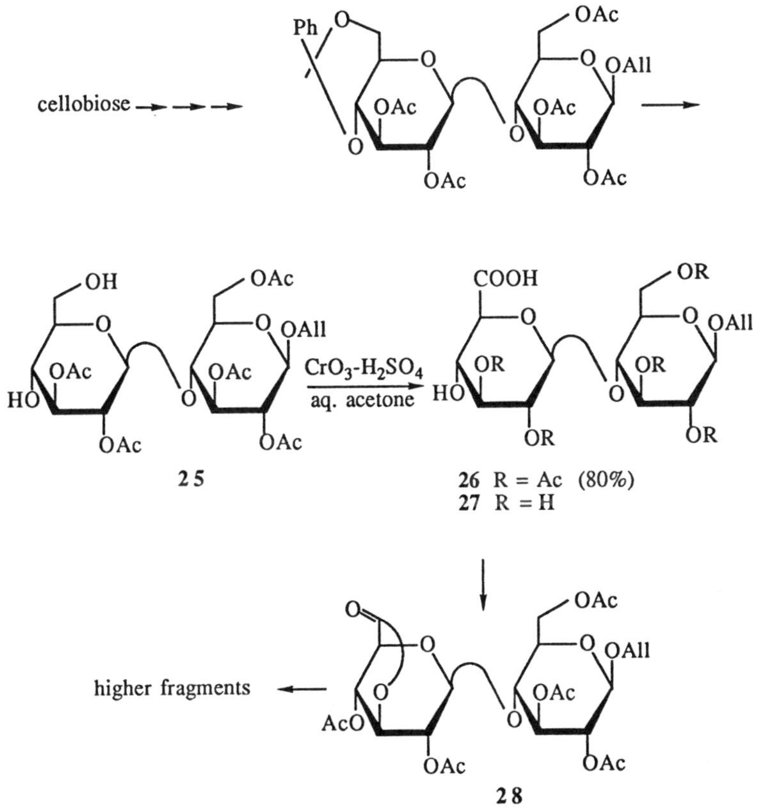

cellobiose → → → →

25

26 R = Ac (80%)
27 R = H

higher fragments ←

28

Synthesis of Monosaccharidic Glycuronamides of Amino Acids. The natural combinations of amino acids which have formed amides with uronic acids described so far are alanine, lysine, serine, or threonine with glucuronic acid (mostly β) or galacturonic acid (mostly α). For the coupling to azidoethyl glycosides of uronic acids partially protected amino acids **29-32** have been used as *tert*-butyl esters with *O-tert*-butyl and *N*ε-*tert*-butyloxycarbonyl protecting groups (*38, 50*). In our opinion, this combination of protecting groups is advantageous (or even optimal) for the coupling step, further transformations, and deprotection in the amino acid moiety, which require acidic condition like a treatment with trifluoroacetic acid. Condensation of the uronic acid azidoethyl glycosides, represented by **33** and amino acid derivatives **29-32** was promoted by ethyl 2-ethoxy-1,2-dihydroquinoline-1-carboxylate (EEDQ) in amine-free DMF. The coupling afforded the protected glycuronamides **34** in a high yield. For conversion of the azidoethyl aglycon into a polymerizable form, catalytic hydrogenation followed by *N*-acryloylation has been used.

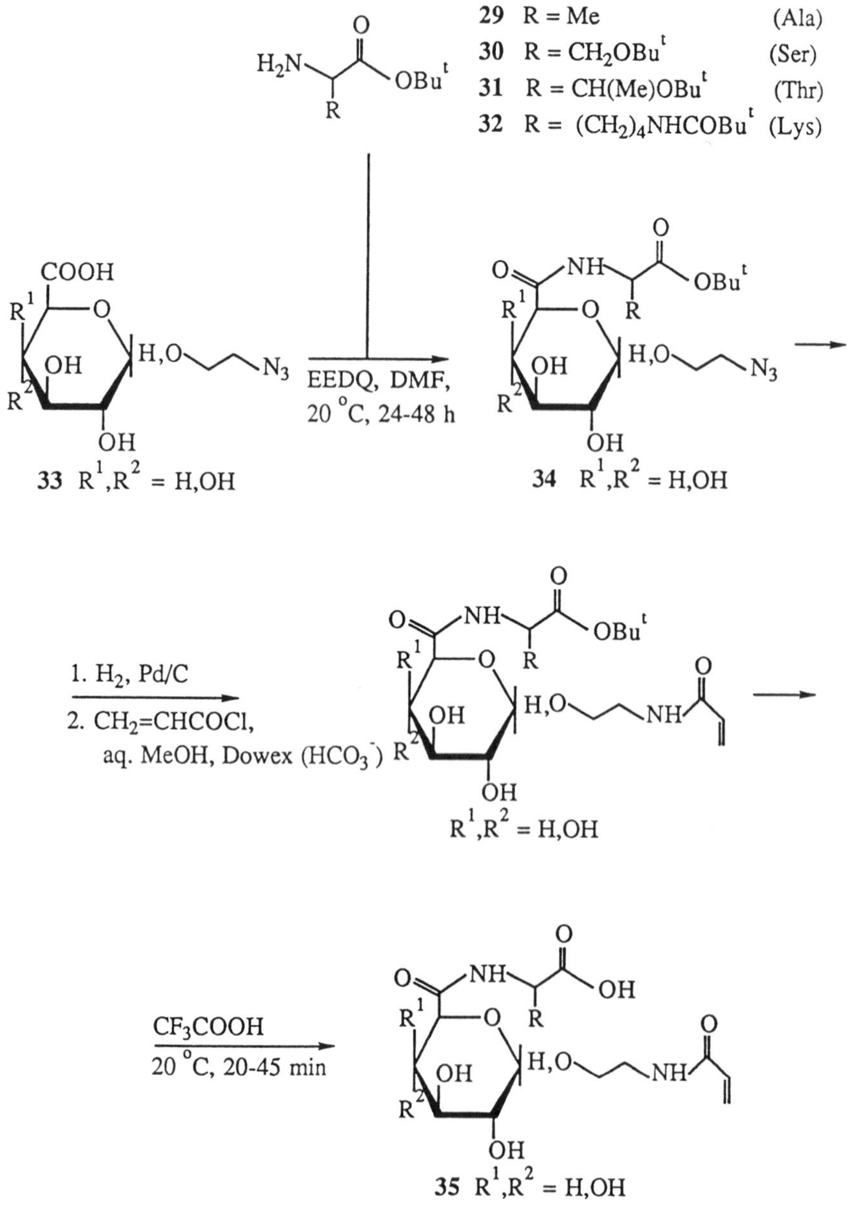

29 R = Me   (Ala)
30 R = CH$_2$OBu^t (Ser)
31 R = CH(Me)OBu^t (Thr)
32 R = (CH$_2$)$_4$NHCOBu^t (Lys)

33 R^1,R^2 = H,OH

EEDQ, DMF, 20 °C, 24-48 h

34 R^1,R^2 = H,OH

R^1,R^2 = H,OH

1. H$_2$, Pd/C
2. CH$_2$=CHCOCl, aq. MeOH, Dowex (HCO$_3^-$)

R^1,R^2 = H,OH

CF$_3$COOH
20 °C, 20-45 min

35 R^1,R^2 = H,OH

After final deprotection with trifluoroacetic acid, the monomeric glycuronamides **35** were transformed by co-polymerisation with acrylamide into a series of polyvalent glycopolymers **36-46** (*38, 50*). Glycopolymers with D-amino acids have also been prepared to test how they bind to antibodies as compared to the natural L-analogs.

$$\begin{array}{cc}
CONH_2 & CONH_2 \\
| & | \\
\end{array}$$

[-(CH_2-CH)_x-CH_2-CH-(CH_2-CH)_y-]_n

|
CONHCH_2CH_2OR

36   R = [6(*N*)-L-Ala]GlcAβ1-

37   R = [6(*N*)-L-Ser]GlcAβ1-

38   R = [6(*N*)-L-Thr]GlcAβ1-

39   R = [6(*N*$^\alpha$)-L-Lys]GlcAβ1-

40   R = [6(*N*)-L-Ala]GalAβ1-

41   R = [6(*N*)-L-Ala]GalAα1-

42   R = [6(*N*)-D-Ala]GalAα1-

43   R = [6(*N*$^\alpha$)-L-Lys]GalAα1-

44   R = [6(*N*$^\alpha$)-D-Lys]GalAα1-

45   R = [6(*N*)-L-Ser]GalAα1-

46   R = [6(*N*)-L-Thr]GalAα1-

Partially protected amino acids and the glucuronic acid derivative of the type **19** discussed in the preceding subsections have been utilized in the syntheses of a part of the main chain (GlcNAcβ1-3[6(*N*$^\alpha$)-L-Lys]GlcAβ1-) in the O-antigen of *P. mirabilis* O27 (*51*) and repeating units of the capsular polysaccharide K54 from *E. coli* O6:K54:H10 (*52*).

In the synthesis of the disaccharide of *P. mirabilis* O27 attempts to use an oxazoline derivative of glucosamine for glycosidation of the azidoethyl glycoside **48** with various promoters (*p*-toluenesulfonic acid in 1,2-dichloroethane or trimethylsilyl triflate in toluene) failed. Only glycosidation of **48** with the glycosyl bromide **47** and promoted by silver triflate gave the protected disaccharide **49** (68%). Deprotection of **49** by the reaction sequence indicated avoided β-elimination in the uronic acid moiety. Coupling of **49** to the lysine derivative **32** gave the target monomer **50**.

We have synthesized all the types of repeating units found in the capsular polysaccharide K54 (*52*) in which the glucuronic acid-rhamnose (**GR**) and rhamnose-glucuronic acid (**RG**) sequences were substituted with each of amino acids (serine and threonine) or not.

The **GR**-sequence was prepared by coupling of glycosyl bromide **51** with the monohydroxy derivative **52** under Helferich condition [Hg(CN)_2, HgBr_2 in acetonitrile] followed by transformations at the anomeric centre in the rhamnose

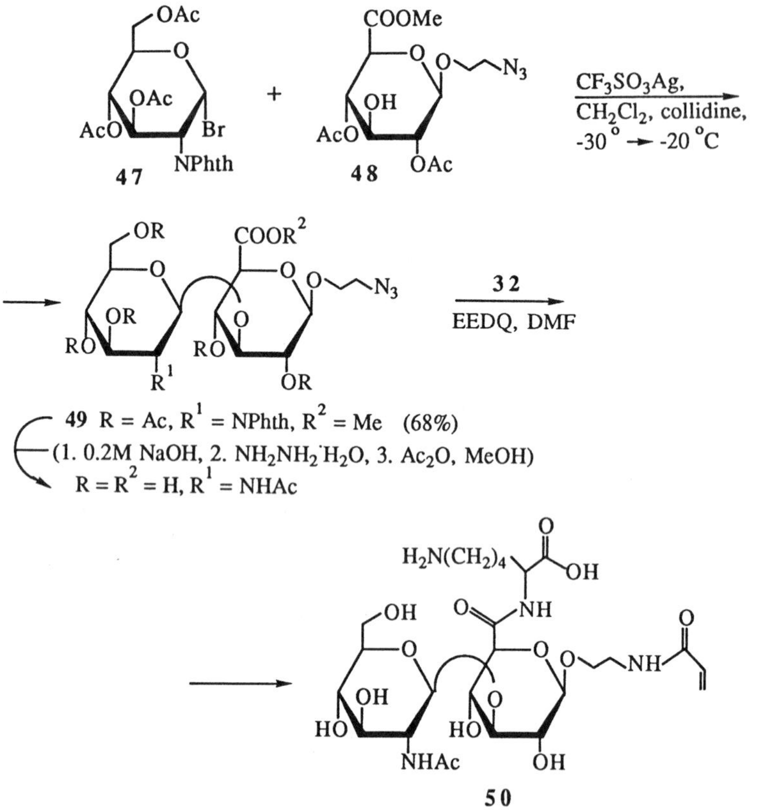
47
48
CF_3SO_3Ag,
CH_2Cl_2, collidine,
-30 ° → -20 °C
32
EEDQ, DMF
49 R = Ac, R¹ = NPhth, R² = Me (68%)
(1. 0.2M NaOH, 2. $NH_2NH_2 \cdot H_2O$, 3. Ac_2O, MeOH)
R = R² = H, R¹ = NHAc

50

residue. The methyl glycoside **53** (R = OMe) was subjected to acetolysis and then brominated. The disaccharide glycosyl bromide **53** (R = Br) was condensed with the *N*-benzyloxycarbonyl protected aminoethanol. The succession of reactions leading to the target **GR**-monomer **54** involved de-*O*-acetylation (MeONa), hydrogenolysis followed by *N*-acryloylation, and finally saponification of the methyl ester group (0.2M NaOH).

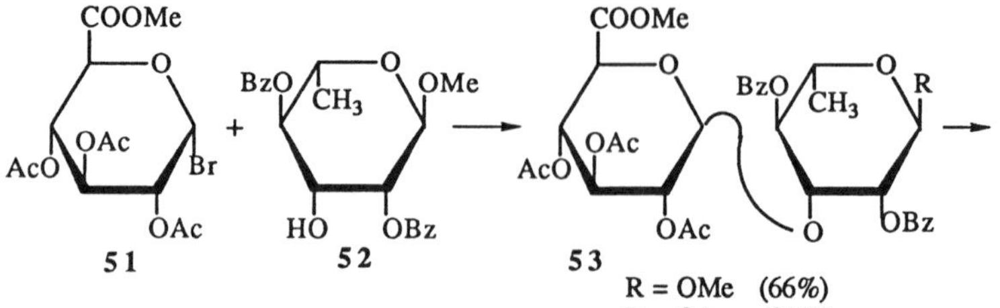
51
52
53

R = OMe (66%)
R = OAc (81%)
R = Br (100%)
R = OCH_2CH_2NHZ (69%)

Glycosidation of the azidoethyl glycoside **48** by acetobromorhamnose using silver triflate as a promoter led to the protected disaccharide **57**. Transformation of the aglycon part in **57** followed by deprotection (saponification with cold 0.14M sodium hydroxide in aqueous methanol) afforded the target **RG**-monomer **58**. Then, according to the procedure discussed above the unsubstituted **GR**- and **RG**-fragments were coupled to the protected serine and threonine derivatives **30** and **31** to give both sequences modified with amino acids (**55, 56** and **59, 60**).

From these target monomers and using co-polymerisation, we have prepared a complete set of glycopolymers **61-66** related to the repeating units of the K54 polysaccharide.

$$\begin{array}{cc} CONH_2 & CONH_2 \\ | & | \end{array}$$

$$[-(CH_2-CH)_x-CH_2-CH-(CH_2-CH)_y-]_n$$

$$|$$

$$CONHCH_2CH_2OR$$

61	R = Rhaα1-3GlcAβ1-	RG-PAA
62	R = Rhaα1-3[6(N)-L-Ser]GlcAβ1-	R(GS)-PA
63	R = Rhaα1-3[6(N)-L-Thr]GlcAβ1-	R(GT)-PAA
64	R = GlcAβ1-3Rhaα1-	GR-PAA
65	R = [6(N)-L-Ser]GlcAβ1-3Rhaα1-	(GS)R-PAA
66	R = [6(N)-L-Thr]GlcAβ1-3Rhaα1-	(GT)R-PAA

Application of OS-PAA conjugates

The key feature of the OS-PAA conjugates is that they contain only one serologically active component as compared to their neoglycoprotein counterparts, and this is the carbohydrate moiety. Therefore, they are very suitable as screening antigens in immunoassays (35) for immunochemical studies (59), for characterisation of epitope specificity of monoclonal antibodies (60, 61), as probes for the analysis of lectins (15, 62), as tools for cell adhesion studies (63) etc.

The OS-PAA conjugates with specificity for the blood group substances have found various application in haemotology e.g. for preparation of universal anti-Rhesus sera by neutralisation of α and β isohaemagglutinins found in human sera using glycopolymers with synthetic A and B trisaccharide ligands (64).

The OS-PAA conjugates were found to be very efficient as coating antigens in ELISA with coating doses of about 0.01 μg/mL, which are low when compared with 1-10 μg/mL required for neoglycoproteins (59).

The OS-PAA conjugates **6-15** have been used as highly specific reagents in EIA for improvement of the serological diagnosis of salmonellosis (65). Antibodies belonging to different immunoglobulin classes (M, G, and A) have been detected in sera from patients with salmonellosis, with the estimation of IgM- and IgG-antibodies being of a higher diagnostic value. The OS-PAA conjugate **16**, modified with lipophilic ligands, has been used in a passive haemagglutination (PH) test for the diagnosis of salmonellosis (30). The OS-PAA conjugates derived from 3,6-di-O-methyl-β-D-glucose and longer synthetic OS-sequences of the glycolipid of *Mycobacterium leprae* have been used in EIA for serological diagnosis of leprosy (17, 25).

The OS-PAA conjugates are useful as research tools in immunochemical studies. For example, the glycopolymers **61-66** were used as coating antigens in ELISA for titration of antisera elicited against the capsular polysaccharides K54 and K96. The latter has the same polysaccharide chain as K54 but is devoid of amino acids. Only the **RG**-sequences showed binding strong enough to be inhibited (66). The interaction of anti-K96 serum with the **RG**-copolymer **61** was inhibited only by glycopolymers with **RG** and related sequences (Fig. 5). The same pattern were observed for the inhibition of the interaction of the anti-K54 serum with **R(GT)-PAA** and **R(GS)-PAA** (Fig. 5).

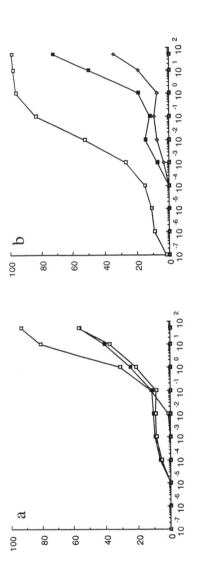

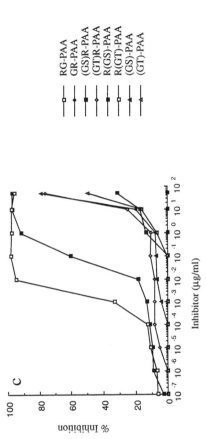

Figure 5. EIA-inhibition by glycopolymers of binding of: a - anti-K96 antiserum with RG-PAA, b - anti-K54 serum with R(GT)-PAA, and c - anti-K54 serum with R(GS)-PAA.

These results can be regarded as an indication that the rhamnosyl residue occupies the terminal, non-reducing position in the K54 and K96 polysaccharides.

Similar results have been obtained with glycopolymers related to the capsular polysaccharide of *S. pneumoniae* type 3 (*47*). As it can be seen from the results of PH (Table II) one of the sequences, namely GlcA-Glc, bound to the anti-type 3 antibodies stronger than the inverted sequence, Glc-GlcA did.

Table II. Passive haemagglutination titration of rabbit anti-S3 antiserum

Titer[a] *in haemagglutination against SRBC coated with*	
capsular polysaccharide of *S. pneumoniae* type 3	1:8000
GlcA-Glc-PAA	1:4000
Glc-GlcA-PAA	1:250

[a]The haemagglutination titer is expressed as the last serum dilution giving clear haemagglutination

A possible explanation is that only the sequence common to the terminal sequence in the natural glycoconjugate and multi-exposed in glycopolymers according to their general design (Figure 2) is recognized by the corresponding antibody in the population of antibodies. These data which allow definition of the chain ends may be of interest in the consideration of biosynthesis of regular polysaccharides.

Acknowledgment. I am grateful to my many colleagues in different countries (Russia, Sweden, India, Poland, and Germany) who are cited in the references for their contributions to this project. The financial support by a grant "Swedish-Russian Scientific Cooperation" from the Royal Swedish Academy of Sciences is thankfully acknowledged.

Literature cited

1. Varki, A. *Glycobiol.* **1993**, *3*, 97-130.
2. Aplin, J. D.; Wriston, J. C., Jr. *CRC Crit. Rev. Biochem.* **1981**, *10*, 259-306.
3. Magnusson, G. In *Protein-Carbohydrate Interactions in Biological Systems*; Lark, D. L., Ed.; FEMS Symp. Series; Academic Press: London, 1986, Vol. 31, pp. 215-228.
4. Stowell, C. P.; Lee, Y. C. *Adv. Carbohydr. Chem. Biochem.* **1980**, *37*, 225-281.
5. Landsteiner, K.; Lampl, H. *Biochem. Z.* **1918**, *86*, 343-394.
6. Roy, R.; Tropper, F. D.; Romanowska, A. *Bioconjugate Chem.* **1992**, *3*, 256-261.
7. Kraska, B.; Mester, L. *Tetrahedron Lett.* **1978**, 4583-4586.
8. Kurita, K.; Hirakawa, N.; Iwakura, Y. *Makromol. Chem.* **1979**, *180*, 855-858.
9. Jann, K.; Westphal, O. In *The Antigens*; Sela, M., Ed.; Academic Press: New York, 1975, Vol. 3, pp. 1-125.

10. Bishop, C. T.; Jennings, H. J. In *The Polysaccharides*; Aspinall, G. O., Ed.; Academic Press: New York, 1982, Vol. 1, pp. 291-330.
11. Andresz, H.; Richter, G. C.; Pfannemüller, B. *Makromol. Chem.* **1978**, *179*, 301-312.
12. Black, W.-A. P.; Dewar, E. T.; Rutherford, D. *J. Chem. Soc.* **1963**, 4433-4439.
13. Kossmehl, G.; Volkheimer, J. *Liebigs Ann. Chem.* **1989**, 1127-1130.
14. Kobayashi, K.; Sumitomo, H.; Itoigawa, T. *Macromolecules* **1987**, *20*, 906-908.
15. Horejsi, V.; Smolek, P.; Kocourek, J. *Biochim. Biophys. Acta* **1978**, *538*, 293-298.
16. Kosma, P.; Schulz, G.; Unger, F. *Carbohydr. Res.* **1988**, *180*, 19-28.
17. Mariño-Albernas, J. R.; Verez-Bencomo, V.; Gonzales-Rodriguez, L.; Perez-Martinez, C. S. *Carbohydr. Res.* **1988**, *183*, 175-182.
18. Paulsen, H.; Höffgen, E. C.; Brenken, M. In *Carbohydrate Antigens*; Garegg, P. J., Lindberg, A. A., Eds.; ACS Symp. Series; ACS Books: Washington, DC, 1993, Vol. 519, pp. 132-145.
19. Chernyak, A. Ya.; Levinsky, A. B.; Tendetnik, Yu. Ya.; Kochetkov, N. K.; Pokrovsky , V. I. *Bioorg. Khim.* **1992**, *18*, 680-688.
20. Byramova, N. E.; Mochalova, L. V.; Belyanchikov, I. M.; Matrosovich, M. N.; Bovin, N. V. *J. Carbohydr. Chem.* **1991**, *10*, 691-700.
21. Kallin, E.; Lönn, H.; Norberg, T.; Elofsson, M. *J. Carbohydr. Chem.* **1989**, *8*, 597-611.
22. Spaltenstein, A.; Whitesides, G. M. *J. Am. Chem. Soc.* **1991**, *113*, 686-687.
23. Kochetkov, N. K.; Dmitriev, B. A.; Chernyak, A. Ya.; Pokrovsky, V. I.; Tendetnik, Yu. Ya. *Dokl. Akad. Nauk SSSR.* **1982**, *263*, 1277-1280.
24. Bovin, N. V.; Zurabyan, S. E.; Khorlin, A. Ya. In *Proc. XII Mendeleyev Congress Pure Appl. Chem. (USSR)*; Nauka Publisher House: Moscow, 1981, p. 120.
25. Bovin, N. V.; Korchagina, E. Yu.; Zemlyanukhina, T. V.; Byramova, N. E.; Galanina, O. E.; Zemlyakov, A. E.; Ivanov, A. E.; Zubov, V. P.; Mochalova, L. V.*Glycoconjugate J.* **1992**, *9*, 7-20.
26. Matrosovich, M. N.; Mochalova, L. V.; Marinina, V. P.; Byramova, N. E. *FEBS Lett.* **1990**, *272*, 209-212.
27. Abramenko, I. V.; Gluzman, D. F.; Korchagina, E. Yu.; Zemlyanukhina, T. V.; Bovin, N. V. *FEBS Lett.* **1992**, *307*, 283-286.
28. Kochetkov, N. K.; Dmitriev, B. A.; Chernyak, A. Ya.; Pokrovsky, V. I.; Tendetnik, Yu. Ya.; Ovcharova, N. M. *USSR Pat.* SU 879,970; *Chem. Abstr.* **1983**, *98*, 17777e.
29. Kochetkov, N. K.; Dmitriev, B. A.; Chernyak, A. Ya.; Levinsky, A. B. *Carbohydr. Res.* **1982**, *110*, c16-c20.
30. Tendetnik, Yu. Ya.; Pokrovsky, V. I.; Chernyak, A. Ya.; Levinsky, A. B.; Kochetkov, N. K. *FEMS Microbiol. Immunol.* **1991**, *76*, 93-98.
31. Weigel, P. H.; Schnaar, R. L.; Roseman, S.; Lee, Y. C. In *Methods Enzymol*; Ginsburg, V., Ed.; Academic Press: New York, 1982, Vol. 83; pp. 294-299.
32. Bovin, N. V.; Ivanova, I. A.; Khorlin, A. Ya. *Bioorg. Khim.* **1985**, *11*, 662-670.
33. Chernyak, A. Ya.; Levinsky, A. B.; Kochetkov, N. K. *Bioorg. Khim.* **1988**, *14*, 1047-1058.
34. Chernyak, A. Ya.; Demidov, I. V.; Kochetkov, N. K. *Bioorg. Khim.* **1989**, *15*, 1673-1685.
35. Pokrovsky, V. V.; Tregub, A. V.; Tendetnik, Yu. Ya.; Pokrovsky, V. I.; Kochetkov, N. K.; Chernyak, A. Ya.; Levinsky, A. B. *Immunologiya.* 1986, 46-50.
36. Chernyak, A. Ya.; Kononov, L. O.; Kochetkov, N. K. *Bioorg. Khim.* **1989**, *15*, 1394-1410.
37. Chernyak, A. Ya.; Sharma G. V. M.; Kononov, L. O.; Radha Krishna P.; Levinsky, A. B.; Kochetkov, N. K.; Rama Rao, A. V. *Carbohydr. Res.* **1992**, *223*. 303-309.

38. Kononov, L. O.; Chernyak, A. Ya.; Kochetkov, N. K. In *VII Eur. Carbohydr. Symp.* Book of Abstracts: Cracow, Poland, 1993, Abstract A093.
39. Sharma G. V. M.; Radha Krishna P. *Carbohydr. Res.* **1993**, *243*, 393-397.
40. Hasegawa, A.; Terada, T.; Ogawa, H.; Kiso, M. *J. Carbohydr. Chem.* **1992**, *11*, 319-331.
41. Lee, R. T.; Lee, Y. C. *Carbohydr. Res.* **1974**, *37*, 193-201.
42. Roy, R.; Tropper, F. *Glycoconjugate J.* **1988**, *5*, 203-206.
43. Roy, R.; Laferrière, C. A. *J. Chem. Soc.,Chem. Commun.* **1990**, 1709-1711.
44. Chernyak, A. Ya., Weintraub, A.; Norberg, T.; Kallin, E. *Glycoconjugate J.* **1990**, *7*, 111-120.
45. Likhosherstov, L. M.; Novikova, O. S.; Derevitskaya, V. A.; Kochetkov, N. K. *Carbohydr. Res.*, **1986**, *146*, c1-c5.
46. Dmitriev, B. A.; Ovchinnikov, M. V.; Lapina, E. B.; Pluzhnikova, G. N.; Lopyrev, I. V.; Chernyak, A. Ya. *Glycoconjugate J.* **1992**, *9*, 168-173.
47. Chernyak, A. Ya.; Antonov, K. V.; Kochetkov, N. K.; Padyukov, L. N.; Tsvetkova, N. V. *Carbohydr. Res.* **1985**, *141*, 199-212.
48. Chernyak, A. Ya.; Antonov, K. V.; Kochetkov, N. K. *Bioorg. Khim.* **1987**, *13*, 958-966.
49. Chernyak, A. Ya.; Antonov, K. V. *Bioorg. Khim.* **1992**, *18*, 716-725.
50. Chernyak, A. Ya.; Sharma G. V. M.; Kononov, L. O.; Radha Krishna P.; Rama Rao, A. V.; Kochetkov, N. K. *Glycoconjugate J.* **1991**, *8*, 82-89.
51. Chernyak, A. Ya.; Kononov, L. O.; Radha Krishna P.; Kochetkov, N. K.; Rama Rao, A. V. *Carbohydr. Res.* **1992**, *225*, 279-289.
52. Chernyak, A. Ya.; Kononov, L. O.; Kochetkov, N. K. *Carbohydr. Res.* **1991**, *216*, 381-398.
53. Chernyak, A. Ya.; Levinsky, A. B.; Dmitriev, B. A.; Kochetkov, N. K. *Carbohydr. Res.* **1984**, *128*, 269-282.
54. Chernyak, A. Ya.; Demidov, I. V.; Karmanova, I. B.; Chernayk, N. V.; Kochetkov, N. K. *Bioorg. Khim.* **1989**, *15*, 111-122.
55. Gromska, W.; Mayer, H. *Eur. J. Biochem.* **1976**, *62*, 391-399.
56. Vinogradov, E. V.; Krajewska-Pietrashik, D.; Kaca, W.; Shashkov, A. S., Knirel, Yu. A.; Kochetkov, N. K. *Eur. J. Biochem.* **1989**, *185*, 645-650.
57. Hoffmann, P.; Jann, B.; Jann, K. *Carbohydr. Res.* **1985**, *139*, 261-271.
58. Chernyak, A. Ya.; Kononov, L. O.; Antonov, K. V. *Izv. Akad. Nauk SSSR, Ser. Khim.* **1988**, 1660-1667.
59. Chernyak, A. Ya., Weintraub, A.; Kochetkov, N. K.; Lindberg, A. A. *Molecular Immunol.* **1993**, *30*, 887-893.
60. Brade, L.; Kosma, P.; Appelmelk, B. J.; Paulsen, H.; Brade, H. *Infect. Immun.* **1987**, *55*, 462-466.
61. Rozalski, A.; Brade, L.; Kuhn, H.-M.; Brade, H.; Kosma, P.; Appelmelk, B. J.; Kusumoto, S.; Paulsen, H. *Carbohydr. Res.* **1989**, *193*, 257-270.
62. Bovin, N. V. In *Lectins and Glycobiology*; Gabius, H.-J., Ed.; Springer Verlag, 1993, pp. 23-30.
63. Weigel, P. H.; Schnaar, R. L.; Kuhlenschmidt, M. K.; Schmell, E.; Lee, R. T.; Lee, Y. C.; Roseman, S. *J. Biol. Chem.* **1979**, *254*, 10830-
64. Chagiashvili, Ts. N.; Zotikov, E. A.; Bovin, N. V.; Korchagina, E. Yu. *Gematologiya Transfusiologiya*, **1992**, 15-16.
65. Tendetnik, Yu. Ya.; Ovcharova, N. M.; Pokrovsky, V. I.; Grudkova, M.; Mali, J.; Chernyak, A. Ya.; Levinsky, A. B.; Kochetkov, N. K. *Immunologiya.* **1987**, 80-82.
66. Chernyak, A. Ya., Weintraub, A.; Kononov, L. O.; Kochetkov, N. K.; Jann, B; Jann, K. In *VII Eur. Carbohydr. Symp.* Book of Abstracts: Cracow, Poland, 1993, Abstract C014

RECEIVED December 3, 1993

Chapter 10

Synthetic Saccharides Can Delineate the Binding of Polysaccharide to Monoclonal Antibodies

Cornelis P. J. Glaudemans, P. Kováč, E. M. Nashed, E. A. Padlan, and S. Rao Arepalli

National Institutes of Health, Building 8, Bethesda, MD 20892

We have prepared the methyl α-glycosides of various isomaltose-related oligo-saccharides that mimic the polysaccharide dextran. The approaches used are described in detail.

The interaction of two monoclonal anti-dextran antibodies has been examined with these ligands. From these studies a detailed picture emerges as to the specific mode of binding between the principal saccharide immuno-determinant and each immunoglobulin.

Specific binding patterns, defined by hydrogen-bonding, between antigenic (poly)saccharides and antibodies, can be elucidated using deoxy- and deoxyfluoro sugars (1). Thus, in a sugar capable of non-covalent binding to a protein, replacement of a hydroxyl group with hydrogen (OH→H) denies any potential for H-bonding at that position. If the hydroxyl is replaced with fluorine (OH→F) donation of a hydrogen bond by the sugar is denied, but reception by fluorine is still possible (2,3). Early work on Concanavalin A (4,5) and D-galactose oxidase (6) first made use of simple deoxy-fluoro derivatives of sugars in order to help define saccharide-protein interactions. Over the years, this laboratory has refined this approach (7-9) and developed a systematic method to map antibody-antigen interactions (1,9). The application (3,10) of this method to two monoclonal antidextran antibodies will be discussed here. Both are antibodies that are specific for the glycosyl terminus (10,11) of the dextran antigen (in an oligo- or polysaccharide, such as A→B→C→D→m, where A, B, C, and D are sugars, either the same or different, and m is any moiety, we name A the glycosyl or upstream terminus, and D the glycoside or downstream terminus). We have previously reported a method that can distinguish antibodies possessing this mode of binding from those that can bind interior segments of the antigenic polysaccharide chain (11). The reader is referred to that report also for more general observations on the structure of polysaccharides and antibodies.

SYNTHETIC PROCEDURES

The syntheses of ligands, which are needed for investigations on the mode of binding of immunoglobulins that show specificity to dextrans, constitute a special challenge to a carbohydrate chemist. This is because the substances required in these studies are methyl α-glycosides of α-linked (1→6)-D-gluco-oligosaccharides and their specifically modified analogs, *i.e.* 1,2-*cis* glycosides. Due to the nature of the glycosidation reaction, such *cis*-glycosides are far more difficult to obtain than the 1,2-*trans* glycosides.

SYNTHESIS OF METHYL α-GLYCOSIDES OF ISOMALTO-OLIGOSACCHARIDES

Although great progress has been made during the last decade in the chemical synthesis of complex oligosaccharides (12-16) no general, high yielding and predictably stereoselective synthesis of α-(1,2-*cis*)-linked oligosaccharides has as yet been developed. The preparation of higher (1→6)-α-D-glucooligosaccharides suffers from additional problems in that the target molecules are poorly crystalline, and the β-linked products, often formed due to the lack of stereospecificity of the coupling reaction, have chromatographic properties similar to the desired α-linked saccharides. Careful verification of the stereochemical purity of the products must therefore be emphasized.

In the glucose series the prerequisite for the formation of an α-(1,2-*cis*)-glycosidic linkage is that the glycosyl donor possesses at C-2 a group that is incapable of neighboring group participation. Examples of functions which have proven useful for this purpose in oligosaccharide syntheses are benzyl- and allyl-ethers. For example, the fully acetylated β-isomaltose (17) **2** (Scheme 1) and methyl α-isomaltoside (18) **5** (Scheme 2) were synthesized in high yields and with excellent stereoselectivities, using the benzyl blocking group at position 2 of the respective D-glucosyl donors **1** and **3**. Other approaches to the

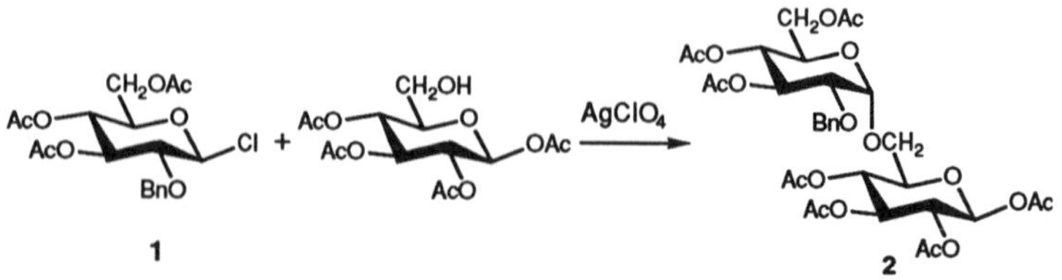

Scheme 1

synthesis of isomaltose and its methyl glycosides have been described (*c.f.* ref. (19) and papers cited therein). Glycosyl halides **1** and **3** can not be used for the synthesis of higher members of the isomalto-oligo

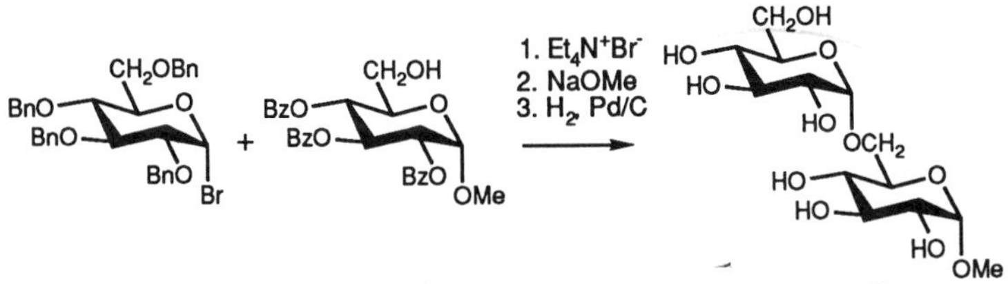

3   **4**   **5**

Scheme 2

saccharides series, since stepwise extension of the isomalto-oligosaccharide chain requires a D-glucosyl donor containing a selectively removable blocking group at *O*-6. A compound of this type, 6-*O*-acetyl-2,3,4-tri-*O*-benzyl-α-D-glucopyranosyl chloride **7**, was first obtained by Pravdic and Keglevic (20) from 1,6-anhydro-β-glucopyranose *via* its 2,3,4-tri-*O*-benzyl derivative **6** (Scheme 3).

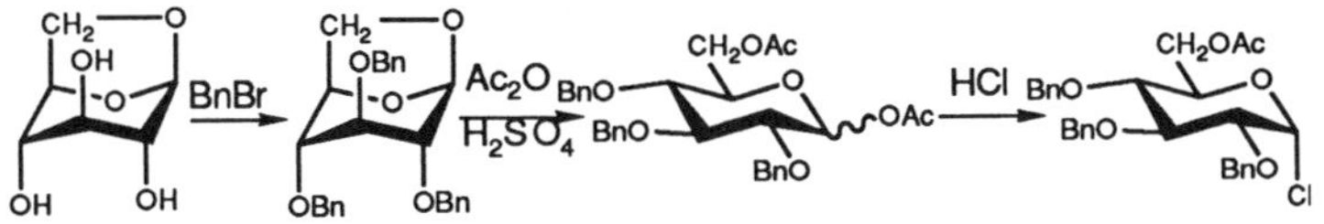

6   **7**

Scheme 3

Halide **7** became readily available (21) by the selective acetolysis of the benzyl group at position 6 of commercially available 2,3,4,6-tetra-*O*-benzyl-D-glucose. (Scheme 4).

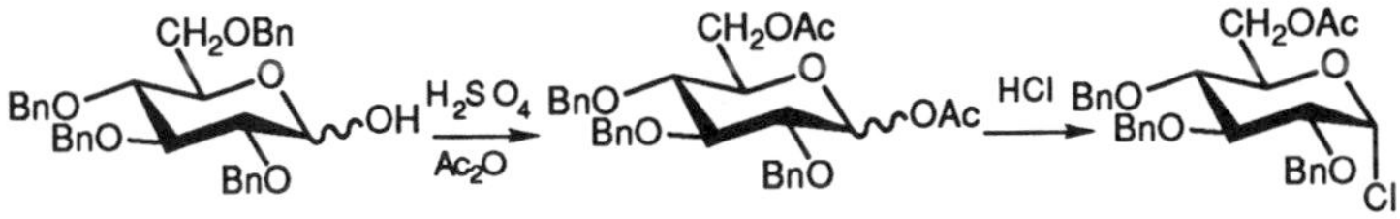

7

Scheme 4

In our approach to the stepwise synthesis of methyl α-glycosides of isomalto-oligosaccharides, we prepared glucosyl chloride **7** by a slight modification of the existing (22) procedure. Next (Scheme 5), **7** was condensed with methyl 2,3,4-tri-*O*-benzyl-α-D-glucopyranoside (19,22) **8** by silver perchlorate-mediated (23) glycosylation, and the fully protected product **9** was deacetylated. Derivative **10** was then hydrogenolysed to give methyl α-isomaltoside **5**.

    To prepare methyl α-isomaltotrioside **12**, compound **10** was submitted to another condensation with **7**, followed by a two-step deprotection. This approach is similar to that of Eby and Schuerch, (24) who have described a stepwise synthesis of fully protected compounds in this series, up to the hexasaccharide. They also prepared the deprotected methyl α-isomaltopentaoside.

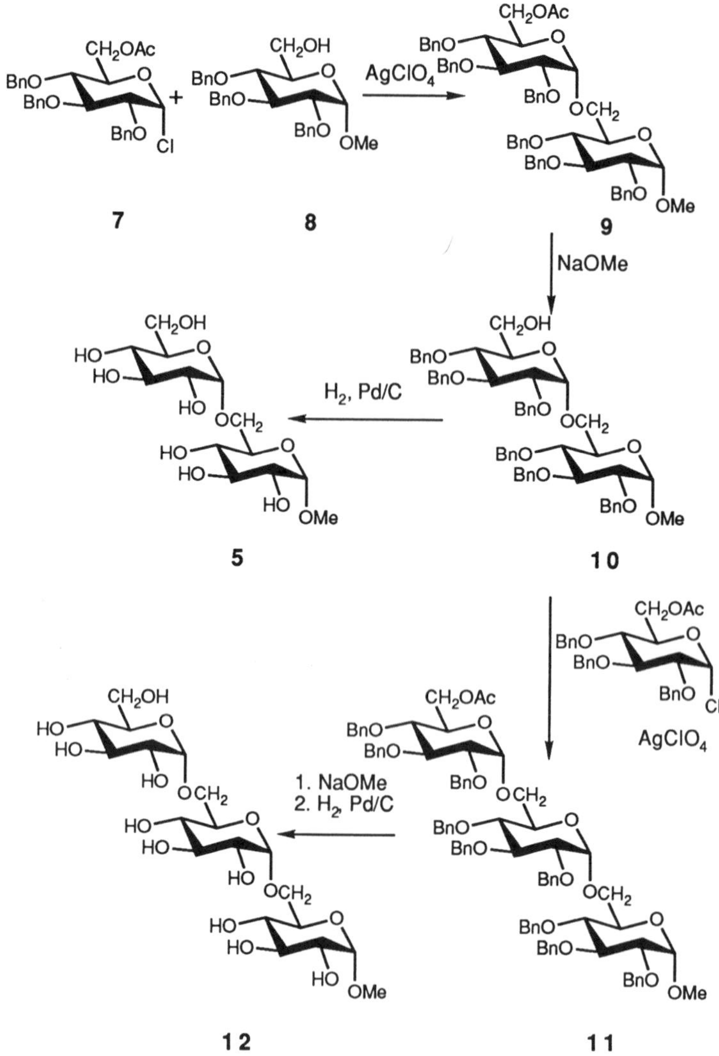

Scheme 5

They also prepared the deprotected methyl α-isomaltopentaoside. The blockwise synthesis of isomalto-tetraose and isomalto-octaose was pioneered by Koto *et al.* (25) who used ethyl $2^1,2^2,3^1,3^2,4^1,4^2$-hexa-*O*-benzyl-6^2-*O*-*p*-nitrobenzoyl-1-thio-α-isomaltoside as the glycosyl donor. The yields of the coupling reactions were from poor to moderate. Nevertheless, they suggested the advantage of the blockwise synthesis of higher oligosaccharides, as compared to a stepwise assembly. We have explored the use of other glycosyl (isomaltosyl) donors for the syntheses of methyl α-glycosides of higher isomalto-oligosaccharides. It appeared to us that an alkyl 6^2-O-acetyl-$2^1,2^2,3^1,3^2,4^1,4^2$-hexa-O-benzyl-1-thio-isomaltoside, such as **16**, could be a versatile compound for this purpose. 1-Thio-glycosides can either be used as glycosyl donors (15) themselves, or they can be readily converted to glycosyl halides by techniques introduced three decades ago (26,27). Compound **16** was previously obtained by Pfäffli *et al.* (28) in 62% yield as the major component of a disaccharide mixture, isolated, formed by condensation of phenyl 2,3,4-tri-O-benzyl-1-thio-β-D-glucopyranoside (**14**) with the chloride **7**.

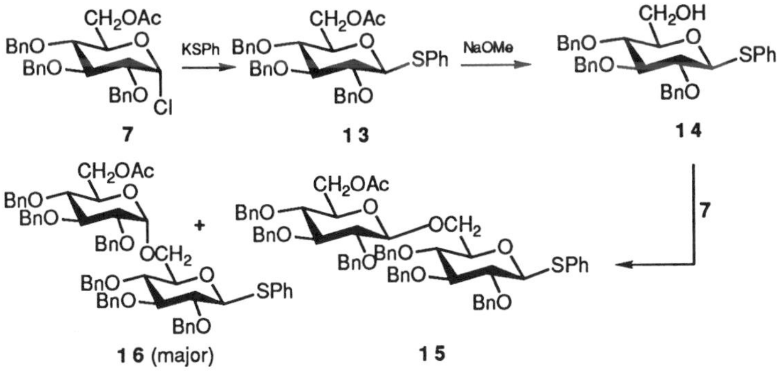

Scheme 6

We obtained the α- and β-linked disaccharides (**15**) and (**16**), in a much higher combined yield (90%), using an improved condensation technique (23). The desired disaccharide (**16**) was again the major

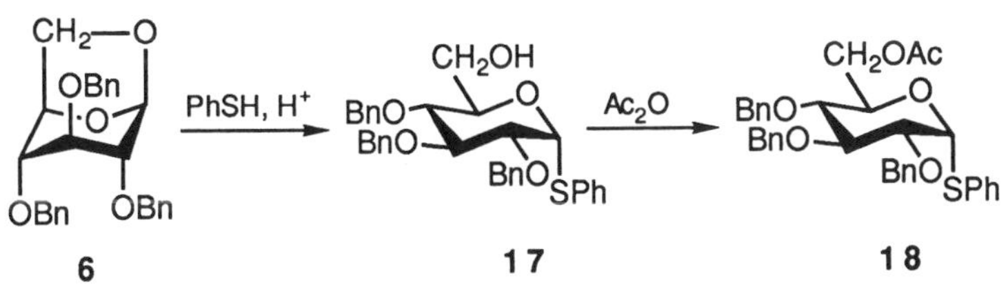

Scheme 7

component of the reaction mixture, but because of separation problems it was difficult to adopt the method to a large scale. A building block in the isomaltose series that was easier to purify turned out to be the

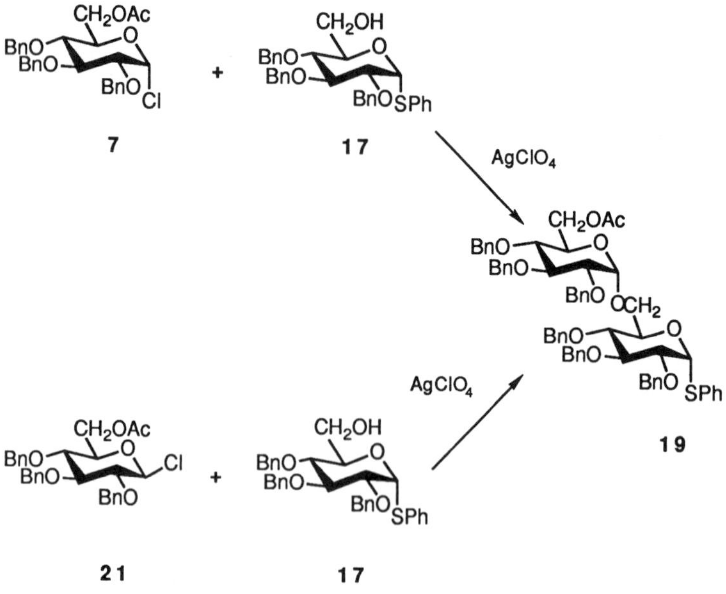

Scheme 8

disaccharide 1-thio-glycoside **19** (Scheme 8). The necessary nucleophile **(17)** was prepared (Scheme 7) from fully benzylated 1,6-anhydro-β-D-glucopyranose (19,29) **(6)** by acid-catalyzed thiophenolysis. Acetylation of **17** readily yielded **18** which was used, together with the isomeric **13**, to study the glycosyl halide-forming reactions of benzylated phenyl 1-thio-glycosides with chlorine.

The reaction of ethyl 2,3,4,6-tetra-O-benzyl-1-thio-α- and β-D-glucopyranosides with bromine in ether was investigated by Weygand and Ziemann (27). Based on optical rotation data, they concluded that the 1-thio-β-D-glycoside reacted with inversion of configuration at C-1 to give 2,3,4,6-tetra-O-benzyl-α-D-glucopyranosyl bromide. With the 1-thio-α-glycoside, the β-glycosyl bromide was formed in a similar fashion, which then anomerized to the thermodynamically more stable α-bromide. We have treated phenyl α- **(18)** and β- **(13)** 1-thioglycosides in carbon tetrachloride with excess (2 moles) of chlorine and monitored the course of the reaction by ¹H-n.m.r. spectroscopy (19). Under these conditions the conversion of either **18** or **13** was practically instantaneous, and no anomerisation of the glycosyl halides was observed over 24 h. While a 1 : 2 mixture of **7** (α) : **21** (β) was formed from the phenyl β-thioglycoside **13**, the reaction of **18** was stereospecific, yielding only 6-O-acetyl-2,3,4-tri-O-benzyl-β-D-glucopyranosyl chloride **(21)** (Scheme 9). Only in the presence of an organic base, such as 2,4,6-trimethylpyridine, did the β-glycosyl

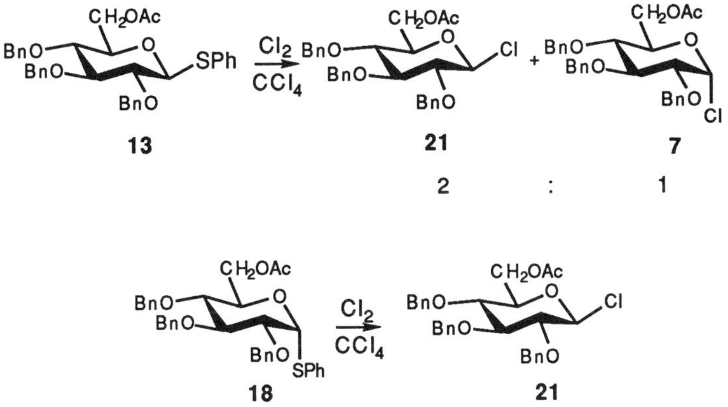

Scheme 9

chloride slowly anomerise (20) to form the thermodynamically more stable α-chloride. This resulted in a 1:1 mixture of **7** and **21** in 24 h. Thus, our observation is at variance with the previous generalization (15) regarding reactions of halogens with 1-thioglycosides.

Compound **21**, when condensed with **17**, produced (19) the disaccharide **19** with the same stereo-selectivity as did halide **7** (Scheme 8). When a solution of the disaccharide 1-thioglycoside **19** in carbon tetrachloride was treated with chlorine (Scheme 10) the corresponding β-glycosyl chloride **22** was again formed rapidly and stereo-specifically, and, once more, no anomerisation was observed

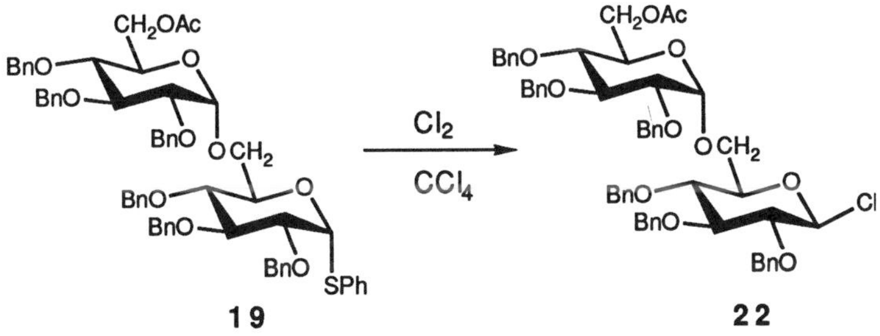

Scheme 10

in the reaction mixture during the 24 h period following its preparation. Fig. 1 shows a ¹H-n.m.r. spectrum of **22** taken at 500 MHz. The proton-signal assignment shown was supported (19) by two-dimensional, ¹H-¹H and ¹H-¹³C correlation spectroscopy.

Crystalline compound **22** is relatively stable and easy-to-handle, and it is a convenient disaccharide building block for making isomalto-oligosaccharides and related substances. We have demonstrated (19)

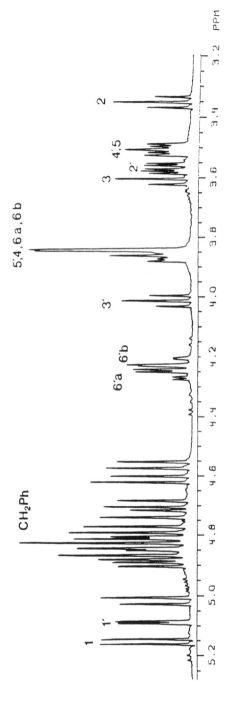

Fig. 1. ^{1}H-n.m.r. spectrum of **22** taken at 500 MHz. Prior to the measurement, the reaction mixture was kept at room temperature for 24 h, concentrated at $40°/0.5$ mmHg with addition of toluene (thrice, to remove all volatiles), and the residue was dissolved in $CDCl_3$. The conversion **19 → 22** was carried out in an n.m.r. tube using 2 molar equivalents of chlorine and CCl_4 as the solvent. The conversion was practically instantaneous. Except for the solvent shift-effect, the spectrum taken 1 min after addition of the reagent (a solution of chlorine in CCl_4) was essentially the same as the one shown.

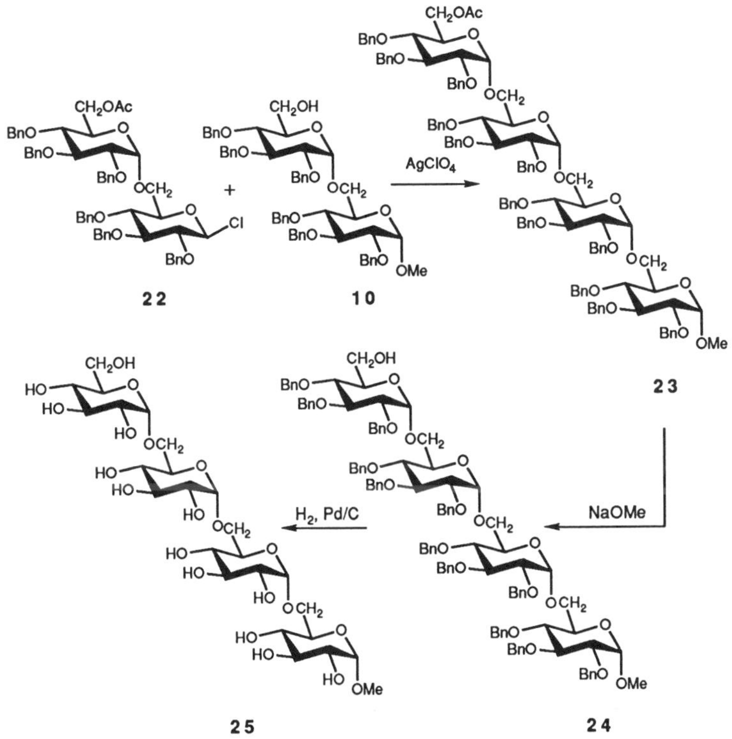

Scheme 11

the versatility of the glycosyl donor **22** by its use in the synthesis of methyl α-glycosides of isomaltotetraose (**25**, Scheme 11) and isomaltopentaose (**27**, Scheme 12). In addition, its versatility is clear in the preparation of isomaltohexaose (**28**, Scheme 13), and isomalto-octaose (**32**, Scheme 14). To make our highest oligosaccharide (**32**) in this series, halide **22** was first condensed (Scheme 15) with the nucleophile **29**, obtained by simple deacetylation of the thioglucoside **19**. Similar to our earlier observations, the resulting tetrasaccharide 1-thioglucoside **30**, was readily, and stereo-specifically converted by chlorine to the corresponding glucosyl chloride **31**.

These results show that (1→6)-α-D-gluco-oligosaccharides can be efficiently synthesized using as glycosyl donors α- or β-glucosyl chlorides of mono- or oligosaccharides that bear a non- participating benzyl group at position 2. Moreover, we have shown that these compounds can be readily obtained from the corresponding phenyl 1-thio-glucosides, as their reaction with chlorine in carbon tetrachloride is fast and simple to perform. Preparation of glucosyl chlorides from 1-thioglucosides in this way has a special advantage when glucosyl donors are to be generated from higher oligosaccharides. For instance

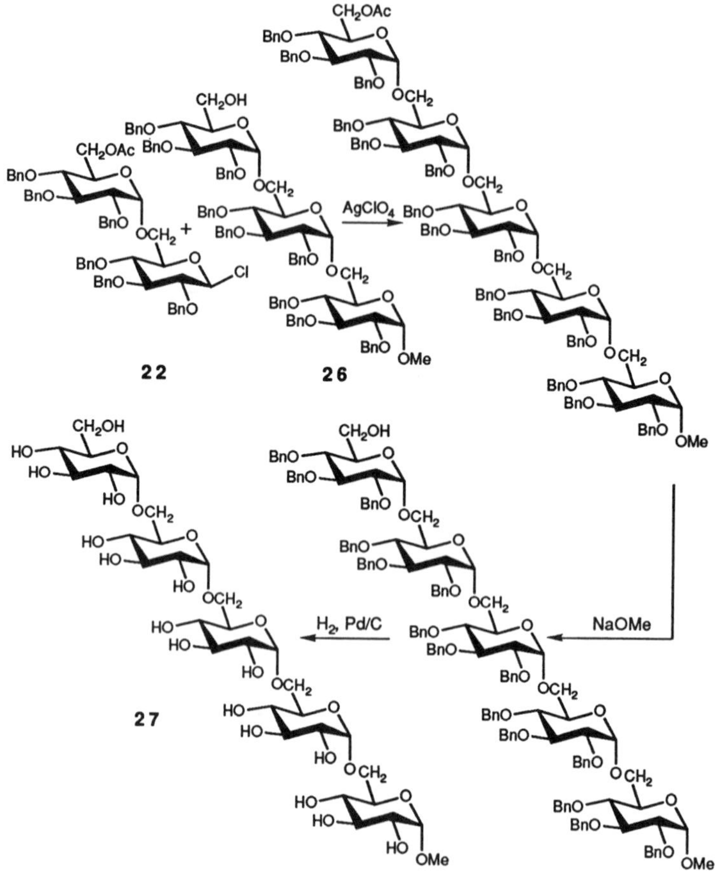

Scheme 12

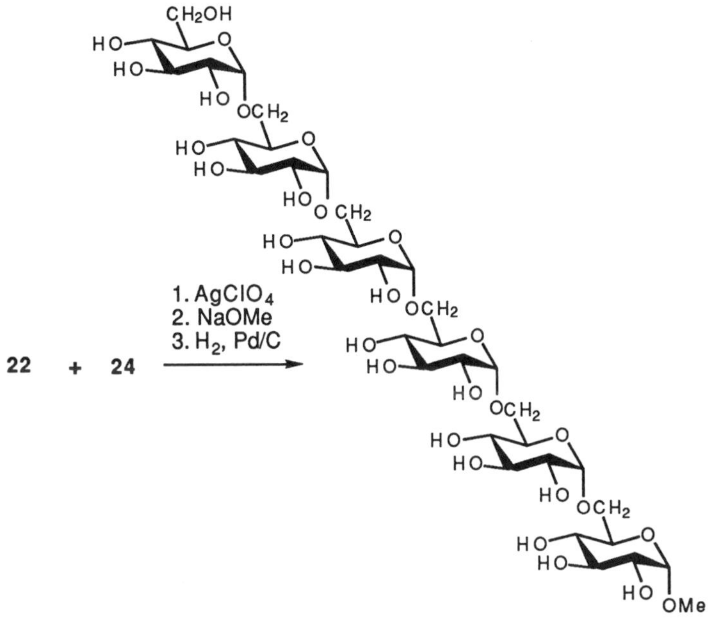

Scheme 13

cleavage of substituents at the anomeric center of sugars with 1,1-dihalogenomethyl methyl ethers (30) or boron trichloride (31) to generate glucosyl chlorides, can be accomplished chemo-selectively, although side reactions do occur (31-33). However, no side reactions were observed during the conversions leading to the glucosyl chlorides **22** and **31,** derived from a di- and a tetra-saccharide 1-thioglucoside, respectively. The use of benzylated glucosyl chlorides as glucosyl donors is much easier than that of their bromo analogs.

Glycosylations mediated by methyl(methylthio)sulfonium triflate (34) $CuBr_2$-Bu_4NBr (35), or alkyl sulfenyl triflates (36) in conjunction with 1-thioglycosides do not seem to give good 1,2-α-(*cis*)-stereo-selectivity and/or tend to be sluggish. Another approach (37) employing the conversion of a disaccharide 1-thioglycoside to the corresponding glucosyl bromide followed by glycosylation resulted in a noticeably low yield (33%) of the trisaccharide, presumably due to side reactions of the reactive glucosyl bromide.

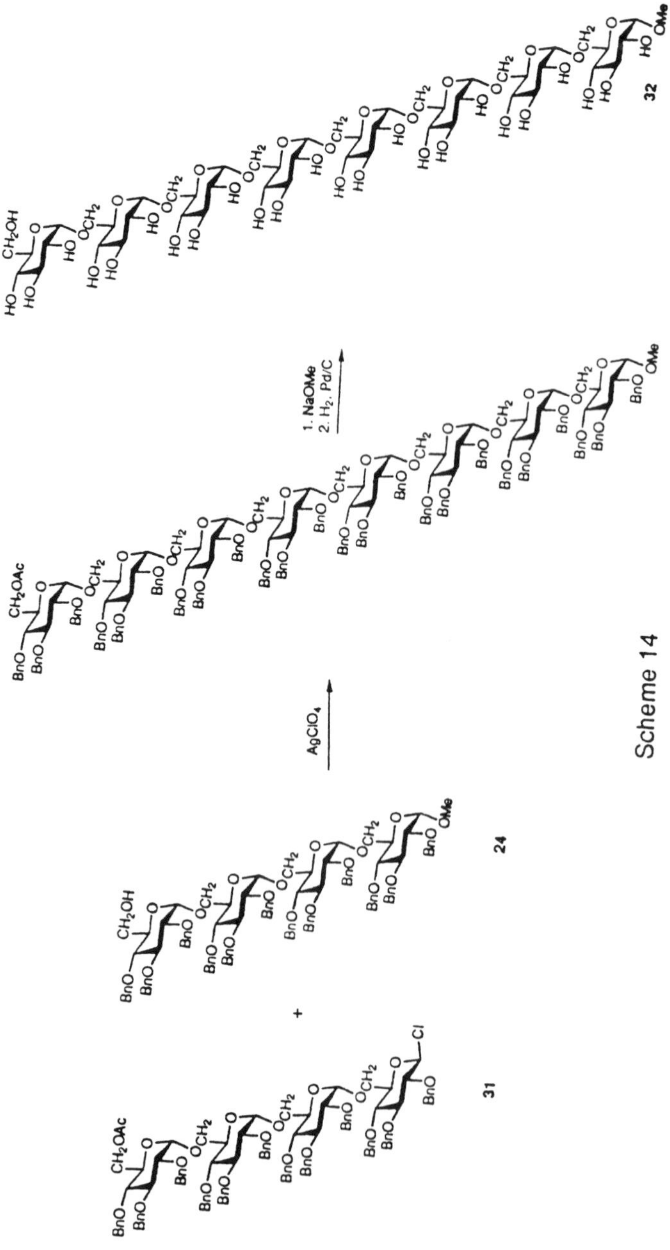

Scheme 14

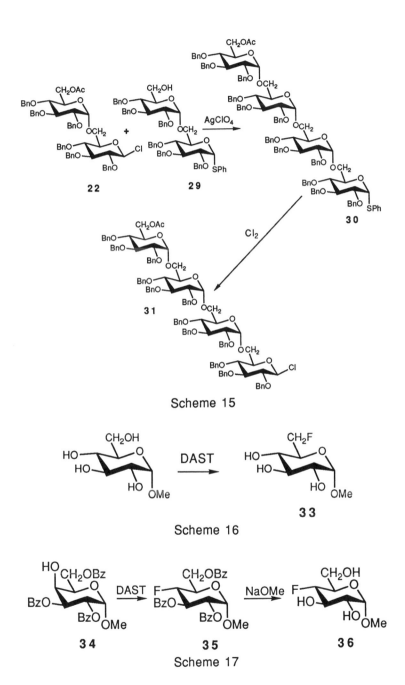

Scheme 15

Scheme 16

Scheme 17

SYNTHESIS OF DEXTRAN-RELATED LIGANDS CONTAINING DEOXYFLUORO GROUPS

The 6-deoxy-6-fluoro derivative **33** was prepared by the procedure of Card (38), who showed that the primary position in methyl α-D-glucopyranoside can be fluorinated selectively (Scheme 16) by treatment with diethylaminosulfur trifluoride (DAST), while treatment of **34** with DAST gave methyl 2,3,6-tri-*O*-benzoyl-4-deoxy-4-fluoro-α-D-glucopyranoside (**35**). We obtained methyl 4-deoxy-4-fluoro-α-D-glucopyranoside (**36**) by debenzoylation of **35** (Scheme 17). Methyl 3-deoxy-3-fluoro-α-D-glucopyranoside (**39**) had previously been obtained (39) in admixture with the β-anomer. When we treated (40) 1,2:5,6-di-*O*-isopropylidene-3-deoxy-3-fluoro-D-glucofuranose (**41**) (Scheme 18) with methanolic hydrogen chloride a mixture of methyl glucosides was formed from which the pure α-anomer could not be isolated. However, when the mixture was acetylated, derivatives **37** and **38** were readily separable, and deacetylation of **38** then gave the desired 3-deoxy-3-fluoro derivative **39**.

Several approaches were investigated to prepare methyl 2-deoxy-2-fluoro-α-D-glucopyranoside (**46**). An attempt to introduce fluorine into position 2 of methyl 3-*O*-benzyl-4,6-*O*-benzylidene-α-D-mannopyranoside (**40**) using DAST as a fluorinating reagent, expected to give the D-*gluco* derivative, was unsuccessful (42). Because of the participation of the axial methoxyl group at C-1 of **40**, 3-*O*-benzyl-4,6-*O*-benzylidene-2-*O*-methyl-α- (**41**) and β- (**42**) -D-glucopyranosyl fluorides were formed as major reaction products, together with methyl 3-*O*-benzyl-4,6-*O*-benzylidene-α-D-*erythro*-hex-2-enopyranoside (**43**) (Scheme 19). Various attempts (40) to prepare the methyl α-glycoside **46** via a protected glycosyl halide, by reacting it with methanol under conditions of the modified Königs-Knorr reaction, failed. Compound **46** was eventually obtained (40) from a mixture of methyl glycosides formed by methanolysis of 1,3,4,6-tetra-*O*-acetyl-2-deoxy-2-fluoro-β-D-glucopyranose (**45**) followed by fractional crystallization. Compound **45** in turn had been obtained by treatment (43) of 1,3,4,6-tetra-*O*-acetyl-β-D-mannopyranose (**44**) with DAST (Scheme 20).

Next we required a series of oligosaccharides containing a terminal 6-deoxy-6-fluoro-α-D-glucopyranosyl group. Two strategies presented themselves for the assembly of the fluorinated glycosides **47**, **51**, and **52** (Schemes 21-23). In the first, fully substituted methyl α-glycosides of isomalto-oligosaccharides containing a selectively removable blocking group at the position to be fluorinated could be synthesized. After partial deblocking, the free primary hydroxyl group at the designated position would be transformed into the fluorodeoxy function by one of the many methods available (45-47). Alternatively, a glycosyl donor derived from 6-deoxy-6-fluoro-D-glucose could be condensed with a suitable derivative of methyl α-D-glucopyranoside. We have used both approaches. In the first approach, compound **10** was treated (44) with DAST, and the resulting, fully protected, fluorinated disaccharide was hydrogenolyzed, to give the desired derivative **47** (Scheme 21). We found during this work (44) that

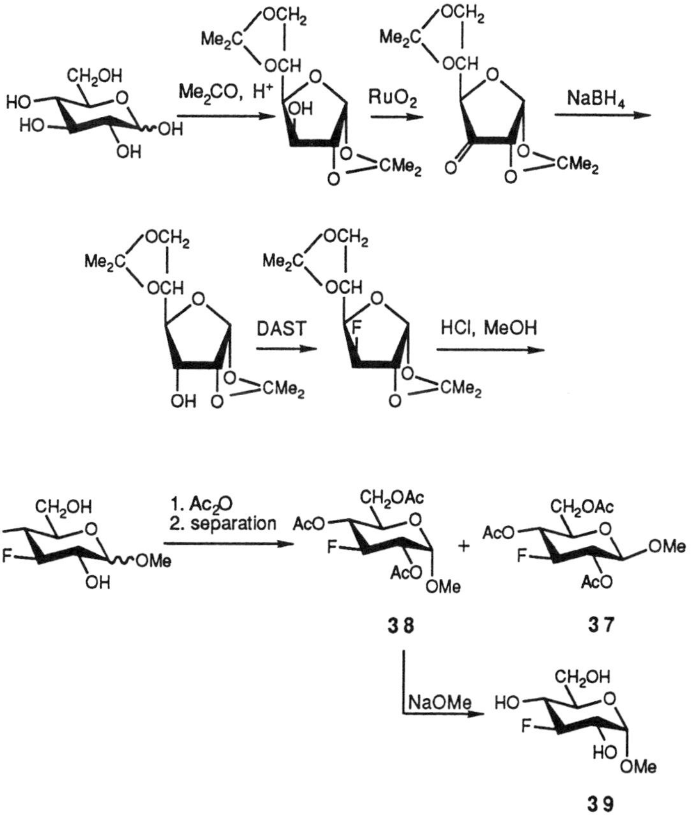

Scheme 18

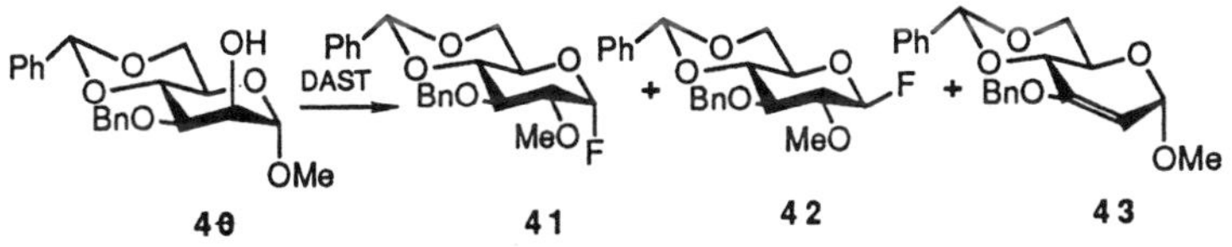

Scheme 19

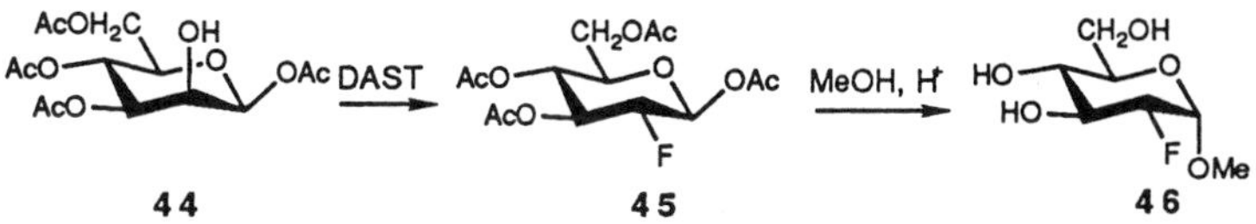

Scheme 20

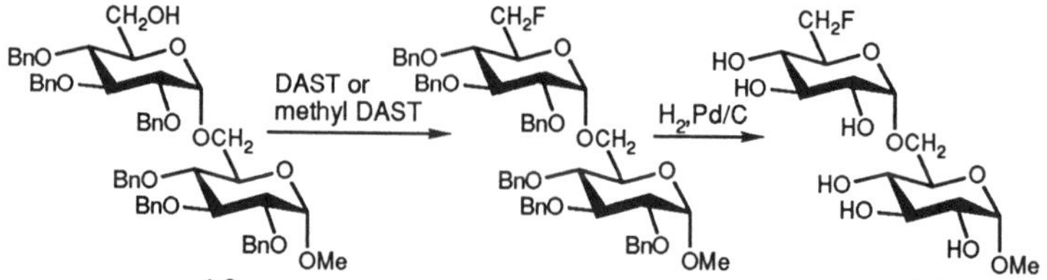

Scheme 21

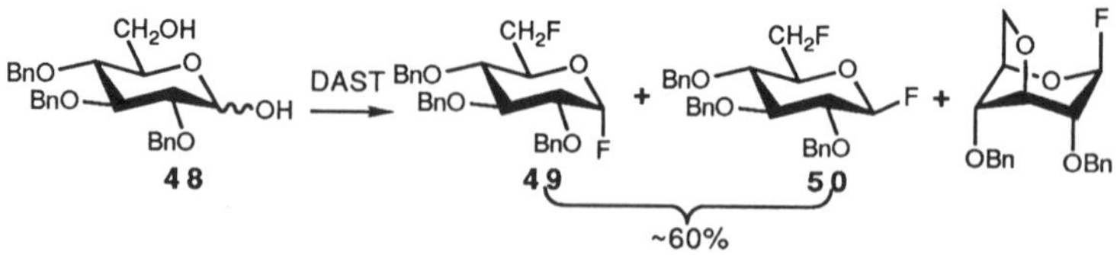

Scheme 22

fluorination could best be accomplished by using dimethylaminosulfur trifluoride (methyl DAST) instead of DAST.

To use the second approach, a 6-deoxy-6-fluoro-α-D-glucosyl donor had to be synthesized. It has been shown (45) that glycosyl fluorides, activated with stannous chloride can be used as glycosyl donors in stereo-controlled α-glycosidation reactions. Successful application of this method prompted us to develope a synthesis starting with 2,3,4-tri-O-benzyl-D-glucose (48) to affect simultaneous C-OH → C-F conversions at both the primary- and the anomeric position using DAST (46). Accordingly (47), 48 gave (Scheme 22) the corresponding glycosyl fluorides 49 and 50 in a combined yield of ~60%. Their condensation (19,44) with nucleophiles 26 and 24 gave the fully substituted, specifically fluorinated isomalto-oligosaccharides. The desired ligands 51 and 52 were obtained then by simple hydrogenolysis (Scheme 23).Other compounds which had to be prepared for our binding studies were methyl α-glycosides of 6-deoxy-6-fluoro derivatives of the four theoretically possible (1→2)- and (1→3)-linked D-glucobioses, i.e. sophorose [β-(1→2)], laminaribiose [β-(1→3)], kojibiose [α-(1→2)], and nigerose [α-(1→3)]. Derivatives (48) 53 and 54 were used as nucleophiles in the coupling reactions leading to these substances. To synthesize methyl 6'-deoxy-6'-fluoro-α-kojibioside (55, Scheme 24) and methyl 6'-deoxy-6'-fluoro-α-nigeroside (56, Scheme 25) we used the same approach as in the synthesis of the fluorinated isomalto-oligosaccharides, namely condensations of a benzylated, 6-fluorinated D-glycosyl chloride promoted with silver perchlorate. The 6-fluorinated glycosyl donor used to assemble methyl 6'-deoxy-6'-fluoro-α-sophoroside (61) and methyl 6'-deoxy-6'-fluoro-α-laminaribioside 62

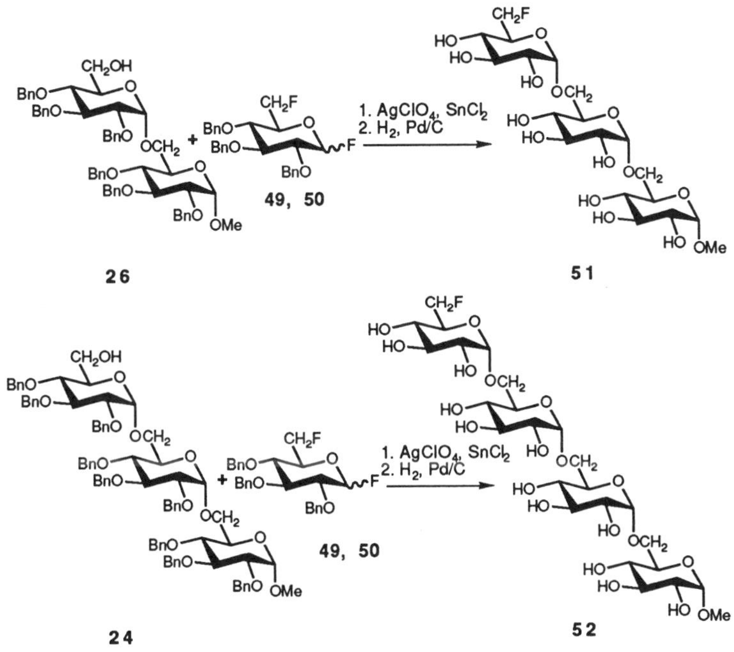

Scheme 23

had to bear a trans-directing participating group at position 2. The benzoylated glucosyl chloride **60**, derived from 6-deoxy-6-fluoro-D-glucose, could be prepared from three different starting materials, as shown in Scheme 26. In the first approach, methyl 2,3,4-tri-*O*-benzoyl-α-D-glucopyranoside, prepared from methyl α-D-glucopyranoside by sequential tritylation, benzoylation and detritylation, was fluorinated using DAST, and the resulting derivative **57** was cleaved (30) with dichloromethyl methyl ether, to give the glucosyl chloride (**60**). The same glucosyl donor was obtained by cleavage (30) of the α- (**58**) or β- (**59**) benzoate derivatives. These were prepared (49) (Scheme 26) by benzoylation of 6-trityl-D-glucose, detritylation, separation of the resulting detritylated α- and β-anomers, and their subsequent fluorination. The target disaccharides **61** and **62** were then prepared as shown in Schemes 27 and 28, respectively.

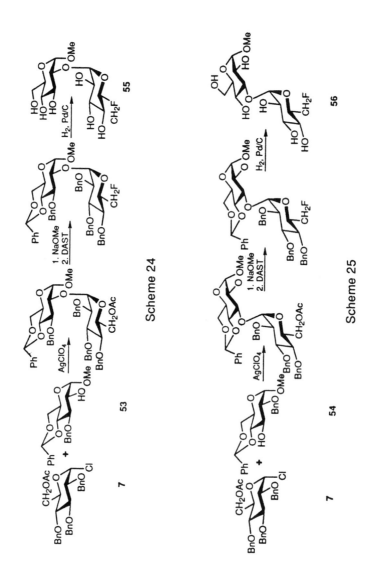

Scheme 24

Scheme 25

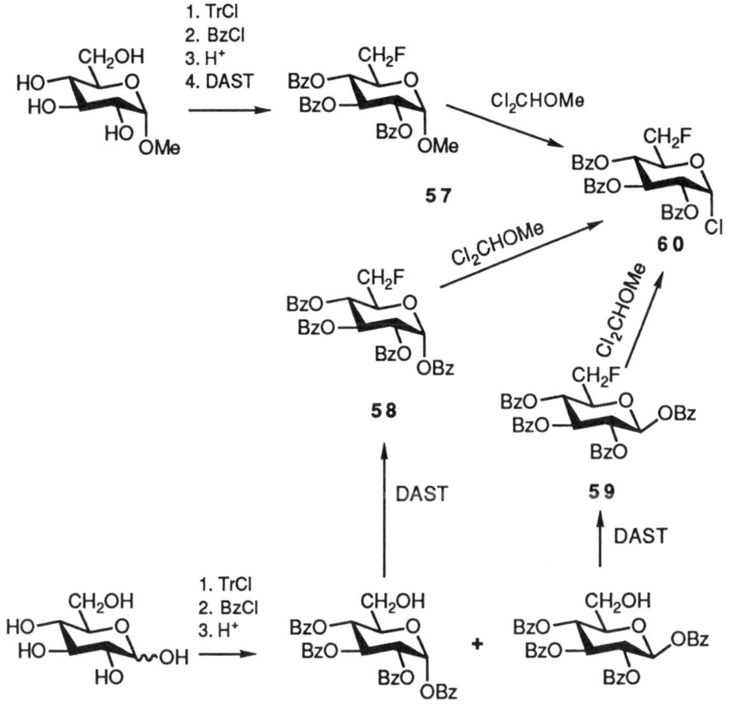

Scheme 26

BINDING STUDIES AND CONCLUSIONS

The binding of the synthetic haptens described above, as well as various known methyl deoxy α-D-glucopyranosides, were studied with two murine monoclonal antibodies. These are W3129 (50,51) and 16.4.12E (52), both IgAs. Their amino acid sequences are known (52,53), and the proposed three dimensional structures for their variable fragment (Fv), comprising the first two domains, have been proposed (10,54).

Our laboratory has frequently used the ligand-induced change of immunoglobulin fluorescence (55,56) to determine the affinity constant, Ka, for the binding of small haptens (56) or polysaccharides to antibodies and to (monovalent) Fab fragments (57). In studying the interaction of haptens with IgA W3129 we could show (3) that the immunoglobulin's affinity is maximal for the hapten methyl α-isomalto-tetraoside (25), the same as that for the pentaoside 27 and the hexaoside 28. That affinity was essentially the same as the one found for the Fab fragment of the immunoglobulin with a synthetic, linear (58), polymeric α(1→6)-linked dextran of molecular mass (Mr) ~36,000

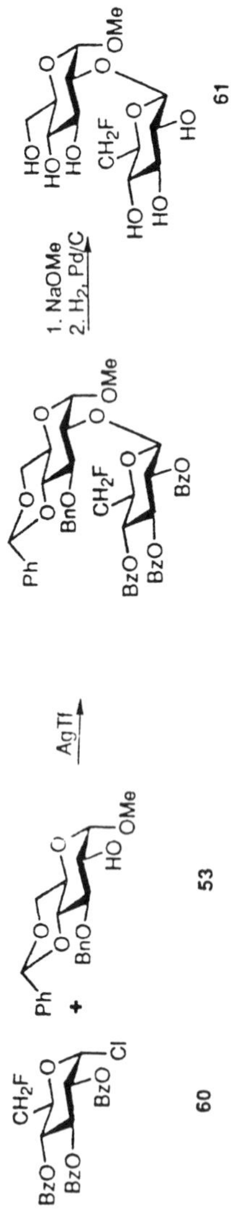

Scheme 27

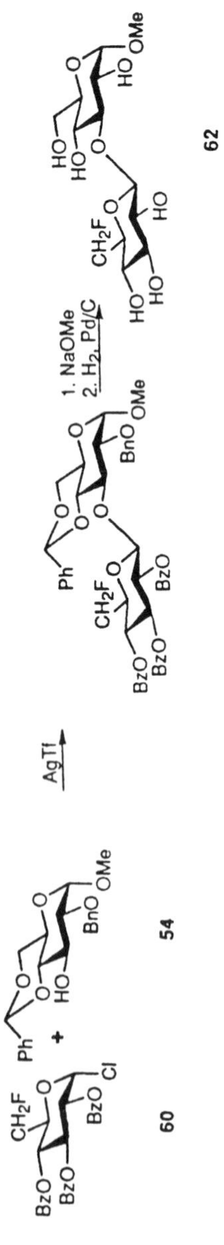

Scheme 28

(59). Thus, only associations in a 1:1 molar ratio can take place between the IgA Fab and that linear dextran polysaccharide (11). In turn that indicates that IgA W3129 binds only to the chain-terminal tetrasaccharide segment of the polysaccharide, and cannot bind to internal sequences of $\alpha(1\rightarrow6)$-linked glucosyl residues. This was confirmed by our laboratory (3) as follows: Methyl α-D-glucopyranoside showed an affinity of 1.8 x 10^3 M^{-1} for the immunoglobulin. Keep in mind that the antibody is capable of maximally binding to four sequential residues of glucose possessing the $\alpha(1\rightarrow6)$-linkage structure. When adding a simple monosaccharide to an antibody whose homologous antigen is a homopolymer of one sugar, that monosaccharide hapten will occupy the antibody subsite that has the highest affinity for that single sugar determinant [It can be calculated (3) that the effect of other subsites can be ignored in those computations]. However, methyl 6-deoxy-6-fluoro α-D-glucopyranoside showed no affinity. Since the only glucosyl residue in an $\alpha(1\rightarrow6)$-linked dextran that carries a free hydroxyl group at C-6 is the non-reducing terminal moiety, that observation shows that the antibody's highest affinity subsite is specific for that terminal residue, and that the C-6 hydroxyl group must be engaged in hydrogen bond donation. This was confirmed by the observation that methyl α-isomaltosyl di-, tri- and tetra-saccharides possessing an upstream 6-deoxy-6-fluoro α-D-glucopyranosyl terminus (47, 51 and 52) showed significantly reduced affinities for the antibody when compared with the affinities of unsubstituted methyl α-isomaltosyl di-, tri- and tetra-saccharides 5, 12 and 25. This is in complete agreement with that terminal saccharide (6-deoxy-6-fluoro-α-D-glucopyranosyl) unit residing, but (essentially) not binding, in the highest affinity subsite. The resulting low affinities for those oligosaccharides were, respectively, 0.5, 2.6, and 4.6 x 10^2 M^{-1} (3).

Methyl 2-deoxy-2-fluoro α-D-glucopyranoside (46) had an affinity for the major subsite that was the same as that of methyl α-D-glucopyranoside, methyl 3-deoxy-3-fluoro α-D-glucopyranoside (39) had a significantly higher affinity (7.7 x 10^3 M^{-1}), while methyl 4-deoxy-4-fluoro α-D-glucopyranoside (36) did not bind at all. This suggests that the IgA W3129 receives a H-bond from OH-4 and OH-6, donates a H-bond to O-3, and that OH-2 is not involved. We also observed that methyl 6'-deoxy-6'-fluoro-laminaribioside (62), as well as methyl 3-O-β-D-galactopyranosyl α-D-glucopyranoside showed binding (1.8 and 2.1 x 10^2 M^{-1} respectively) with IgA W3129. All this indicates that the OH-3 of the glucose residue in the major affinity subsite can tolerate substitution by a large moiety that by itself does not bind.

Our laboratory also studied a hybridoma monoclonal antibody (16.4.12E) obtained by Matsuda and Kabat by immunization with a conjugate antigen of isomaltohexaose linked to keyhole limpet hemocyanin (52). This IgA antibody has a reported amino acid sequence in the V_L and V_H regions that is sufficiently similar to that of IgA W3129 (53,54) so that one might expect it also to be specific for the non-reducing terminal epitope of an $\alpha(1\rightarrow6)$-linked dextran. Corroboration of the end-binding property of 16.4.12E came when others employed Enzyme Linked ImmunoSorbent Assay (ELISA). It was found (52) that the antibody 16.4.12E did not show detectable binding to a linear, synthetic $\alpha(1\rightarrow6)$-linked dextran (58), but did so to a heavily branched dextran possessing 39% terminal glycosyl end groups, 26% $\alpha(1\rightarrow6)$-linked backbone and 35% $\alpha(1\rightarrow6)$-linked backbone with $\alpha(1\rightarrow2)$-linked linked branches (60). In the ELISA test, the amount of IgA binding to the linear dextran is too small to be dectable (1:1 on a molar basis if the antibody binds the glycosyl terminus only), while with the heavily branched dextran, 39 moles of antibodies could bind per dextran molecule, and these would be detected more easily.

In binding studies our laboratory showed (10) that with IgA 16.4.12E (52) there is a dramatic decrease in binding in going from methyl α-D-glucopyranoside (Ka 4.5 x 10^3 M^{-1}) to methyl 6-deoxy-6-fluoro-α-D-glucopyranoside (33, Ka 0.3 x 10^2 M^{-1}). Again, methyl α-isomaltosyl di-, tri- and tetra-saccharides possessing an upstream 6-deoxy-6-fluoro α-D-glucopyranosyl terminus (47, 51 and 52) showed significantly reduced affinities for the antibody when compared with the affinities of unsubstituted methyl α-isomaltosyl di-, tri- and tetra-saccharides 5, 12 and 25. Here too, that suggests that the terminal, upstream 6-deoxy-6-fluoro α-D-glucopyranosyl unit resides in the major subsite but shows only very weak binding. The resulting affinities for those oligosaccharides in this case were 3.6, 6.3 and 37 x 10^2 M^{-1} respectively. It is possible, but unlikely that an antibody would be selected for on the basis of an affinity for internal residues of the dextran that is as low as 3.7 x 10^3 M^{-1}. Thus it shows the crucial role of the major affinity for the terminal glusosyl unit in the selection of IgA 16.4.12E.

Here the methyl 3-deoxy-3-fluoro-α-D-glucopyranoside 39 had a lesser affinity than methyl α-D-glucopyranoside had, indicating that for this IgA the position of the glycosyl group in the major subsite is different from that in IgA W3129. Methyl 2-deoxy-2-fluoro-α-D-glucopyranoside (46) bound reasonably well (Ka 0.8 x 10^2 M^{-1}), but methyl 4-deoxy-4-fluoro-α-D-glucopyranoside (36) did not bind at all. These data show that the major subsite of 16.4.12E is H-bonded to OH-6 and OH-4 of the methyl α-D-glucopyranoside hapten (donation by the sugar), and that subsite represents therefore the one binding the terminal glucosyl group of a $\alpha(1\rightarrow6)$dextran. The H-bond from the protein to OH-3 of the terminal glucosyl group present in W3129, is evidently absent in IgA 16.4.12E. The spacial orientation of the terminal glucosyl group in the major subsite also differs from that in IgA W3129. Here the sugar moiety must have its OH-3 group located close to the protein subsite's van der Waals surface, as neither methyl 6'-deoxy-6'-fluoro-laminaribioside (62), nor methyl 3-O-β-D-galacto-pyranosyl α-D-glucopyranoside showed any binding.

A model for the three dimensional structure of the Fv of IgA W3129 has been published (54), and our laboratory proposed a model for the F_V of IgA 16.4.12E based on the analogy with W3129 (10). From the binding affinities of the large number of our synthetic haptens mimicking the determinant of the dextran chain it is possible to propose a detailed mode of binding of that polysaccharide to these two antibodies.

Begining with IgA W3129, we observed that methyl α-D-glucopyranoside, as well as all α-(1→6)-linked glucopyranosyl oligosaccharides perturbed the tryptophanyl (TRP) fluorescence of the immunoglobulin to the same extent, i.e. the protein subsite with the major affinity for its' glucosyl residue harbored the perturbable tryptophanyl residues. In the published model (54) there is a pronounced, deep cavity at the interface of the H and the L chain in the Fv, and in its wall it posseses the only solvent exposed TRP residue at position 33H of the F_V (In the amino acid numbering, H stands for the heavy-, and L for the light-chain, and the numbering system used is that of Kabat *et al.* (61)). Therefore, that cavity appears to be the subsite with the major affinity for its glucosyl residue. Neither methyl 6-deoxy-α-D-glucopyranoside nor methyl 6-deoxy-6-fluoro-α-D-glucopyranoside bind to IgA W3129. These two ligands both lack the 6-hydroxyl group, and in the dextran the upstream terminal glucosyl group is the only sugar residue in the entire antigen which posseses a 6-hydroxyl group. This suggests that this cavity is the subsite for the dextran's glycosyl terminus. Thus that terminal sugar donates a hydrogen bond to the protein. Methyl 3-deoxy-3-fluoro-α-D-glucopyranoside (**39**) shows strong binding, while the corresponding 3-deoxy derivative does not. This indicates that the 3-OH group can receive a H-bond from the protein (Fluorine is electronegative, and *can* receive a hydrogen bond (2,3)). Histidine 93L can donate such a H-bond. Since neither the 4-deoxy-4-fluoro nor the 6-deoxy-6-fluoro derivatives (**36** and **33**) bind, H-bond donation by the 4- and 6-hydroxy groups of the terminal glucosyl moiety appears critical. There are two amino acid residues that can receive a H-bond: Glutamic acid 50H and the carbonyl group of Serine 92L. Methyl α-D-glucopyranoside is placed in the subsite interacting with these residues. Since we observed that substitution at position 3 in that hapten with a large moiety still allows binding, that position would have to project out of the subsite, rather than to its spacially restricted rear. Finally, it must be kept in mind that methyl α-D-glucopyranoside represents the chain-terminal residue of the dextran antigen, i.e., the remainder of the polysaccharide chain must be able to exit from the subsite. That means that the CH_3 group (which represents the CH_2 group of the penultimate glucose of the dextran) must face outwards toward the solvent. Simultaneous submission to these restraints allows a good fit into the subsite (Fig. 2a). When the van der Waals surface of the glucoside is displayed, this fit is remarkable for its fine tolerance (Fig. 2b).

For the IgA 16.4.12E, we found that the ligand-induced protein fluorescence change was exceptionally high. This would be expected if multiple TRP residues would be perturbed upon binding of ligands. In our published model of IgA 16.4.12E (10) there are three such TRP residues, 33H, 100^{A}H and 96L, located in or near the pronounced cavity at the H/L interface of the F_V. For this immunoglobulin also, methyl α-D-glucopyranoside and all subsequent oligo-saccharides in

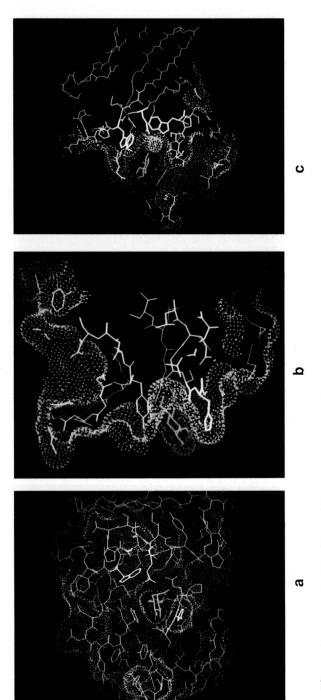

Figure 2a. IgA W3129 with methyl α-D-glucopyranoside, representing the terminal glucosyl residue of a α(1,6)-dextran, located in its major affinity subsite. Note the white outline of TRP 33H, GLU 50H, HIS 93L, and the carbonyl of SER 92L. Also note the OH-3 and the OCH₃ projecting out of the subsite. For details see text.

Figure 2b. IgA W3129 with the methyl α-D-glucopyranoside in the major subsite as above, showing its van der Waals surface.

Figure 2c. IgA 16.4.12E with methyl α-D-glucopyranoside, representing the terminal glucoside residue of a α(1,6)-dextran, in its major subsite. Two of the three TRP residues (33H and 96L) are visible, the third (100^AH) is outside the periphery of the projection. Note that the glucoside is turned over relative to the position it had in W3129. Here its OH-3 projects to the back of the subsite, while the OCH₃ still projects out toward the solvent. For details see text.

the methyl α-glycosides of the α-(1→6)-series (**5, 12, 25, 27** and **28**) cause nearly the same TRP fluorescence change on binding. Again this indicates that the subsite possessing the major affinity for a glucosyl residue holds the perturbable TRP residues. Thus, we propose that cavity to be the major affinity subsite. We have placed methyl α-D-glucopyranoside in there (Fig. 2c), subject to the following restrictions: Since that moiety again represents the glucosyl antigenic terminus, and its OCH_3 group mimicks the CH_2 of the penultimate glucosyl residue, that group is positioned as exiting the subsite towards the solvent. Hydroxyl groups 4 and 6 donate H-bonding, and were placed to interact with the carboxyl groups of Glycine 96H and 98H. The hydroxyl groups 2 and 3 were placed in the cavity near a water molecule that was H-bonded to the ε-oxygen of Glutamic acid 50H, based on the very weak binding of methyl 2-deoxy-α-D-glucopyranoside and the corresponding 3-deoxy-3-fluoro glycoside **39**, as well as the lack of binding of methyl 3-deoxy-α-D-glucopyranoside. All H-bonding proposed was of the correct required distance, and others (62) have shown that network H-bonds involving water are quite common in saccharide-protein interactions. In the case of the binding of the glucosyl residue in the major affinity subsite of IgA 16.4.12E there is one great difference, when compared to IgA W3129: since methyl 3-O-β-D-galactopyranosyl α-D-glucopyranoside a derivative of methyl α-D-glucopyranoside bearing a large substitution at position 3, does not bind, position 3 is likely to be in a spacially hindered environment. Thus the 3-OH is turned to the back of the cavity. Subject to those restraints, Figure 2c shows the mode of binding.

To recapitulate: First, we have shown that 1,2-cis glucosides can be easily prepared by silver perchlorate mediated condensation using benzylated glucosyl chlorides as donors. These can be generated efficiently by treatment of parent 1-thioglucosides with chlorine. Second, it can be seen that in both W3129 and 16.4.12E the maximally binding determinant is the tetrasaccharide. In W3129 the subsite affinities diminish in order of their location (3), while in l6.4.12E, the fourth subsite has a slightly higher affinity than that of the third subsite (10). Also it should be noted that a synthetically modified ligand can have a higher affinity than the natural one, comparing for instance the affinities of methyl α-D-glucopyranoside (Ka 1.8 x 10^3 M^{-1}) and methyl 3-deoxy-3-fluoro α-D-glucopyranoside (Ka 7.7 x 10^3 M^{-1}) for IgA W3129.

It must also be remembered that affinity measurements of ligands and antibodies by microcalorimetry can be more informative with respect to obtaining additional thermodynamic parameters. In the method employed by us, only the affinity constsnt is obtained. The advantage of the method used by us is that it requires far less antibody.

LITERATURE CITED
1.   Glaudemans, C. P. J.; Kovac, P. *ACS Symposium Series,* 1988, **374,** 78.
2.   Murray-Rust, P.; Stallings, W. C.; Monti, C. T.; Preston, R. K.; Glusker, J. P. *J. Amer. Chem. Soc.,* 1983, **105,** 3206.
3.   Glaudemans, C. P. J.; Kovac, P.; Rao, A. S. *Carbohydr. Res.,* 1989, **190,** 267.
4.   Poretz, R. D.; Goldstein, I. J. *Biochemistry,* 1970, **9,** 2890.
5.   Goldstein, I. J.; Reichert, C. M.; Misaki, A. 1., 234, 283. *Ann. N.Y. Acad. Sc.,* 1974, 283.
6.   Nam Shin, J. A.; Maradufu, A.; Marion, J.; Perlin, A. S. *Carbohydr. Res.,* 1980, **84,** 328.
7.   Ittah, Y.; Glaudemans, C. P. J. *Carbohydr. Res.,* 1981, **95,** 189.
8.   Glaudemans, C. P. J.; Kovac, P.; Rasmussen, P. *Biochemistry,* 1984, **23,** 6732.
9.   Glaudemans, C. P. J. *Chem. Rev.,* 1991, **91,** 25.
10.   Nashed, E. M.; Perdomo, G. R.; Padlan, E. A.; Kovac, P.; Matsuda, T.; Kabat, E. A.; Glaudemans, C. P. J. *J. Biol. Chem.,* 1990, **265,** 20699.
11.   Glaudemans, C. P. J.; Bennett, L. G.; Bhattacharjee, A. K.; Nashed, E. M.; Ziegler, T. *A. C. S. Symposium Series,* 1993, **519,** 72.
12.   Schmidt, R. R. *Angew. Chem. Int. Ed. Engl.,* 1986, **25,** 212.
13.   Flowers, H. *Methods in Enzymology,* 1987, **138,** 359.
14.   Paulsen, H. *Angew. Chem. Int. Ed. Engl.,* 1990, **29,** 823.
15.   Fügedi, P.; Garegg, P. J.; Lönn, H.; Norberg, T. *Glycoconjugate J.,* 1987, **4,** 97.
16.   Toshima, K.; Tatsuta, K. *Chem. Rev.,* 1993, **93,** 1503.
17.   Igarashi, K.; Irisawa, J.; Honma, T. *Carbohydr. Res.,* 1975, **39,** 341.
18.   Forsgren, M.; Jansson, P. E.; Kenne, L. *J. Chem. Soc. Perkin Trans. 1,* 1985, 2383.
19.   Kovac, P.; Lerner, L. *Carbohydr. Res.,* 1988, **184,** 87.
20.   Pravdic, N.; Keglevic, D. *Tetrahedron,* 1965, **21,** 1897.
21.   Eby, R.; Sondheimer, S. J.; Schuerch, C. *Carbohydr. Res.,* 1979, **73,** 273.
22.   Kovac, P.; Alfödi, J.; Kosik, M. *Chem. Zvesti,* 1974, **28,** 820.
23.   Igarashi, K.; Irisawa, J.; Honma, T. *Carbohydr. Res.,* 1975, **39,** 213.
24.   Eby, R.; Schuerch, C. *Carbohydr. Res.,* 1976, **50,** 203.
25.   Koto, S.; Uchida, T.; Zen, S. *Bull. Chem. Soc. Jpn.,* 1973, **46,** 2520.
26.   Weygand, F.; Ziemann, H. *Ann.,* 1962, **657,** 179.
27.   Wolfrom, M. L.; Groebke, W. *J. Org. Chem.,* 1963, **28,** 2986.
28.   Pfaffli, P. J.; Hixson, S. H.; Anderson, L. *Carbohydr. Res.,* 1972, **23,** 195.
29.   Zemplen, G.; Csuros, Z.; Angyal, S. *Chem. Ber.,* 1937, **70,** 1848.
30.   Gross, H.; Farkas, I.; Bognár, R. *Z. Chem.,* 1978, **18,** 201.
31.   Perdomo, G. R.; Krepinsky, J. *Tetrahedron Lett.,* 1987, **28,** 5595.
32.   Kovac, P.; Whittaker, N. F.; Glaudemans, C. P. J. *J. Carbohydr. Chem.,* 1985, **4,** 243.

33. Kovac, P.; Taylor, R. B. *J. Org. Chem.,* 1985, **50**, 5323.
34. Andersson, F.; Fugedi, P.; Garegg, P. J.; Nashed, M. *Tetrahedron Lett.,* 1986, **27**, 3919.
35. Sato, S.; Mori, M.; Ito, Y.; Ogawa, T. *Carbohydr. Res.,* 1986, **155**, C6.
36. Dasgupta, F.; Garegg, P. J. *Carbohydr. Res.,* 1988, **177**, C13.
37. Norberg, T.; Ritzen, H. *Glycoconjugate J.,* 1986, **3**, 135.
38. Card, P. *J. Org. Chem.,* 1983, **48**, 393.
39. Kent, P. W.; Dwek, R. A.; Taylor, N. F. *Tetrahedron,* 1971, **27**, 3887.
40. Kovac, P.; Yeh, H. J. C.; Glaudemans, C. P. J. *Carbohydr. Res.,* 1987, **169**, 23.
41. Klemm, G. H.; Kaufman, R. J.; Sidhu, R. S. *Tetrahedron Lett.,* 1982, **23**, 2927.
42. Kovac, P.; Yeh, H. J. C.; Jung, G. L.; Glaudemans, C. P. J. *J. Carbohydr. Chem.,* 1986, **5**, 499.
43. Kovac, P. *Carbohydr. Res.,* 1986, **153**, 168.
44. Kovac, P.; Sklenar, V.; Glaudemans, C. P. J. *Carbohydr. Res.,* 1988, **175**, 201.
45. Mukayama, T.; Murai, Y.; Shoda, S. *Chem. Lett.,* 1981, 431.
46. Card, P. *J. Carbohydr. Chem.,* 1985, **4**, 451.
47. Kovac, P.; Yeh, H. J. C.; Jung, G. L.; Glaudemans, C. P. J. *J. Carbohydr. Chem.,* 1987, **6**, 423.
48. Ogawa, T.; Kaburagi, T. *Carbohydr. Res.,* 1982, **103**, 53.
49. Kovac, P.; Glaudemans, C. P. J. *J. Carbohydr. Chem.,* 1988, **7**, 317.
50. Weigert, M.; Raschke, W. C.; Carson, D.; Cohn, M. *J. Exp. Med.,* 1974, **139**, 137.
51. Cisar, J.; Kabat, E. A.; Liao, J.; Potter, M. *J. Exp. Med.,* 1974, **139**, 159.
52. Matsuda, T.; Kabat, E. A. *J. Immunol.,* 1989, **142**, 863.
53. Borden, P.; Kabat, E. A. *Proc. Natl. Acad. Sci. U. S. A.,* 1987, **84**, 2440.
54. Padlan, E. A.; Kabat, E. A. *Proc. Natl. Acad. Sci.,* 1988, **85**, 6885.
55. Jolley, M. E.; Rudikoff, S., Potter, M.; Glaudemans, C. P. J. *Biochemistry,* 1973, **12**, 3039.
56. Jolley, M. E.; Glaudemans, C. P. J. *Carbohydr. Res.,* 1974, **33**, 377.
57. Glaudemans, C. P. J.; Manjula, B. N.; Bennett, L. G.; Bishop, C. T. *Immunochemistry,* 1977, **14**, 675.
58. Ruckel, E. R.; Schuerch, C. *J. Amer. Chem. Soc.,* 1966, **88**, 2605.
59. Bennet, L. G.; Glaudemans, C. P. J. *Carbohydr. Res.,* 1979, **72**, 315.
60. Seymour, F. R.; Slodki, M. E.; Plattner, R. D.; Jeanes, A. *Carbohydr. Res.,* 1977, **53**, 153.
61. Kabat, E. A.; Wu, T. T.; Perry, H. M.; Gottesman, K. S.; Foeller, C. *Sequences of Proteins of Immunological Interest, U. S. Dept. of Health and Human Services. NIH Publication 91-3242, NIH, Bethesda MD 20892.,* 1991,
62. Quicho, F. A.; Vyas, N. K. *Nature,* 1984, **310**, 381.

RECEIVED December 27, 1993

Chapter 11

Synthesis of Sialo-oligosaccharides and Their Ceramide Derivatives as Tools for Elucidation of Biologic Functions of Gangliosides

Akira Hasegawa

Department of Applied Bioorganic Chemistry, Gifu University, Gifu 501–11, Japan

A facile, regio- and α-stereoselective glycoside synthesis of sialic acids by use of the methyl or phenyl 2-thioglycoside of sialic acids as the glycosyl donor and the suitably protected galactose and lactose acceptors in acetonitrile under kinetically controlled conditions is described. This procedure is effectively applied to the systematic synthesis of sialyloligosaccharides such as the sialyl Lewis X epitope and its analogs, and their ceramide derivatives. These compounds can be used to elucidate the structural features of sialyl oligosaccharides necessary for selectin recognition.

Sialic acid-containing oligosaccharides are the important constituents of gangliosides and glycoproteins. Biologically, these membrane components are considered to be responsible for many primary physiological activities (*1-5*) and the sialyl oligosaccharide moieties of these glycoconjugates are exposed as ligands to the external environment, capable of expressing biological functions which are harmonious to these chemical structures. An approach towards the systematic understanding of structure-function relationships of the sialo-oligosaccharides necessitates efficient regio- and stereoselective synthetic routes, affording various sialo-oligosaccharides, their derivatives and analogs.

The focal point in the synthesis of sialo-oligosaccharides has been the stereoselective α-glycosylation of sialic acid with various sugar residues. Recently, we have developed (*6-9*) a facile regio- and α-stereoselective glycosylation of sialic acids using the 2-thioglycosides of sialic acids as the glycosyl donors and the suitably protected galactose and lactose acceptors with dimethyl(methylthio)sulfonium triflate (DMTST) (*10*) or *N*-iodosuccinimide (NIS) (*11-12*) as the glycosyl promoter in acetonitrile under kinetically controlled conditions. The α-glycosides of sialic acids thus obtained have been effectively employed as the building blocks for the systematic synthesis of sialo-glycoconjugates. In the first part of this article we describe the efficient method for the α-glycoside synthesis of sialic acids. A systematic synthesis of the sialyl Le^x oligosaccharide, various types of analogs, and their ceramide derivatives, which are useful for determining the structural requirements necessary for selectin recognition is then described.

0097–6156/94/0560–0184$08.00/0

Regio- and α-Stereoselective α-Glycoside Synthesis of Sialic Acids

Sialic acids are as constituents of glycoproteins and gangliosides of cell membranes, and they play important roles in many biological processes. The naturally occurring sialo compounds contain sialic acids with an α-glycosidic linkage, except for CMP-*N*-acetylneuraminic acid.

Kuhn *et al.* (*13*) were the first to attempt the glycosylation of sialic acid with sugar derivatives, employing the 2-chloro derivative of *N*-acetylneuraminic acid (Neu5Ac) as the glycosyl donor. With the 2-halo-derivatives of Neu5Ac, the yield and stereoselectivity of the glycosides are generally poor, especially of those with the secondary hydroxyl groups of sugar derivatives. Particularly annoying is the competitive elimination due to the deoxy center at position 3, resulting of the 2,3-dehydro derivative. The bottleneck in these reactions has been controlling the formation of this elimination product by the selection of suitably designed glycosyl donors and glycosyl promoters. In addition, achieving α-glycosides in high yield is very difficult because the β-glycosides are thermodynamically favored.

Recently, several new efforts have been developed towards obtaining mainly α-glycosides, using the 2-halo-3-substituted Neu5Ac derivatives (*14-15*), Neu5Ac phosphites (*16-17*), S-glycosyl xanthates (*18-19*) of Neu5Ac, and the 2-thioglycosides of sialic acids (*6-9*).

A facile sialylation of suitably protected sugar derivatives with thioglycosides derivatives of sialic acids by use of thiophilic promoters will be described herein. With the wide utility of thioglycosides in oligosaccharide synthesis, preliminary attempts (*6*) using the methyl α-2-thioglycoside (*20*) or the 1:1 anomeric mixture (*7*) of Neu5Ac as a suitable glycosyl donor and DMTST [dimethyl(methylthio)sulfonium triflate] as a promoter and various alcohols as the acceptors, indicated that the reactions conducted in acetonitrile under kinetically controlled conditions afforded predominantly α-glycosides in high yields. This method was successfully extended to the synthesis of sialyl α(2→3)- and sialyl α(2→6)-sugar derivatives in high yields (50~70%) which is, perhaps, the best among those reported so far. Iodonium-ion-promoted glycosylations (*11-12*) are also attractive for oligosaccharide syntheses. We have examined its application to the sialylation involving the 2-thioglycoside of sialic acids and reported (*9*) the comparative reactivities of DMTST and NIS/TfOH in acetonitrile or dichloromethane, with the objective of obtaining predominant α-glycoside (Table 1; Figure 1). Notably, with DMTST in acetonitrile, secondary hydroxyls were glycosylated to give exclusively the α-configuration (40~50%), while an anomeric mixture (α:β ~4:1; 65%~90%) was formed with primary hydroxyls. With NIS/TfOH in acetonitrile, on the other hand, even hindered primary and unreactive secondary hydroxyls were glycosylated in high yields (~70%), but the increased amount of the β-glycoside was formed in some cases. Further, the reactions in dichloromethane with either DMTST or NIS/TfOH showed poor stereoselectivity.

Based on the aforementioned results, a reasonable reaction mechanism can be postulated as follows (Figure 2). A less reactive acceptor nucleophile and stable donor anomeric-intermediate (d) are the two probable factors leading to the formation of α-glycoside of sialic acids. On the other hand, the reactive alcohol can attack other intermediates along with (d), consequently increasing the amounts of the β-glycosides are formed. In addition, using dichloromethane, the nucleophile reacts with the intermediates (a), (b) and (c) to give an anomeric mixture nonstereoselectively.

General glycosylation procedure: *2-(Trimethylsilyl)ethyl O-(methyl 5-acetamido-4,7,8,9-tetra-O-acetyl-3,5-dideoxy-D-glycero-α-D-galacto-2-nonulopyranosylonate)-(2→3)-6-O-benzoyl-β-D-galactopyranoside* (**10**).

  (a) **DMTST-promoted glycosylation by use of the methyl 2-thioglycoside (1a) of Neu5Ac.** To a solution of **1a** (2.7 g, 5.2 mmol) and **2**

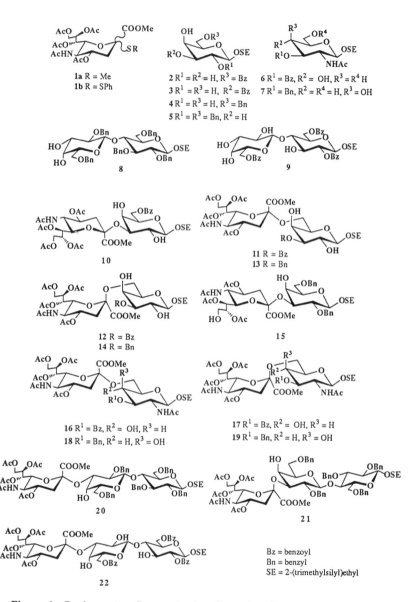

Figure 1   Regio- and α-Stereoselective Glycoside Synthesis of Sialic Acids

Table I. DMTST*- and NIS-TfOH**-Promoted Glycosylation
by use of 2-Thioglycosides 1 of Neu5Ac

Acceptor	Donor	Promoter	Solvent	Product	Yield*** (%)	
					α	β
2	1a	DMTST	CH₃CN	10	52	0
2	1a	NIS	CH₃CN	10	61	0
2	1b	NIS	CH₃CN	10	70	0
3	1a	DMTST	CH₃CN	11	70	0
3	1a	NIS	CH₃CN	11	59	0
3	1a	NIS	CH₂Cl₂	11,12	49	25
4	1a	DMTST	CH₃CN	13,14	50	15
4	1a	NIS	CH₃CN	13,14	51	26
4	1a	NIS	CH₂Cl₂	13,14	43	45
5	1b	NIS	CH₃CN	15	70	0
6	1a	DMTST	CH₃CN	16,17	71	20
7	1a	DMTST	CH₃CN	18,19	63	14
8	1a	DMTST	CH₃CN	20,21	30	8
8	1a	NIS	CH₃CN	20,21	59	10
9	1a	DMTST	CH₃CN	22	47	0
9	1b	NIS	CH₃CN	22	55	0

*   Reactions were performed at -15°C
**   Reactions were performed at -35 ~ -40°C
*** Isolated yield

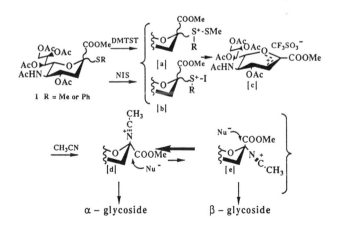

α – glycoside   β – glycoside

Figure 2   The Reaction Mechanism Suggested for α-Predominant
Glycoside Formation of Neu5Ac

(1.0 g, 2.6 mmol) in dry acetonitrile (20 mL) were added powdered molecular sieves-3Å (MS-3Å; 3 g), and the suspension was stirred for 5 h at room temperature, then cooled to -40°C. To the cooled suspension DMTST (6.53 g; 62% DMTST by weight, 3.0 equiv. relative to the donor) and MS-3Å were added, and the mixture was stirred for 17 h at -15°C. Methanol (1 mL) was added to the mixture, and it was neutralized with triethylamine. The solids were filtered off and washed thoroughly with CH_2Cl_2, and the combined filtrate and washings were concentrated. The residue was dissolved in CH_2Cl_2 and the solution was successively washed with M Na_2CO_3 and water, dried (Na_2SO_4), and evaporated. Column chromatography (1:1 EtOAc-hexane) of the residue on silica gel gave **10** (1.16 g, 52%) as an amorphous mass, $[\alpha]_D$ -6.0° (c 2.0 $CHCl_3$).

(b) **NIS-TfOH-promoted glycosylation by use of the methyl 2-thioglycoside (1a) of Neu5Ac.** To a solution of **1a** (1.67 g, 3.2 mmol) and **2** (730 mg, 1.9 mmol) in dry acetonitrile (15 mL) powdered MS-3Å (2.3 g) were added. The mixture was stirred for 5 h at room temperature and then cooled to -40°C. To the cooled mixture were added NIS (1.44 g, 6.4 mmol) and TfOH (96 mg, 0.64 mmol), and the mixture was stirred for 2 h at -40°C. A similar work-up as described above gave **10** (995 mg, 61%).

(c) **NIS-TfOH-promoted glycosylation by use of the phenyl 2-thioglycoside of Neu5Ac.** To a solution of **1b** (10.7 g, 18.4 mmol) and **2** (3.84 g, 10 mmol) in dry acetonitrile (50 mL) and CH_2Cl_2 (5 mL) were added powdered MS-3Å (20 g), and the mixture was stirred overnight at room temperature, then cooled to -40°C. To the cooled mixture were added NIS (8.28 g, 36.8 mmol) and TfOH (540 mg, 3.6 mmol), and the mixture was stirred for 2.5 h at -35°C. A similar work-up as described above gave **10** (6.0 g, 70%).

2-(Trimethylsilyl)ethyl O-(methyl 5-acetamido-4,7,8,9-tetra-O-acetyl-3,5-dideoxy-D-glycero-α-D-galacto-2-nonulopyranosylonate)-(2→6)-3-O-benzoyl-β-D-galactopyranoside (**11**). To a solution of **1a** (2.7 g, 5.2 mmol) and **3** (1.0 g, 2.6 mmol) in dry acetonitrile (20 mL) were added powdered MS-3Å (3 g), and the mixture was stirred for 6 h at room temperature, then cooled to -15°C. A mixture (6.53 g, 62% DMTST by weight) of DMTST and MS-3Å was added to the mixture, and this was stirred for 17 h at -15°C. Processing as described for the synthesis of **10**, and chromatography (silica gel) using 1:1 EtOAc-hexane as eluent afforded **11** (1.56 g, 70%) as an amorphous mass, $[\alpha]_D$ -6.4° (c 0.4, $CHCl_3$).

Synthesis of sialyl Lewis X and its analogs

Sialyl Lewis X (sLe[x]) was first isolated (21) from the human kidney and found to be widespread as a tumor-associated ganglioside antigen. Recently, it has been demonstrated (22-32) that the selectin family, such as E-selectin (endothelial leukocyte adhesion molecule-1, ELAM-1), P-selectin (granule membrane protein, GMP-140), and L-selectin (leukocyte adhesion molecule-1), recognizes the sLe[x] determinant, α-Neu5Ac-(2→3)-β-D-Gal-(1→4)-[α-L-Fuc-(1→3)]-β-GlcNAc. This sequence is found as the terminal carbohydrate structure in both cell membrane glycolipids and glycoproteins. In view of these new findings, it is of interest to elucidate the structural requirements for the expression of such biological activities which are related to cell-cell adhesion, tumor-metastasis, inflammation, and thrombosis.

Synthesis of Sialyl Le[x] and its Position Isomer. Sialyl Le[x] (33) and its sialyl α(2→6) positional isomer (34) with regard to the substitution of the Gal residue by Neu5Ac were synthesized according to our established method. As shown in Figure 3, the trisaccharide acceptor **25** (35) was first coupled with **23** or **24** in the presence of DMTST in benzene to give the desired α-tetrasaccharides (**26**) in 86~95% yields. Reductive ring opening of the benzylidene acetal in **26** with sodium cyanoborohydride-hydrogen chloride (36) gave **27**, which, on glycosylation with **28**, afforded the

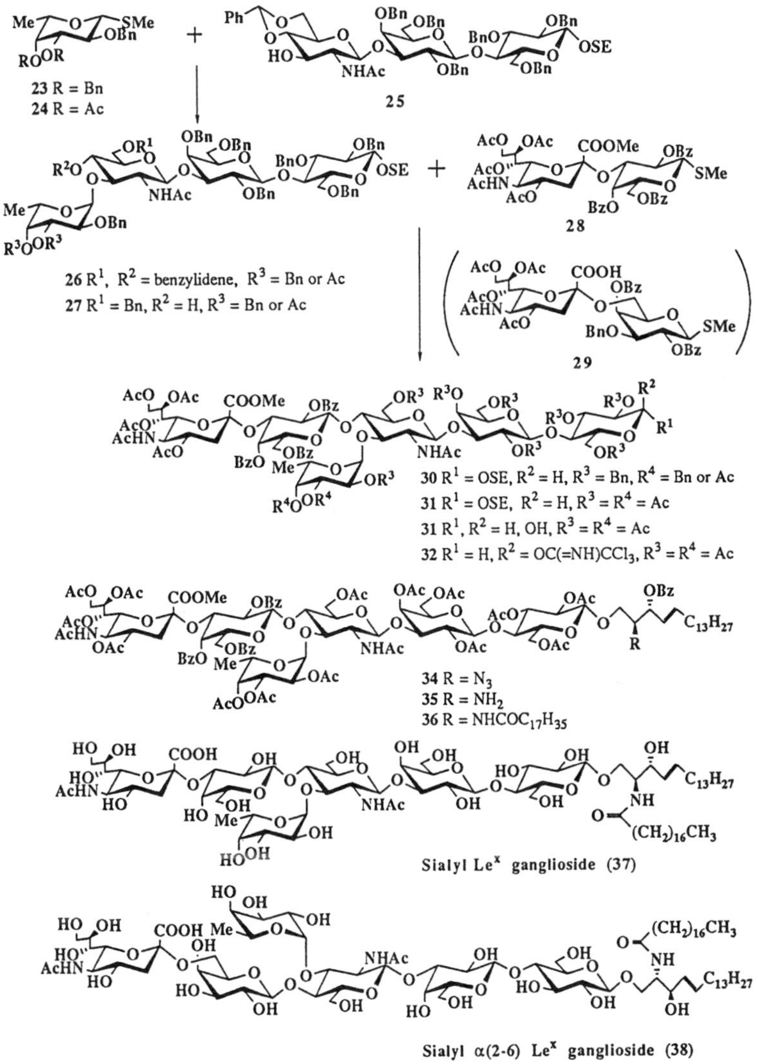

23 R = Bn
24 R = Ac

25

26 R^1, R^2 = benzylidene, R^3 = Bn or Ac
27 R^1 = Bn, R^2 = H, R^3 = Bn or Ac

28

29

30 R^1 = OSE, R^2 = H, R^3 = Bn, R^4 = Bn or Ac
31 R^1 = OSE, R^2 = H, R^3 = R^4 = Ac
31 R^1, R^2 = H, OH, R^3 = R^4 = Ac
32 R^1 = H, R^2 = OC(=NH)CCl$_3$, R^3 = R^4 = Ac

34 R = N$_3$
35 R = NH$_2$
36 R = NHCOC$_{17}$H$_{35}$

Sialyl Le^x ganglioside (37)

Sialyl α(2-6) Le^x ganglioside (38)

Sialyl Le^x oligosaccharide (39)

Figure 3   Synthesis of sLe^x, sLe^x Ganglioside and its Position Isomer

hexasaccharide (41%). When the tetrasaccharide **27** containing the 3,4-di-*O*-acetyl-2-*O*-benzyl-L-fucose moiety was used as acceptor, the yield of **30** was increased to 70%. Hydrogenolytic removal of the benzyl groups and subsequent acetylation afforded **31** (81%). Selective removal of the 2-(trimethylsilyl)ethyl group in **31** using trifluoroacetic acid (*37*), and subsequent treatment (*38*) with CCl$_3$CN in the presence of 1,8-diazabicyclo[5.4.0]undec-7-ene (DBU) gave the α-imidate **33** (91%). The final glycosylation of (2*S*,3*R*,4*E*)-2-azido-3-*O*-benzoyl-4-octadecene-1,3-diol (*39*) with **33** thus obtained in the presence of boron trifluoride etherate, afforded only the expected β-glycoside **34** (56%), which was transformed *via* selective reduction of the azido group, coupling of the amine **35** with octadecanoic acid, *O*-deacylation, and saponification of the methyl ester group, into the title sialyl Le^x ganglioside **37** in high yield. Similarly, by the coupling of **27** with sialyl α(2→6)-galactose donor **29** (*41*) and subsequent reactions as described for the synthesis of **37**, a positional isomer (**38**) of sialyl Le^x, was synthesized. Sialyl Le^x oligosaccharide (**39**) (69%) was also synthesized from **32** *via* *O*-(tetrahydropyran-2-yl)ation, *O*-deacylation, saponification of the methyl ester group, and hydrolysis of the tetrahydropyranyl group.

  2-(Trimethylsilyl)ethyl O-(methyl 5-acetamido-4,7,8,9-tetra-O-acetyl-3,5-dideoxy-D-glycero-α-D-galacto-2-nonulopyranosylonate)-(2→3)-O-(2,4,6-tri-O-benzoyl-β-D-galactopyranosyl)-(1→4)-O-[3,4-di-O-acetyl-2-O-benzyl-α-L-fucopyranosyl)-(1→3)]-O-(2-acetamido-6-O-benzyl-2-deoxy-β-D-glucopyranosyl)-(1→3)-O-(2,4,6-tri-O-benzyl-β-D-galactopyranosyl)-(1→4)-2,3,6-tri-O-benzyl-β-D-glucopyranoside (Protected sialyl Le^x oligosaccharide) (**30**) To a solution of **27** (1.25 g, 0.78 mmol) and **28** (1.1 g, 1.1 mmol) in dry CH$_2$Cl$_2$ were added powdered MS-4Å (5 g), and the mixture was stirred overnight at room temperature, then cooled to 0°C. A mixture (1.14 g; 50% DMTST by weight) of DMTST and MS-4Å (500 mg) was added, and the stirring was continued for 48 h at 0°C. Methanol (1 mL) and Et$_3$N (0.5 mL) were added to the mixture, and the solids were filtered off and washed with CH$_2$Cl$_2$. The filtrate and washings were combined and washed with water, dried (Na$_2$SO$_4$), and concentrated. Column chromatography (50:1 CHCl$_3$-MeOH) of the residue on silica gel gave **30** (1.4 g, 70%) as an amorphous mass, [α]$_D$ -23.2° (*c* 1.2, CHCl$_3$).

  Sialyl Le^x oligosaccharide (**39**) To a solution of **32** (150 mg, 0.07 mmol) in *p*-dioxane (3 mL) were added 3,4-dihydro-2*H*-pyran (0.5 mL) and *p*-toluenesulfonic acid monohydrate (20 mg), and the mixture was stirred for 5 h at room temperature, then neutralized with Amberlite IR-410 (HO⁻) resin and concentrated. To a solution of the residue in MeOH (5 mL) was added NaOCH$_3$ (30 mg), and the mixture was stirred for 2 days at room temperature, water (0.5 mL) was added to the mixture, and it was stirred for 16 h at room temperature, and neutralized with Amberlite IR-120 (H⁺) resin, then concentrated. A solution of the residue in aq. 60% AcOH (10 mL) was heated for 2 h at 45°C and concentrated. Column chromatography (1:1 MeOH-H$_2$O) of the residue on Sephadex LH-20 (50 g) gave **39** (60 mg, 69%) as an amorphous mass, [α]$_D$ -20.3° (*c* 0.26, H$_2$O; equil.).

Synthesis of the deoxy-fucose-containing sialyl Le^x gangliosides

The synthesis of the sialyl Le^x oligosaccharides containing 2-, 3-, and 4-deoxy-fucose moieties is described next. These were utilized to clarify the structural requirements on the fucose moiety necessary for selectin recognition. For the synthesis of the target sLe^x analogs, we employed the methyl 1-thioglycosides **40-42** of the appropriate deoxy-L-fucose derivatives as the glycosyl donors (Hasegawa, A.; Ando, T.; Kato, M.; Ishida, H.; Kiso, M. *Carbohydr. Res.*, in press) and 2-(trimethylsilyl)ethyl *O*-(2-acetamido-4,6-*O*-benzylidene-2-deoxy-β-D-glucopyranosyl)-(1→4)-2,4,6-tri-*O*-benzyl-β-D-galactopyranoside (*42*) as a suitably protected glycosyl acceptor (Figure 4).

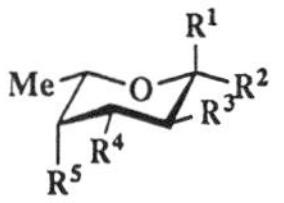

	R¹	R²	R³	R⁴	R⁵
40	H , SMe		H	OBz	OBz
41	H	SMe	OBn	H	OBn
42	H	SMe	OBn	OBn	H

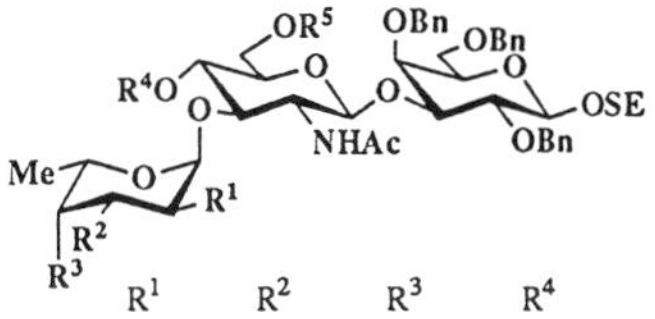

	R¹	R²	R³	R⁴	R⁵
43	H	OBz	OBz	benzylidene	
44	H	OBz	OBz	H	Bn
45	OBn	H	OBn	benzylidene	
46	OBn	H	OBn	H	Bn
47	OBn	OBn	H	benzylidene	
48	OBn	OBn	H	H	Bn

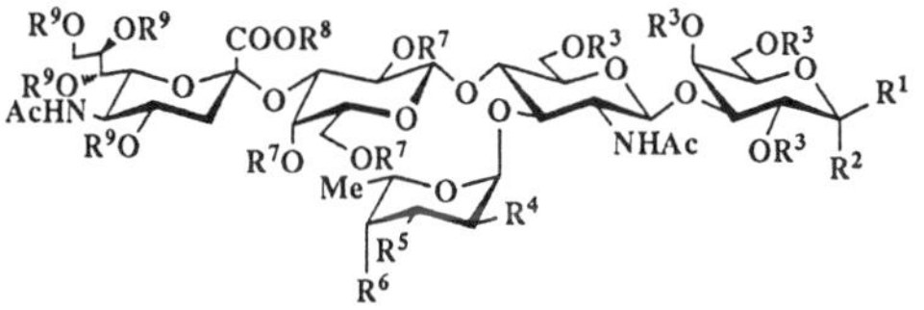

	R¹	R²	R³	R⁴	R⁵	R⁶	R⁷	R⁸	R⁹
49	OSE	H	Bn	H	OBz	OBz	Bz	Me	Ac
50	OSE	H	Ac	H	OBz	OBz	Bz	Me	Ac
51	H	OC(=NH)CCl₃	Ac	H	OBz	OBz	Bz	Me	Ac
52	OSE	H	H	H	OH	OH	H	H	H
53	OSE	H	Bn	OBn	H	OBn	Bz	Me	Ac
54	OSE	H	Ac	OAc	H	OAc	Bz	Me	Ac
55	H	OC(=NH)CCl₃	Ac	OAc	H	OAc	Bz	Me	Ac
56	OSE	H	H	OH	H	OH	H	H	H
57	OSE	H	Bn	OBn	OBn	H	Bz	Me	Ac
58	OSE	H	Ac	OAc	OAc	H	Bz	Me	Ac
59	H	OC(=NH)CCl₃	Ac	OAc	OAc	H	Bz	Me	Ac
60	OSE	H	H	OH	OH	H	H	H	H

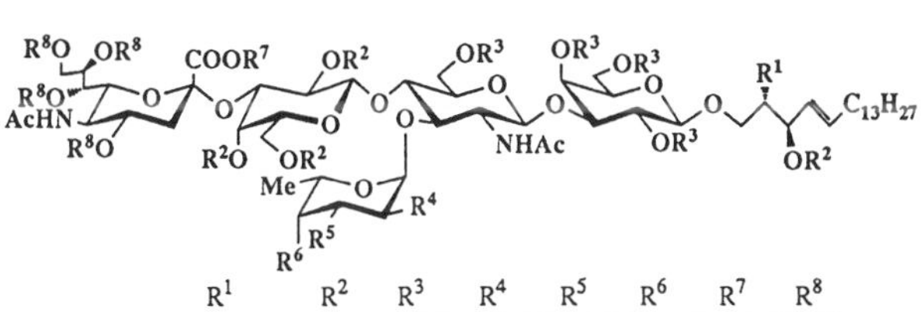

	R¹	R²	R³	R⁴	R⁵	R⁶	R⁷	R⁸
61	N₃	Bz	Ac	H	OBz	OBz	Me	Ac
62	NHCOC₁₇H₃₅	H	H	H	OH	OH	H	H
63	N₃	Bz	Ac	OAc	H	OAc	Me	Ac
64	NHCOC₁₇H₃₅	H	H	OH	H	OH	H	H
65	N₃	Bz	Ac	OAc	OAc	H	Me	Ac
66	NHCOC₁₇H₃₅	H	H	OH	OH	H	H	H

Figure 4   Synthesis of the Deoxy-Fucose-Containing sLeˣ Gangliosides

Glycosylation of the acceptor with **40-42** in benzene for 10 h at 5-10°C, using DMTST as a promoter, gave exclusively the α-glycosides **43** (86%), **45** (82%), and **47** (57%), respectively. These were converted by reductive ring-opening of the benzylidene acetal with sodium cyanoborohydride, into the glycosyl acceptors **44** (97%), **46** (80%), and **48** (87%). DMTST-promoted glycosylation of **44, 46** and **48** with the sialyl α(2→3)-Gal donor **28** afforded the desired pentasaccharides **49, 53** and **57**, which were converted *via* reductive removal of the benzyl groups, *O*-acetylation, selective removal of the 2-(trimethylsilyl)ethyl group, and subsequent imidate formation, into the corresponding α-trichloroacetimidates **51, 55,** and **59** respectively in good yields. Glycosylation of (2*S*,3*R*,4*E*)-2-azido-3-*O*-benzoyl-4-octadecene-1,3-diol with **51, 55,** or **59** thus obtained afforded only the expected β-glycosides **61** (85%), **63** (64%), and **65** (60%) respectively, which were transformed in good yields, *via* selective reduction of the azido group, coupling with octadecanoic acid, *O*-deacetylation, and de-esterification, into the target gangliosides **62** (55%), **64** (97%), and **66** (82%). The sialyl Le^x oligosaccharide analogs **52, 56,** and **60** were obtained in good yields, by *O*-deacylation of **50, 54,** or **58,** and subsequent hydrolysis of the methyl ester group.

Synthesis of the chemically modified sialic acid-containing sialyl Le^x ganglioside analogs

The synthesis of sLe^x ganglioside analogs containing the C7-Neu5Ac, C8-Neu5Ac and 8-epi-Neu5Ac, to explore the structural requirements of the sialic acid moiety for selectin recognition, was completed as follows.

Methyl *O*-(methyl 5-acetamido-4,7-di-*O*-acetyl-3,5-dideoxy-β-L-*arabino*-2-heptulopyranosylonate)-(2→3)-, methyl *O*-(methyl 5-acetamido-4,7,8-tri-*O*-acetyl-3,5-dideoxy-α-D-*galacto*-2-octuropyranosylonate)-(2→3)-, and methyl *O*-(methyl 5-acetamido-4,7,8,9-tetra-*O*-acetyl-3,5-dideoxy-L-*glycero*-β-D-*galacto*-2-nonulopyranosylonate)-(2→3)-2,4,6-tri-*O*-benzoyl-1-thio-β-D-galactopyranosides (**77, 81,** and **85**) were selected as the glycosyl donors, and 2-(trimethylsilyl)ethyl *O*-(2,3,4-tri-*O*-benzyl-α-L-fucopyranosyl)-(1→3)-*O*-(2-acetamido-6-*O*-benzyl-2-deoxy-β-D-glucopyranosyl)-(1→3)-2,4,6-tri-*O*-benzyl-β-D-galactopyranoside (*42*) as the glycosyl acceptor (Figure 5). Compounds **77, 81,** and **85** were prepared by glycosylation of **2** with the 2-thioglycosides **70, 72,** and **73** (*43*), using NIS-TfOH in acetonitrile for 2 h at -35°C, followed by *O*-benzoylation, selective transformation of the 2-(trimethylsilyl)ethyl group to acetyl by treatment (*37*) with boron trifluoride etherate, and introduction (*7*) of the methylthio group with methylthio(trimethyl)silane.

Glycosylation of the trisaccharide acceptor with these donors (1.45 equiv. with respect to the acceptor), in dichloromethane for 48 h at 5°C in the presence of DMTST, yielded the expected β-glycosides, **86** (53%), **90** (53%), and **94** (47%), respectively. Catalytic hydrogenolysis (10% Pd-C) in ethanol-acetic acid for 3 days at 45°C of the benzyl groups in **86, 90,** or **94,** and subsequent *O*-acetylation gave the per-*O*-acyl derivatives, **87** (85%), **91** (87%), and **95** (83%) after column chromatography. Compounds **87, 91,** and **95** were converted into the corresponding α-trichloroacetimidates, **89, 93,** and **97,** according to the method described for the synthesis of **33,** and these, when coupled with the azido-sphingosine derivative, gave the required β-glycosides, **98** (68%), **101** (43%), and **104** (46%). Finally, these were transformed *via* selective reduction of the azido group, condensation with octadecanoic acid, *O*-deacylation and de-esterification into the target sLe^x ganglioside analogs **100, 103,** and **106** in good yields, according to our established method (*44*).

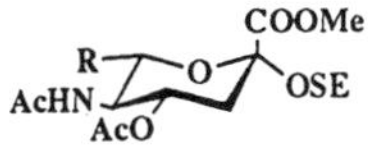

67 R = X¹ ... (structures)

68 R = Y¹

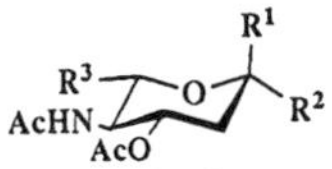

	R¹, R²	R³
69	OAc, COOMe	X¹
70	SPh, COOMe	X¹
71	OAc, COOMe	Y¹
72	SPh, COOMe	Y¹

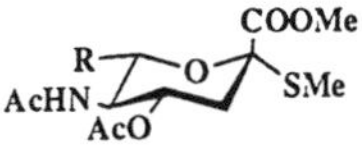

R = Z¹
73

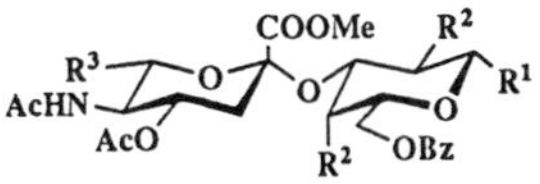

	R¹	R²	R³
74	OSE	OH	X¹
75	OSE	OBz	X¹
76	OAc	OBz	X¹
77	SMe	OBz	X¹
78	OSE	OH	Y¹
79	OSE	OBz	Y¹
80	OAc	OBz	Y¹
81	SMe	OBz	Y¹
82	OSE	OH	Z¹
83	OSE	OBz	Z¹
84	OAc	OBz	Z¹
85	SMe	OBz	Z¹

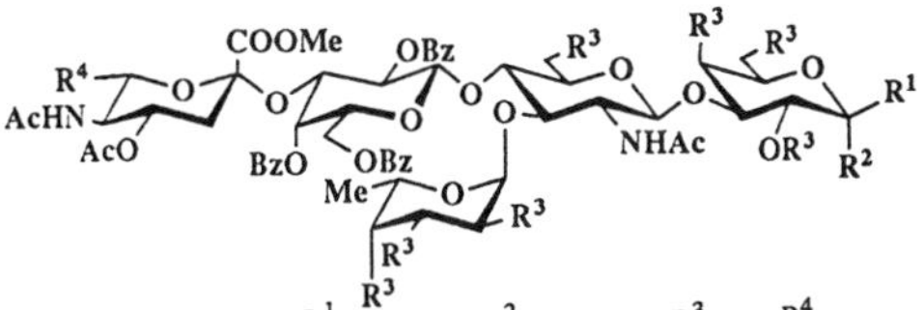

	R¹	R²	R³	R⁴
86	OSE	H	OBn	X¹
87	OSE	H	OAc	X¹
88		H, OH	OAc	X¹
89	H	OC(=NH)CCl₃	OAc	X¹
90	OSE	H	OBn	Y¹
91	OSE	H	OAc	Y¹
92		H, OH	OAc	Y¹
93	H	OC(=NH)CCl₃	OAc	Y¹
94	OSE	H	OBn	Z¹
95	OSE	H	OAc	Z¹
96		H, OH	OAc	Z¹
97	H	OC(=NH)CCl₃	OAc	Z¹

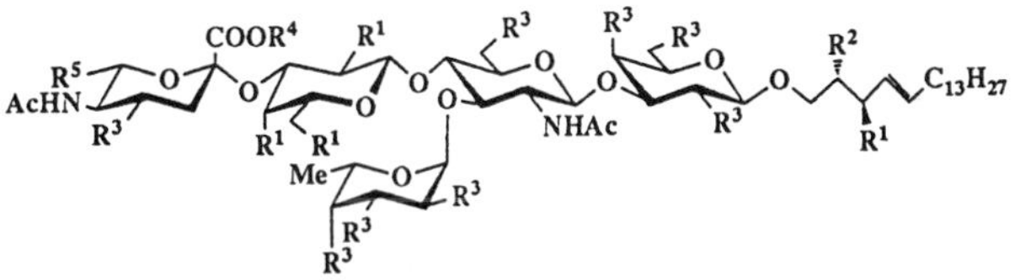

	R¹	R²	R³	R⁴	R⁵
98	OBz	N₃	OAc	Me	X¹
99	OBz	NHCOC₁₇H₃₅	OAc	Me	X¹
100	OH	NHCOC₁₇H₃₅	OH	H	X²
101	OBz	N₃	OAc	Me	Y¹
102	OBz	NHCOC₁₇H₃₅	OAc	Me	Y¹
103	OH	NHCOC₁₇H₃₅	OH	H	Y²
104	OBz	N₃	OAc	Me	Z¹
105	OBz	NHCOC₁₇H₃₅	OAc	Me	Z¹
106	OH	NHCOC₁₇H₃₅	OH	H	Z²

X¹ = CH₂OAc   Y¹ = HCOAc/CH₂OAc   Z¹ = HCOAc/AcOCH/CH₂OAc   X² = CH₂OH   Y² = HCOH/CH₂OH   Z² = HCOH/HOCH/CH₂OH

Figure 5   Synthesis of the Modified Neu5Ac-Containing sLeˣ Ganglioside Analogs

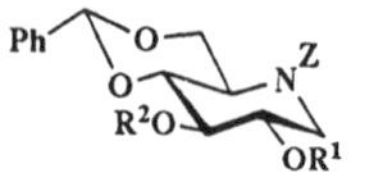

	R^1	R^2
107	H	H
108	H	ClAc
109	Ac	H

Z = benzyloxycarbonyl
ClAc = chloroacetyl

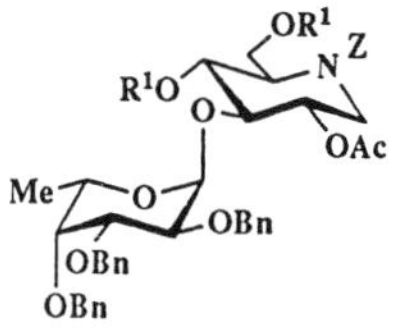

	R^1	R^2
110	benzylidene	
111	H	Bn

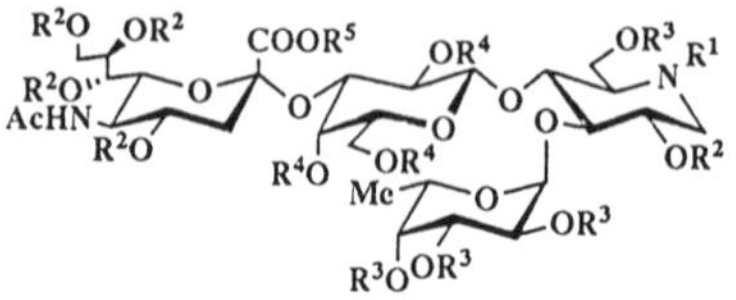

Sialyl Le^x analog

	R^1	R^2	R^3	R^4	R^5
112	Z	Ac	Bn	Bz	Me
113	Me	Ac	H	Bz	Me
114	Me	H	H	H	H

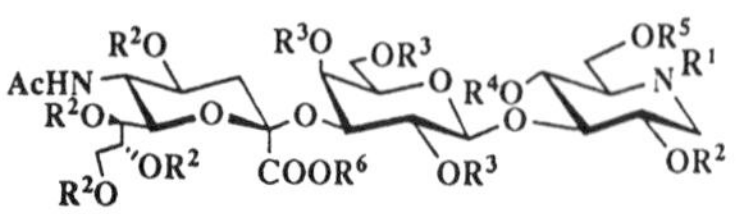

	R^1	R^2	R^3	R^4	R^5	R^6
115	Z	Ac	Bz	benzylidene		Me
116	Z	Ac	Bz	H	Bn	Me

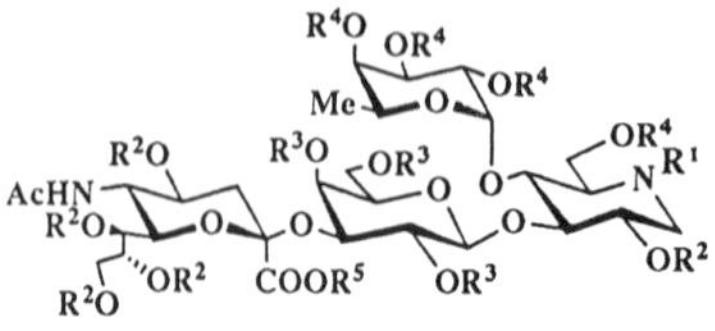

Sialyl Le^a analog

	R^1	R^2	R^3	R^4	R^5
117	Z	Ac	Bz	Bn	Me
118	Me	Ac	Bz	H	Me
119	Me	H	H	H	H

Figure 6   Synthesis of 1-Deoxynojirimycin-Containing sLe^a and sLe^x Ganglioside Analogs

Synthesis of 1-deoxynojirimycin-containing sLe^a and sLe^x analogs.

Synthesis (Kiso, M.; Furui, H.; Ando, K.; Hasegawa, A. *J. Carbohydr. Chem.* in press) of sLe^a and sLe^x analogs, in which *N*-acetylglucosamine residue in sLe^a and sLe^x oligosaccharide epitopes is replaced by 1-deoxynojirimycin derivative, is described (Figure 6). Glycosylation of **109** (obtained from 4,6-*O*-benzylidene-*N*-benzyloxycarbonyl-1,5-dideoxy-1,5-imino-D-glucitol (*45*) (**107**) *via* selective 3-*O*-chloroacetylation, 2-*O*-acetylation of **108**, and dechloroacetylation with aq. pyridine) with the fucose donor **23** in the presence of DMTST in benzene for 2 h gave **110** (92%) as an amorphous mass. Reductive ring-opening of the benzylidene acetal in **110** gave the disaccharide glycosyl acceptor **111** (81%). The glycosylation of **111** with the sialyl galactose donor **28** using NIS-TfOH in dichloromethane gave the desired tetrasaccharide **112** (61%), which was transformed *via* catalytic hydrogenolysis of the benzyl and benzyloxycarbonyl groups using Pd-black in 1:1 methanol-acetic acid, *O*-deacylation and saponification of the methyl ester group, into the sLe^x analog **114** in good yield. On the other hand, glycosylation of **109** with the sialyl α(2→3)Gal donor **28** using NIS-TfOH as described for **112** gave **115** (90%). Reductive ring opening of the benzylidene group in **115** gave **116** which was then coupled with **23** in the presence of DMTST (as described for **110**) to afford the desired tetrasaccharide **117** (89%). Compound **117** was easily transformed *via* hydrogenolysis of the benzyl and benzyloxycarbonyl groups, *O*-deacylation, and de-esterification, into the target sLe^a analog **119**, quantitatively.

As described in this article, the facile stereo-controlled glycosylations of sialic acids with the suitably protected sugar residues are now feasible using either DMTST or NIS-TfOH in acetonitrile under kinetically controlled conditions. By using this procedure, a variety of sialyl oligosaccharides and their ceramide derivatives (ganglioside) and analogs have systematically been synthesized. These molecules will be used to define the structural requirements necessary for selectin recognition.

Acknowledgments

This work was supported in part by a Grant-in-Aid (No. 04250102 and No. 03660132) for Scientific Research on Priority Areas from the Ministry of Education, Science and Culture of Japan. The author would like to express his gratitude to various colleagues cited in the references.

Literature Cited

1. Wiegandt, H. In *Glycolipids*; Wiegandt, H. Ed.; New Comprehensive Biochemistry 10; Elsevier: Amsterdam, New York, Oxford, 1985; pp 199-260.
2. Reutter, W., Köttgen, E., Bauer, C., Gerok, W. In *Biological Significance of Sialic Acids*; Schauer, R., Ed.; Cell Biology Monographs 10; Springer-Verlag: Wien, New York, 1983; pp 263-305.
3. Furukawa, K., Kobata, A., in *Cell Surface Carbohydrates-Their Involvement in Cell Adhesion*; Ogura, H., Hasegawa, A., Suami, T., Eds.; Carbohydrates-Synthetic methods and Applications in Medicinal Chemistry; Kodansha-VCH: Tokyo, Weinheim, New York, Cambridge, Basel, 1992; pp 369-384.
4. Hakomori, S. *Sci. Am.* **1986**, *254*, pp 32.
5. Tsuji, S.; Yamakawa, T.; Tanaka, M.; Nagai, Y. *J. Neurochem.* **1988**, *50*, pp 414.

6. Murase, T.; Ishida, H.; Kiso, M.; Hasegawa, A. *Carbohydr. Res.* **1988**, *184*, pp c1.
7. Hasegawa, A; Ohki, H.; Nagahama, T.; Ishida, H.; Kiso, M. *Carbohydr. Res.* **1991**, *212*, pp 277.
8. Hasegawa, A; Ogawa, M.; Ishida, H.; Kiso, M. *J. Carbohydr. Chem.* **1990**, *9*, pp 393.
9. Hasegawa, A.; Nagahama, T.; Ohki, H.; Hotta, K.; Ishida, H.; Kiso, M. *J. Carbohydr. Chem.* **1991**, *10*, pp 493.
10. Fügedi, P.; Garegg, P. J. *Carbohydr. Res.* **1986**, *184*, pp c1.
11. Konradsson, P.; Udodong, U. S.; Fraser-Reid, B. *Tetrahedron Lett.* **1990**, *31*, pp 4313.
12. Veeneman, G. H.; van Leeuwen, S. H.; van Boom, J. H. *Tetrahedron Lett.* **1990**, *31*, pp 1331.
13. Kuhn, R.; Lutz, P.; MacDonald, D. L. *Chem. Ber.* **1966**, *99*, pp 611.
14. Ogawa, T.; Itoh, Y. *Tetrahedron Lett.* **1987**, *28*, pp 6221.
15. Kondo, T.; Abe, H.; Goto, T. *Chem. Lett.* **1988**, pp 1657.
16. Martin, T. J.; Schmidt, R. R. *Tetrahedron Lett.* **1992**, *33*, pp 6123.
17. Sim, M. M.; Kondo, H.; Wong, C. H. *J. Am. Chem. Soc.* **1993**, *115*, pp 2260.
18. Lönn, H.; Stenvall, K. *Tetrahedron Lett.* **1992**, *33*, pp 115.
19. Marra, A.; Sinaÿ, P. *Carbohydr. Res.* **1990**, *195*, pp 303.
20. Hasegawa, A.; Nakamura, J.; Kiso, M. *J. Carbohydr. Chem.* **1986**, *5*, pp 11.
21. Pauvala, H. *J. Biol. Chem.* **1976**, *251*, pp 7517.
22. Phillips, M. L.; Nudelman, E.; Gaeta, F. C. A.; Perez, M.; Singhal, A. K.; Hakomori, S.; Paulson, J. C. *Science* **1990**, *25*, pp 1130.
23. Walz, G.; Aruffo, A.; Kolanus, W.; Bevilacqua, M.; Seed, B. *Science* **1990**, *250*, pp 1132.
24. Tyrell, D.; James, P.; Rao, N.; Foxall, C.; Abbas, S.; Dasgupta, F.; Nashed, M.; Hasegawa, A.; Kiso, M.; Asa, D.; Kidd, J.; Brandley, B. K. *Proc. Natl. Acad. Sci. USA* **1991**, *88*, pp 10372.
25. Polly, M. J.; Phillips, M. L.; Wayner, E.; Nudelman, E; Singhal, A. K.; Hakomori, S.; Paulson, J. C. *Proc. Natl. Acad. Sci. USA* **1991**, *88*, pp 6224.
26. Tanaka, A.; Ohmori, K.; Takahashi, N.; Tsuyuoka, K.; Yago, A.; Hasegawa, A.; Kannagi, R. *Biochem. Biophys. Res. Commun.* **1991**, *179*, pp 713.
27. Kotovuori, P.; Tontti, E.; Pigott, R.; Stepherd, M.; Kiso, M.; Hasegawa, A.; Renkonen, R.; Nortamo, P.; Altieri, D. C.; Gamberg, C. G. *Glycobiology* **1993**, *3*, pp 131.
28. Larkin, M.; Ahern, T. J.; Stoll, M. S.; Shaffer, M.; Sako, D.; O'Brien, J.; Yuen, C. T.; Lawson, A. M.; Childs, R. A.; Barone, K. M.; Langer-Safer, P. R.; Hasegawa, A.; Kiso, M.; Larsen, G. R.; Feizi, T. *J. Biol. Chem.* **1992**, *269*, pp 13661.
29. Hanisch, F. G.; Hanski, C.; Hasegawa, A. *Cancer Res.* **1992**, *52*, pp 3138.
30. Foxall, C.; Watson, S. R.; Dowbenko, D.; Fennie, C.; Lasky, L, A.; Kiso, M.; Hasegawa, A.; Asa, D.; Brandley, B. K. *J. Cell Biol.* **1992**, *117*, pp 895.
31. Tanaka, A.; Ohmori, K.; Yoneda, T.; Tsuyuoka, K.; Hasegawa, A.; Kiso, M.; Kannagi, R. *Cancer Res.* **1993**, *53*, pp 354.
32. Ohmori, K.; Tanaka, A.; Yoneda, T.; Buma, Y.; Hirashima, K.; Tsuyuoka, K.; Hasegawa, A.; Kannnagi, R. *Blood* **1993**, *81*, pp 101.
33. Kameyama, A.; Ishida, H.; Kiso, M.; Hasegawa, A. *Carbohydr. Res.* **1991**, *209*, pp c1; *ibid. J. Carbohydr. Chem.* **1991**, *10*, pp 549.
34. Kameyama, A.; Ishida, H.; Kiso, M.; Hasegawa, A. *J. Carbohydr. Chem.* **1991**, *10*, pp 729.
35. Kameyama, A.; Ishida, H.; Kiso, M.; Hasegawa, A. *Carbohydr. Res.* **1990**, *193*, pp 269.
36. Garegg, P. J.; Hultberg, H.; Wallin, S. *Carbohydr. Res.* **1982**, *108*, pp 97.

37. Jansson, K.; Ahlfors, S.; Frejd, T.; Kihlberg, J.; Gagnusson, G.; Dahmen, J.; Noori, G.; Stenvall, K. *J. Org. Chem.* **1988**, *53*, pp 5629.
38. Schmidt, R. R.; Michel, J. *Angew. Chem. Int. Ed. Engl.* **1980**, *19*, 731.
39. Ito, Y.; Kiso, M.; Hasegawa, A. *J. Carbohydr. Chem.* **1989**, *8*, pp 285.
40. Schmidt, R. R.; Zimmermann, P. *Angew. Chem. Int. Ed. Engl.* **1986**, *25*, pp 725.
41. Hasegawa, A.; Hotta, K.; Kameyama, A.; Ishida, H.; Kiso, M. *J. Carbohydr. Chem.* **1991**, *10*, 439.
42. Hasegawa, A.; Ando, T.; Kameyama, A.; Kiso, M. *Carbohydr. Res.* **1992**, *230*, pp c1; *ibid. J. Carbohydr. Chem.* **1992**, *11*, pp 645.
43. Yoshida, M.; Uchimura, A.; Kiso, M.; Hasegawa, A. *Glycoconjugate J.* **1993**, *10*, pp 3.
44. Murase. T.; Ishida, H.; Kiso, M.; Hasegawa, A. *Carbohydr. Res.* **1989**, *188*, pp 71.
45. Furui, H.; Kiso, M.; Hasegawa, A. *Carbohydr. Res.* **1991**, *229*, pp c1.

RECEIVED November 17, 1993

Chapter 12

Photolabile, Spacer-Modified Oligosaccharides for Regioselective Chemical Modificaion and Probing of Receptor Binding Sites

Jochen Lehmann, Stefan Petry, and Markus Schmidt-Schuchardt

Institut für Organische Chemie und Biochemie, Albert-Ludwigs-Universität Freiburg, Albertstrasse 21, D−79104 Freiburg, Germany

During the past decade, biochemistry has undergone a gradual development from investigations of covalent molecular transformations, which occur in metabolism for instance - to those of noncovalent molecular interactions, which can be summarized under the heading "recognition phenomena". The last, unsolved basic problems in molecular biology fall into this category. Organic chemistry, which is almost exclusively preoccupied with the making and breaking of covalent bonds, has, at least in part, developed into an auxiliary science for biology. Numerous examples exist where naturally occuring compounds have been modified synthetically in order to study changes induced in a biological system, and to compare them with the natural processes. One aim, which is the topic of this chapter, is to "freeze" reversible receptor ligand interactions by covalently linking the interacting molecules.

Molecular recognition in biology is based on noncovalent interaction between receptors and ligands. Regardless of whether a ligand has to be transported through a membrane or in body fluids, chemically transformed by an enzyme, tightly bound by an antibody or a lectin at a cell surface, the process always begins with the formation of a complex between the two complementary systems, resulting in the liberation of binding energy. By definition, the receptor is a protein or a conjugate thereof. Ligands are as diverse as chemical compounds can be. Contacts are made between the different functionalities which are present in a protein structure and complementary structural elements in the ligand. Binding forces are usually subdivided into hydrophobic interactions, hydrogen bonds, charge transfer interactions and salt-bridges.

Among the many different types of ligands, carbohydrates are a special case. The investigation of protein carbohydrate interactions has been stimulated by the important role which glycoconjugates apparently play in biological recog-

0097−6156/94/0560−0198$10.70/0

nition phenomena (*1*). Almost daily, new and important scientific results on the function of carbohydrates in organizing cellular events are reported. Carbohydrates seem to be predestined by their very structure to pass on "chemical information". The recognition is usually based on hydrogen bond formation and Van der Waals interactions between a semi-rigid receptor binding site and the ligand structure, where specificity can of course never be absolute. The "fit" between a given receptor and its carbohydrate ligand is normally quantified by the equilibrium constant of the complex and its components, which can be measured in many different ways. The amount of binding energy liberated on combining ligand and receptor is naturally determined by the biological role the complex formation plays. Binding constants can differ by many orders of magnitude, from the mM- to the pM-range, depending on whether the ligand is normally present in high concentration (serum glucose) or very low levels, such as hormones. Since specific interaction is of biological relevance, one of the aims of glycobiology is altering the specificity of a receptor by artificially changing the structure of the binding site.

Recently it was demonstrated that a calcium-dependent animal lectin which specifically binds terminal D-mannose-, D-glucose- and L-fucose groups in glycoconjugates by recognizing a (-) *syn*-clinal, vicinal diol, can be converted to a mutant protein by exchange of two amino acids. The new protein has a low affinity for D-Man, D-Glc or L-Fuc but now recognizes D-Gal (*2*). Knowing the shape and structure of the catalytic site in an enzyme may also give a clue as to how the activity or specificity of the protein can be altered by exchanging, adding or taking away amino acids in the primary structure. A new line of research deals with the preparation of monoclonal antibodies with catalytic activity (*3*). The first antibody capable of cleaving glycosides was prepared by immunizing mice with an analogue of a glycosyl cation, a so-called "transition state analogue" (*4*). The catalytic antibody had only a moderate rate enhancing effect on glycoside bond cleavage in comparison with real glycoside hydrolases, however, knowledge of the antibody's binding site can be used to optimize catalytic efficiency by protein engineering (*5*). Protein engineering may also be a means to interconvert substrate specificity and reaction specificity in enzymes. Structural similarities between four types of alpha-glucan transforming enzymes, alpha-amylase, pullulanase, cyclodextrin glucanotransferase and branching enzyme, indicate a high degree of relationship, where changes in specificity can be achieved by only slight structural mutations (*6*).

For modulation of receptor propertie it is essential to have a detailed knowledge of the protein structure, especially the binding site. Most carbohydrate binding proteins recognize extended structures of carbohydrates (*7*), therefore, the binding site or active site of an enzyme is not restricted to only one or two points of interaction. Generally a whole area of complementary contact between the receptor and the carbohydrate ligand is seen. The binding area of the receptor is shaped to fit the surface of the ligand and equipped with the right functional groups for optimal, but not necessarily strong, noncovalent interaction. With the progress in X-ray crystallographic analyse many tertiary structures of carbohydrate binding proteins can be solved with a sufficiently high degree of resolution. However, only a limited number of receptor-ligand complexes have been obtained as single crystals suitable for the analysis of their three-dimensional structures by this method. The maltodextrin-binding protein (MBP), a periplasmic component of *E. coli*, has been crystallized as its maltose--complex. The three-dimensional structure shows that hydrogen bonds from the charged sidechains are involved in fixing the ligand (*8*) (Figure 1).

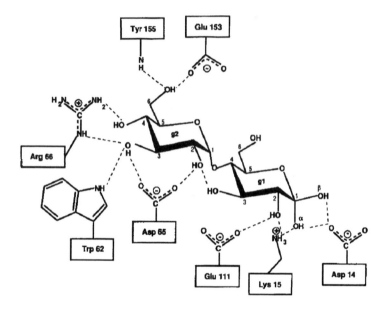

Figure 1   Amino acids of MBP involved in fixing the ligand in the binding site.

The binding area in this case consists of two globular domains, which form a deep groove which is closed at one end. From the numbering of the amino acids in the MBP, it is apparent that the binding area is usually composed of segments in the peptide chain, which lie far apart. There are also binding areas, like in antibodies, made up by separate peptide chains.

An alternative method for structural analysis of binding areas other than by X-ray diffraction or possibly by n.m.r.-spectroscopy of ligand-receptor complexes is photoaffinity labelling with suitable ligands.

The Method of Photoaffinity Labelling and the Choice of a Photolabile Group

Affinity labelling is the chemical modification of amino acid residues in the binding area by reactive ligands. If these ligands can be generated by irradiation when specifically bound to a receptor, the chemical modification is referred to as photoaffinity labelling (*9*). The reactive ligand should covalently modify all structural elements which are close to the binding area, even chemically inert parts such as aliphatic hydrocarbons. Only carbenes and nitrenes are capable of such indiscriminate reactivity. There are a number of chemically fairly stable functional groups from which the reactive species can be photogenerated *in situ*.

Several additional criteria have to be considered when selecting the right photo-labile group for a ligand:
1. The group must not significantly alter the hydrophilic properties of the ligand.
2. It must not interfere with formation of the ligand-receptor complex due to bulkiness.
3. Photogeneration of the reactive ligand should not change the native structure of the receptor.
4. The lifetime of the reactive species should be shorter than the rate of dissociation of the ligand-receptor complex.

Aromatic azides, usually nitrated, are probably the most widely used groups in photoaffinity reagents. Their application for labelling carbohydrate binding receptors is, however, limited due to the drastic physical changes such a hydrophobic residue induces in a normally very hydrophilic ligand. The physical bulk of the substituted aromatic can also distort the natural shape of a ligand. Numerous examples are known of so-called pseudo-photoaffinity labelling by massive, nonspecific covalent modification of the receptor. This pseudo-affinity labelling is also supported by the relatively long lifetime of aromatic nitrenes.

Aliphatic azides can be easily prepared from carbohydrates via nucleophilic displacement of a sulfonate ester group by azide, but can only be photolyzed by high energy u.v.-light of 280-300 nm. This, and the relatively long time (appr. 20 min.) needed for photolysis can cause severe damage to the protein structure. For example, if alpha-amylase is irradiated at 300 nm for 15 min. enzyme activity is totally lost and the enzyme begins to precipitate, but irradiation at 350 nm does not affect enzyme activity (*10*) (Figure 2).

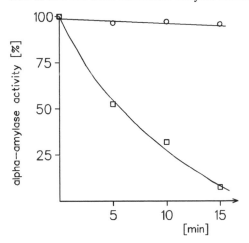

Figure 2   Enzymic activity of alpha-amylase photolysed in solution at 300 nm (□) e.g. 350 nm (○).

Sometimes it is not necessary to keep the native tertiary structure of a receptor intact, and in this case azides may be useful as photoaffinity reagents. There are cases known where an azido group in a carbohydrate can totally prevent binding to a receptor due to steric hindrance (11).

Carbenes can be photogenerated from α-diazoesters, which show a moderate chemical stability. In fact, the first photoaffinity labelling was performed with p-nitrophenyl diazoacetate and chymotrypsin (12). The introduction of electron withdrawing groups can partially overcome the inherent disadvantages of these compounds but their bulk can also be a problem. Even the smallest group of this type, a 2-diazo-3,3,3-trifluoropropionate residue, when attached to a carbohydrate structure, can interfere with binding to the receptor.

Carbenes may also be generated from diazirines. These latter compounds have ideal properties (13). They absorb light between 340 and 380 nm and decay at a rapid rate ($t_{1/2}$ = 1-4 min.). Diazirines are chemically stable under most conditions, although temperatures above 70° are harmful, as are catalytic reduction. Size is usually not a problem with diazirines. The heterocyclic three-membered ring occupies slightly more space than a hydroxymethylene group; in comparison, the azido group extends significantly into the surrounding area.

If the diazirine is adjacent to a primary or secondary hydroxy group, products resulting from intramolecular rearrangement predominate, resulting in a loss of labelling efficiency (14). Ideally, the diazirine ring is placed into a short alkyl chain or at the 6-position of a hexopyranose derivative, allowing the desired intermolecular insertion or addition reactions to take place.

The Principle of Spacer-Modified Oligosaccharides (SMOs)

In order to analyze in detail a large binding site by photoaffinity labelling, suitable ligands are needed which can regioselectively modify defined locations in this area. A direct and comparatively simple way to obtain such compounds would be to introduce substituents into the natural ligand which carry a photolabile group, by regiospecific acylation or alkylation. Numerous reagents for this purpose are commercially available.

The margin, however, between repulsion of such an activated ligand due to steric hindrance (B) and ineffective labelling due to lack of contact with the binding area (C) is narrow (Figure 3).

A   B   C

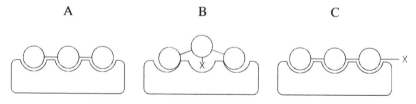

Figure 3   A: optimal binding of native ligand **G3** (G refers to a glycosyl unit) to its receptor, B: reduced binding due to steric hindrance, C: reduced photoaffinity labelling due to lack of contact.

Glycoside hydrolases usually accept structural changes in the aglycone of a substrate yet rarely *O*-alkylation or -acylation of the glycone (*15*). On the other hand, the hexose transporting protein in erythrocyte membrane will reject 1'-methoxy-3'-azibutyl ß-D-glucopyranoside **1** with a photolabile aglycone, but 4-(2'-diazo-3',3',3'-trifluoropropionyl)-D-glucose **2** (*11*) binds strongly to this same enzyme.

For labelling of the active site of ß-D-galactosidase from *E. coli*, three enzyme resistant photoaffinity reagents were synthesised, the free sugar 6-azi-6-deoxy-D-galactose **3** (*11*) and the S- and C-glycosides 3'-azibutyl ß-D-thiogalactoside **4** (*16*) and 3,7-anhydro-2-azi-1,2-dideoxy-D-glycero-L-mannooctitol **5** (*17*).

As shown in Table 1, compound **3** is not bound to the active site at all. The thioglycoside **4** has an excellent affinity but is totally ineffective as labelling reagent for the enzyme's active site. Only the C-glycoside **5** showed an acceptable inhibition constant and is effective as a photoaffinity reagent (*17*).

Table 1. Inhibition and Deactivation of ß-Galactosidase by Different Photolabile Compounds.

Compound	K_I [mM]	% Inactivation[a]
3	-.-	-.-
4	0.07	-.-
5	0.75	40%

[a] Irreversible inactivation of the enzyme on irradiation, Inhibitor concentration 15 mM.

Even more critical can be the selection of labelling reagents for highly specific oligosaccharide binding proteins. Essentially a photolabile group can be placed at the ends, either the aglyconic or the glyconic end of the recognized oligosaccharide structure. This, from a preparative point of view, is relatively easy. For the reasons already discussed, there are of course limitations.

In a three subsite binding area of a hypothetical receptor, a linear trisaccharide segment **G3** is optimally bound (Figure 3). The complex **A** is shown in
the schematic presentation.

Under the assumption that each subsite supplies the same amount of binding
energy, the **G2** ligands may have sufficient affinity in all modes **AI, AII** and **BI,
BII**. The photolabile group is placed in a sterically unobtrusive sidearm. For
obvious reasons, both **G2** ligands in **AI, AII** and **BI, BII** will either be relatively
unspecific concerning the position of chemical modification or relatively ineffective because contact is reduced when the sidearm is placed outside the actual
binding area (Figure 4).

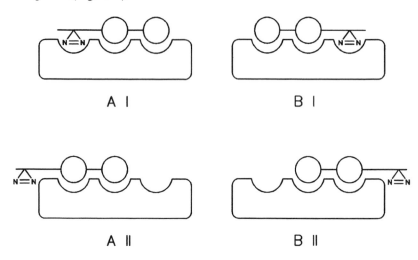

Figure 4   Two **G2** units with differently fixed sidearms carrying the photolabile group in
             different but thermodynamically equivalent binding modes (**AI, AII**) or (**BI,
             BII**).

With structure **G1-G1** in complex **C**, only one strong binding mode is
possible. The photolabile group is in position for highly regioselective labelling,
and the spacer carrying the group has lost its mobility due to the binding areas
at both ends (Figure 5). We call structures such as **G1-G1** spacer-modified
oligosaccharides (SMOs).

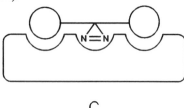

Figure 5   **G1-G1** unit, with spacer carrying the photolabile group. Only one stable binding
             mode is possible.

In an SMO, one or more monosaccharides of an oligosaccharide structure are replaced by a flexible spacer of the same length. This type of spacer is ideally suited to carry the diazirino ring for photoaffinity labelling. As long as the flanking saccharide structures supply sufficient affinity, the photolabile unit can be placed at different positions along the binding area for regioselective chemical modification.

The principle of SMO was first investigated with a maltotriose analogue **6** consisting of two α-D-glucopyranosyl units (1→4)-linked by an acyclic spacer, which replaces the center glucose (*18*). This compound binds specifically to alpha-amylase and also to the maltodextrin binding protein.

6

This compound can be regarded as an α-glucoside or a 4-*O*-alkyl-glucoside, and the affinity of the enzyme for **6** is astonishing. Neither end of this molecule binds on its own to the receptors. This indicates a significant cooperative effect, comparable to the "clustering effect" described by Y. C. Lee (*19*). A series of sulfur linked SMOs mimicking maltotriose have remarkable affinities to pancreatic alpha-amylase (*20*) (Table 2) and serve as good example of the usefulness of these structurally "reduced" ligands as probes for the corresponding receptor sites.

Probing the Binding Site of Pancreatic Alpha-Amylase with three Isomeric, Photolabile SMOs

As part of our investigations of the structures of active sites of glycoside hydrolases, porcine pancreatic alpha-amylase (a gift of Prof. Dr. G. Marchis-Mouren, Marseille) has been a major subject. The enzyme is a typical *endo*-glucanase and is an essential catalyst for starch degradation in mammals. It consists of one polypeptide chain of 496 amino acids (*21*). Its three-dimensional structure has been investigated by X-ray diffraction (*22*) and appears to be consistent with earlier enzyme kinetic investigations. These findings indicate a binding area (**A, B, C, D** and **E**) which can accommodate five (1→4)-α-linked D-glucopyranosyl units (Figure 6).

Table 2.   Ligands that Bind to the Active Site of Alpha-Amylase.

K_I [mM]	Ligand
9	
17	
11/16	
3	

Figure 6   Schematic presentation of the binding area of pancreatic alpha-amylase. ↑ indicates the active site.

Recent investigations indicate the participation of approximately 17 amino acids, which are scattered along the peptide chain, to be involved in forming the binding site with Asp-197 and Asp-300 the most likely candidates for catalytically active groups (*22*). For chemical modification of the five subsite binding area with photolabile SMOs in different, specific places, the ligands used should consist of two carbohydrate parts (one glyconic and one aglyconic end) of one to three monosaccharide units each, linked by a spacer carrying the diazirine as a replacement for one monosaccharide unit (Figure 7).

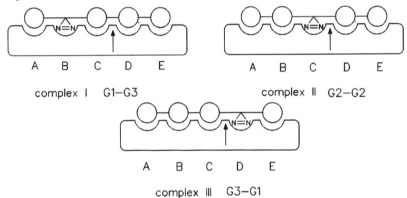

Figure 7   Each complex **I**, **II**, and **III** preferentially exists in the described thermodynamically favoured binding mode.

The monosaccharides have to be α-(1→4)-linked, and the SMOs should be resistant to enzymic degradation to prevent a decrease in their binding affinity during the labelling process. Of the three complexes in Figure 7, **I** would probably be susceptible to cleavage by the enzyme unless one of the glucosyl units occupying subsites **C** and **D** are slightly modified. It is known that the hydroxymethyl group is essential for cleavage, since an α-(1→4)-linked tetrasaccharide **7** with two center xylopyranosyl units mimicking maltotetraose is a stable competitive inhibitor of pancreas alpha-amylase (*23*).

8

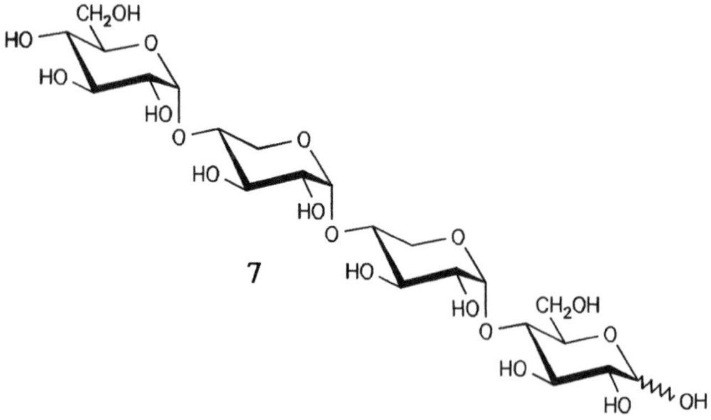

Syntheses of Enzyme Resistant SMOs that Mimic Maltopentaose

Of different possibilities to introduce a 6-membered spacer (length of an α-$(1\rightarrow4)$-linked glucopyranosyl unit including an anomeric oxygen) into a malto-dextrin chain, the most economic is the preparation of an α-D-glucopyranoside carrying as an aglycone the spacer to be (Figure 8).

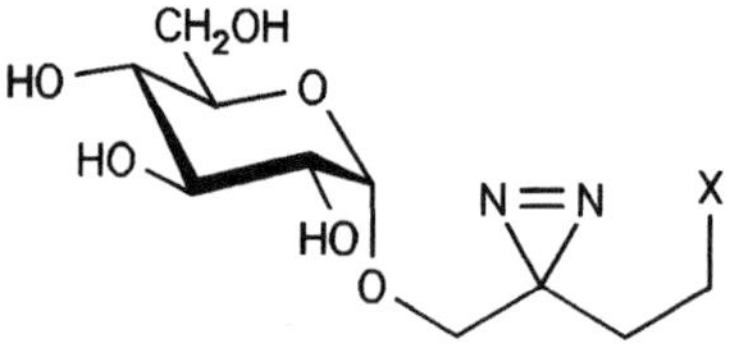

Figure 8   Glucopyranoside carrying the spacer as an aglycone (x: reactive group).

Any D-glucopyranoside can be converted to a homologous series of maltodex-trine-glycosides by enzymic transglucosylation with cyclomaltohexaose (α-CD) and cyclodextrine glucanotransferase (CGTase) (24). The transglucosylation will be carried out as the last step in assembling the SMO before the final purifica-tion. The difficult part in the chemical synthesis is coupling the aglyconic side of the SMO to the spacer. A versatile synthon for coupling nucleophiles stereo- and regioselectively to the 4-position of a pyranosyl ring is methyl 3,4-anhydro-6-deoxy-ß-L-arabino-hex-5-eno-pyranoside 8 (25).

If group X of the spacer (Figure 8) can be converted to a thiol, epoxide opening occurs smoothly and with good yields (26). The double bond in 8 has the ad-ditional advantage of modifying the future D-glucopyranosyl residue in such a way that when incorporated into an oligosaccharide chain, it prevents cleavage by alpha-amylase when it binds to either subsite **C** or subsite **D** (Figure 7).

The oligomeric homologues of 8, 21, and 22 are prepared from the correspon-ding methyl α-maltoside 9 or -maltotrioside 10 in an analogous way (26) (Figure 9). Each of these can be coupled with the glyconic part 23, carrying the spacer (27) (Figure 10).

Compound **26** is already a maltopentaose analogue, the smaller molecules **24** and **25** have still to be glucosylated with CGTase (*24*) and α-CD. The useless higher homologues can be trimmed down with beta-amylase (*27*) (Figure 11). The final separation and purification of homologues is carried out by h.p.l.c. to yield the three SMOs **26, 27,** and **28** (*27*).

Radiolabelling

The detection of a chemical modification in a protein requires a radioactive photoaffinity label. Again, there are several possibilities for introduction of a radioisotope into the different SMOs. The only structural element common to all three synthetic pathways is the glycoside **29** (*27*). Because it is always advisable to have the label directly attached to that part of the ligand that is involved in the chemical modification, the label, ($[^{3}H]$) is introduced into the primary carbon end of the aglycone by selective oxidation-reduction (Figure 12). This prevents the radiolabel from being lost due a secondary chemical reaction, such as hydrolysis.

The rest of the synthesis is carried out as described.

The unsaturated hexosylresidue appears to bind as well as, if not better, than the intact D-glucopyranosyl moiety to the active subsite. It also prevents enzymatic hydrolysis by the enzyme. When compounds **28** and **33a/b** are incubated for extended times (> 1h) with extremely high enzyme concentrations, only a small amount of maltose is cleaved off the glyconic end; the rest of the molecule **G1-G1** is mainly unchanged and inhibitor **G3-G1** can be detected. This fact must be interpreted by the formation of very little, unfavourable, yet productive binding. The small but reproducible difference in the observed K_I s between the diastereomeric azides **33a** and **33b** (*26*) (Table 3) is a clear indication for interaction of the spacer and binding site. This is a necessary condition for successful photoaffinity labelling.

Competitive Inhibition of Alpha-Amylase by the SMOs

Enzyme activity of alpha-amylase can be determined by a number of different assays, most of which are relatively complicated (*28*). Using 4-nitrophenyl α-maltotrioside and relatively high enzyme concentrations allows direct optical determination of enzyme activity (*29*). This method is not very sensitive, however, due to the slow rate of hydrolysis of this glycoside. Inhibition constants for compounds **26, 27,** and **28** were determined by Dixon plot (*27, 30*) and are listed in Table 3 along with those for the analogous azides **31, 32, 33a** and **33b** (*26*). Although not really comparable, the inhibition constants of all SMOs are equal and even lower than the K_M for the substrate maltopentaose (1.1 mM).

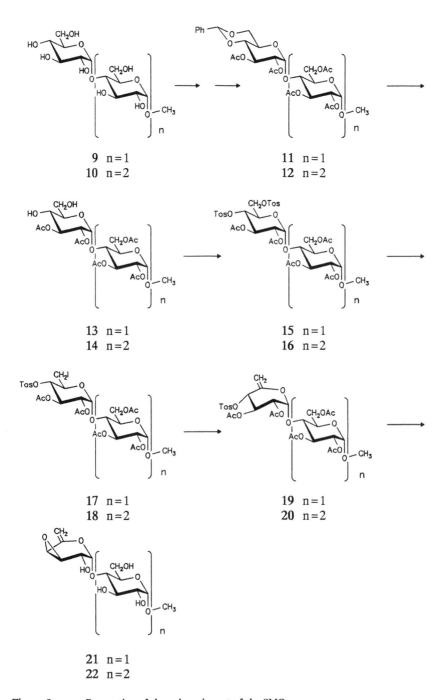

9 n = 1
10 n = 2

11 n = 1
12 n = 2

13 n = 1
14 n = 2

15 n = 1
16 n = 2

17 n = 1
18 n = 2

19 n = 1
20 n = 2

21 n = 1
22 n = 2

Figure 9   Preparation of the aglyconic part of the SMOs.

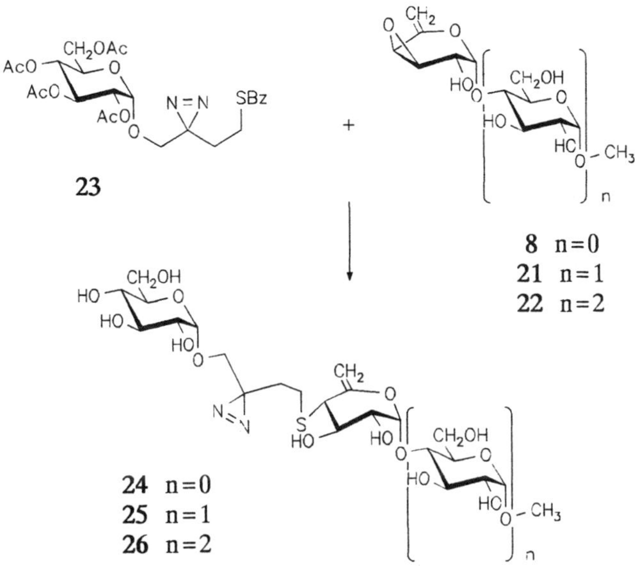

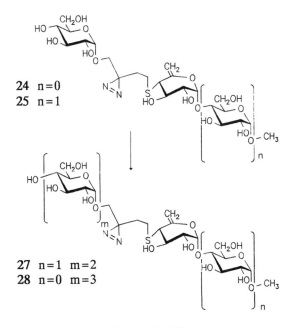

Figure 10   Preparation of the SMOs by coupling of the glyconic and aglyconic precursors.

Figure 11   Enzymic preparation of isomeric SMOs.

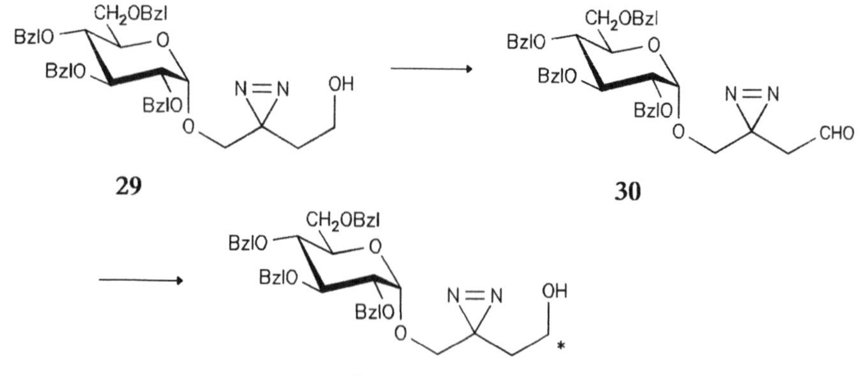

29

30

29*

* Position of the radioactive label

Figure 12   Radiolabelling of compound **29** by selective oxidation-reduction.

Table 3.   Competitive Inhibition of PPA by Different SMOs.

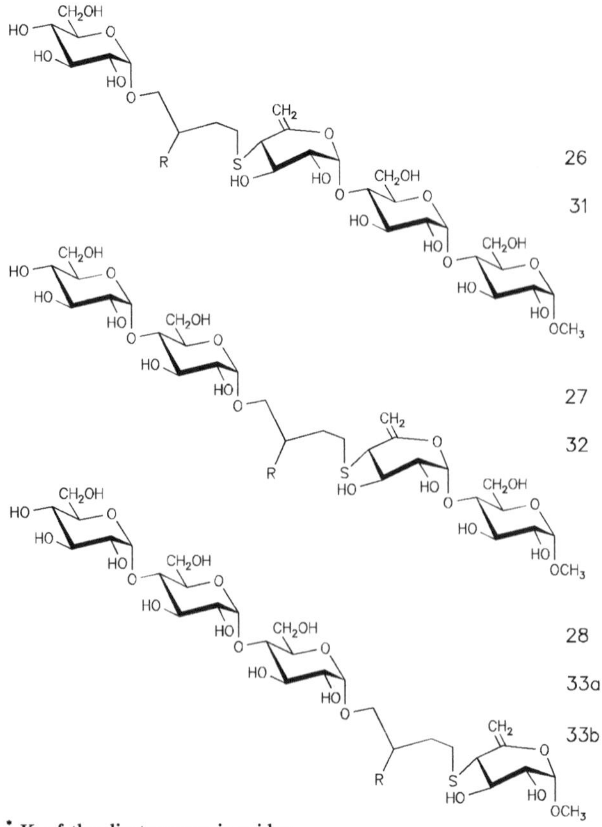

		K_i [mM]
26	R = (diazirine)	0.15
31	R = N_3	0.15*
27	R = (diazirine)	2.1
32	R = N_3	2.4*
28	R = (diazirine)	2.5
33a	R = N_3	1.75
33b	R = N_3	0.92

* K_I of the diastereomeric azides

Irreversible Inhibition of Alpha-Amylase by Photoaffinity Labelling

The SMOs **26**, **27**, and **28** decay when irradiated at 350 nm at $t_{1/2}$ of appr. 2 minutes; generally, photolysis is carried out for 10 minutes or less. Incubation mixtures are placed into a quartz tube which can be cooled by ventilation and rotated at about 30 rpm so that the solution is spread as a film on the inside of the reaction vessel. All experiments are carried out also in the presence of a very high concentration of the reversible inhibitor maltotriose. The solutions after irradiation are clear and colourless and, after thorough dialysis, can be directly used to assay enzyme activity. The results are shown in Figure 13 (*27*).

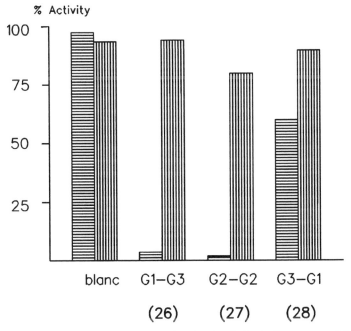

Figure 13   Remaining alpha-amylase activity after irradiation in the presence of SMOs **26**, **27**, and **28** without(═) and with (‖) added maltotriose.

All three spacer-modified tetrasaccharides irreversibly inhibit the enzyme by blocking its active site. In the presence of maltotriose very little deactivation takes place. The deactivation by photoaffinity labelling with either compound **26** or **27** was extraordinarily high and indicates good contact of the spacer with the receptor subsite. The third compound **28** leads to about 40% deactivation, which is still quite efficient. Figure 14 shows the Lineweaver-Burk plot for one photo-deactivated enzyme preparation and a blank, which was prepared in the same way as the samples but did not contain the photolabile ligand. The K_M has not changed, which is an indication for the highly specific deactivation due to blockage of the binding site specifically. The remaining unmodified alpha-amylase molecules are as active as those in the blank experiment (*27*).

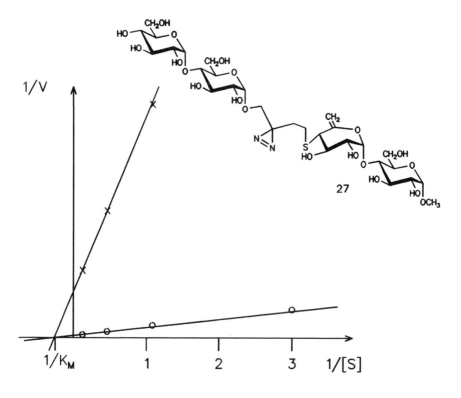

Figure 14   O: Hydrolysis of 4-nitrophenyl α-maltotrioside
                 X: Hydrolytic activity of the same enzyme photoaffinty labelled by
                 compound 27.

Radiolabelling the Binding Site of Alpha-Amylase. Isolation and Analysis of Modified Peptides

The measurements of deactivation are supported by the covalent incorporation of radioactivity into the protein. Each of the three [³H]-labelled ligands **G1-G3** (**26***), **G2-G2** (**27***) and **G3-G1** (**28***) sucessfully radiolabelled alpha-amylase. The **G1-G3**-, **G2-G2**- and **G3-G1**-treated proteins were digested by trypsin and the peptides separated by h.p.l.c., using a water/acetonitrile gradient containing trifluoroacetic acid (5 mM).

Splitting the eluate allowed a crude determination of radioactivity in a flow counting chamber. The elution profile of the peptide mixture (measured at 220 nm) and radioactivity both superimposed to give the approximate location of the modified peptides. By using very small fraction volumes in the area containing the radioactivity, greater resolution can be obtained, as measured by a liquid scintillation counter. By this procedure, single peptides carrying the radiolabel could be isolated. If there is doubt as to the purity of the sample different conditions for the h.p.l.c. must be applied. If h.p.l.c. of the peptide again shows only one signal, usually with an altered retention time, one can assume the peptide in question to be pure.

Labelling alpha-amylase with **G1-G3**[*] gave two labelled peptides, **G1-G3 I** and **G1-G3 II** . The protein labelled with **G2-G2**[*] yielded three labelled peptides, **G2-G2 I**, **G2-G2 II** and **G2-G2 III**. **G1-G3 I**, **G2-G2 I** and **G1-G3 II**, **G2-G2 II** have the same amino acid sequence. PPA labelled with **G3-G1**[*] gave only one major labelled peptide **G3-G1 I** (Table 4).

Table 4.   Photoaffinity Labelling of PPA by Different Radioactive SMOs.

Radioactive photoaffinity label	% Radioactivity covalently attached to protein in SDS-gel[*]	μCi/mg Protein	Yield of radioactivity of isolated pure tryptic peptides	
G1-G3	35	14	G1-G3I	20%
			G1-G3II	6%
G2-G2	38	15.5	G2-G2I	12%
			G2-G2II	3%
			G2-G2III	50%
G3-G1	25	9.8	G3-G1I	30%

[*] 100% radioactivity would be a stoichiometric modification (1 ligand/molecule PPA)

The amino acid sequences of most peptides (Figure 15) correspond well with the amino acid composition of the groove area suspected to be the binding- and active site of alpha-amylase.

A)

57   61
—Pro --------- Arg—

62   68
—Tyr --------- Lys—

93   124
—Ile   --------- Arg—

159   172
—Asp ---------Lys—

195   200
—Ile   --------- Lys—

228   243
—Pro --------- Lys—

258   267
—Tyr --------- Arg—

292   318
—Ala --------- Arg—

B)

      ↓ 334   337 ↓
Tyr── Gly–Phe–Thr–Arg ──Val

C)

       ↓ 269   273 ↓
Lys ──Trp–Ser–Gly–Glu–Lys── Met

D)

       ↓ 262   267 ↓
Lys── Leu–Gly–Thr–Val–Val–Arg──Lys

E)

      ↓ 31   35 ↓
Arg──Tyr–Leu–Gly–Pro–Lys──Gly

Figure 15   A) Amino acid sequences of tryptic peptides containing residues which appear to interact with the bound substrate.
B) Amino acid sequence of G1-G3I and G2-G2I.
C) Amino acid sequence of G1-G3II and G2-G2II.
D) Amino acid sequence of G2-G2III.
E) Amino acid sequence of G3-G1I.

Can Galactosyltransferase (Gal-T) Recognize two GlcNAc-Residues at the same Time?

ß-(1→4)-D-Galactosyltransferase (UDP-galactose: 2-acetamido-2-deoxy-ß-D-glu-copyranosyl-glycopeptide galactose-ß-(1→4)-transferase [EC 2.4.1.38]) is an important enzyme of glycoconjugate anabolism. In mammals, both a soluble (serum and milk) and a membrane associated form (mainly Golgi-membrane) are known. Both have the same kinetic parameters and it is very likely that their active sites are similar if not identical (*31*), the soluble form is probably produced by proteolytic cleavage of the membrane enzyme (*32*).

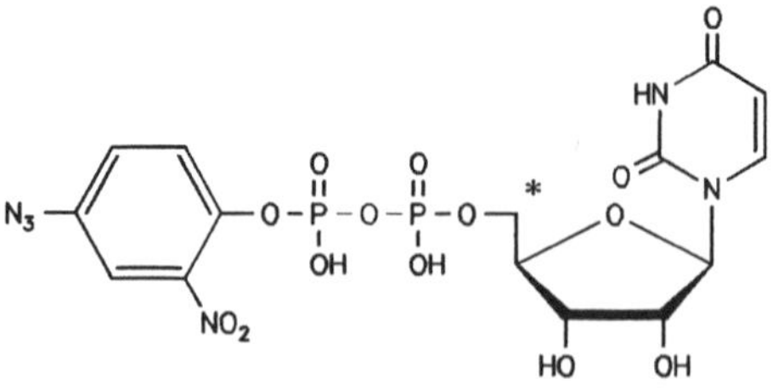

34

The binding site for the donor substrate UDP-Gal could be defined by photoaffinity labelling, using the aromatic nucleotide **34** (*33*). The sequence homology of corresponding peptides between human and bovine Gal-T is more than 90% (*34*).

Obviously there must be, besides a donor-, also an acceptor binding site. One of the natural acceptors is the biantennary N-linked core heptasaccharide **35**.

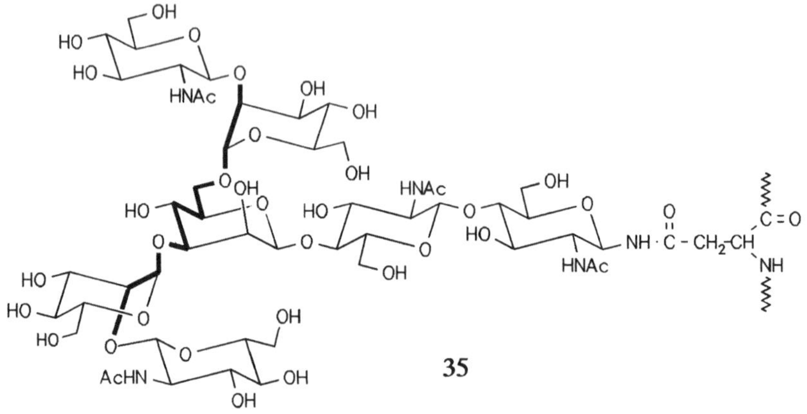

35

An analogous glycopeptide **36** has been used by Rao and Mendicino to determine a K_M for the first (0.25 mM) and the second (2.0 mM) galactosylation (*35*). Apparently the first galactosylation, which yields the branched isomers **37a** or **37b** is favoured over the second to give the endproduct **38**.

Man–GlcNAc
/
Hc–IgG–Asn–GlcNAc–GlcNAc–Man
↑   \
Fuc   Man–GlcNAc

36

Man–GlcNAc–Gal
/
Hc–IgG–Asn–GlcNAc–GlcNAc–Man
↑   \
Fuc   Man–GlcNAc

37

A significant difference in rate constants for the first galactosylation was found by H. Schachter *et al.* (*36*) favouring the GlcNAc attached to the (1→3)-linked mannose. This pronounced "branch-specificity" is hard to explain without assuming an acceptor binding area, which is capable of differentiating between the two structurally different glyconic ends of oligosaccharide **35**.

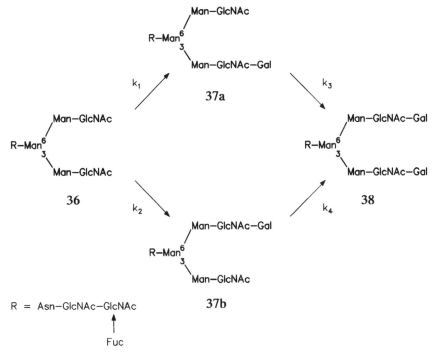

R = Asn–GlcNAc–GlcNAc
↑
Fuc

From these data, we postulate two subsites, designated **1** and **2**, for the two terminal GlcNAc groups. Site **1** is also the site of galactosylation (Figure 16). We also assume there is binding interaction between the enzyme and acceptor along the mannotriose unit **C-D-E** (Figure 17).

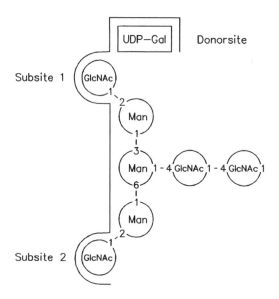

Figure 16   Preferred orientation of the acceptorsubstrate **35** in the binding area of Gal-T

Such a model site would explain the apparent branch-specificity and also the highly reduced rate for the second galactosylation; formation of the enzyme-ligand complex is less probable with one GlcNAc already blocked. Recognition of both terminal GlcNAc groups simultaneously would be a condition for optimal reactivity.

Syntheses of Mono- and Biantennary Acceptors for Gal-T

Total syntheses of compounds like the heptasaccharide **35**, as described by Paulsen *et al.* (*37*), are difficult and time consuming. Often much simpler structures can be used for kinetic investigations as well as covalent modifications of receptor binding sites. If the end groups **A** and **B** in the acceptor substrate **35** are the actual anchoring groups when the ligand and Gal-T form a complex for galactosylation, then the only structural requirement for an optimal acceptor substrate would be a spacer to keep the two GlcNAc residues at the right distance. The shortest connection between **A** and **B** following the covalent bonds through the mannosyl residues **C**, **D** and **E** is a chain of 10 atoms not including the glycosidic oxygen atoms of the GlcNAc residues.

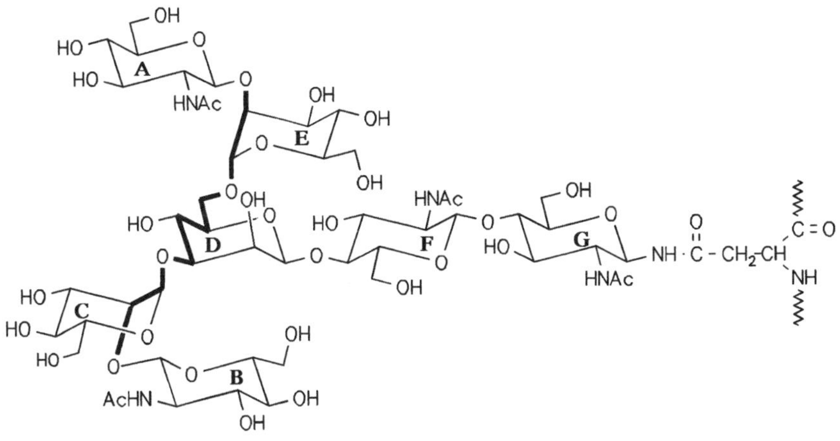

Figure 17 One of the natural acceptorsubstrates of Golgi-Gal-T is the core-oligosaccharide of N-glycoproteins.

The simplest ligand for a two subsite acceptor binding area of Gal-T would consist of an aliphatic, acyclic, 10-membered spacer connecting two GlcNAc residues "head on" (*38*). The spacer is sufficiently flexible to adopt a conformation following the line of the three central mannose residues in structure 35. It can also accommodate photolabile groups in different positions of the chain and possibly also a "handle" replacing the chitobiose unit **F-G**. This unit, which is linked to the protein via asparagine, has been shown by Hindsgaul *et al.* to be non-relevant for the recognition by Gal-T (*39*). Although we call these structures spacer-modified oligosaccharides, because the spacer replaces carbohydrate units of the same longitudinal extension, the compounds resemble ligands prepared by Y. C. Lee *et al.*, which are placed as determinants in neoglycoproteins (shorthand synthesis) for the binding to the hepatic Gal/GalNAc receptor (*40*). Such ligands are usually much easier to synthesize than the complete, natural oligosaccharide structures.

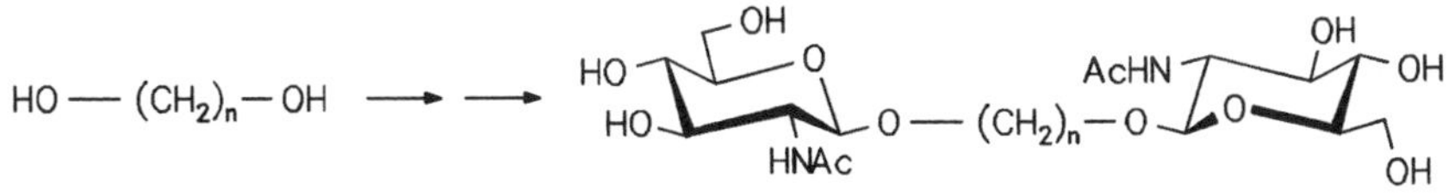

39 n = 12	**43** n = 12	
40 n = 10	**44** n = 10	
41 n = 8	**45** n = 8	
42 n = 6	**46** n = 6	

Four commercially available α-ω-alkanediols (**39**, **40**, **41**, and **42**) were each reacted with 1,3,4,6-tetra-*O*-acetyl-2-deoxy-2-phthalimido-β-D-glucopyranose in the presence of $SnCl_4$ under anhydrous conditions according to a method described by Hanessian *et al.* (*41*). Yields of the condensations were good and after deprotection and re-N-acetylation of the products four spacer-modified "nonreducing" disaccharides **43**, **44**, **45**, and **46** could be obtained (*42*). The compounds are crystalline and turned out to be acceptor substrates for Gal-T. Results of the kinetic data for these compounds will be discussed in the next chapter.

SMOs for the Regioselective Labelling of the Receptor Binding Area

Enzyme assays described in the next chapter clearly demonstrate the superiority of compound **44** with the C_{10}-spacer as acceptor substrate.
Superimposing structure **35** and compound **44**, the sections corresponding to the spacer can clearly be seen in the natural structure (Figure 18).

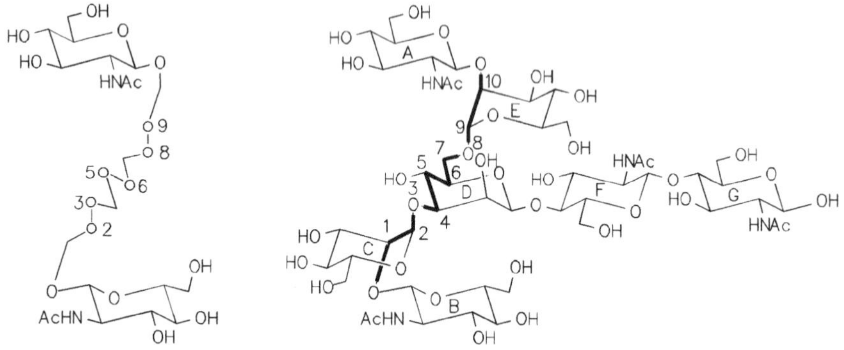

44   **35**

Figure 18   Compound **44** as a superimposable section of structure **35**.
o position of azi group.

For practical reasons the positions 2, 3, and 5 were chosen as sites for the photolabile diazirine. This would also cover the positions 6, 8, and 9 with respect to the receptor binding area, because of the pseudo-symmetry (diazirine instead of methylene should not be recognized by the enzyme) of the ligand. Placing the photolabile groups in such a position should make a regioselective chemical modification of distinct positions in the receptor area possible and give an indication of the closeness between spacer and protein surface.

Three decanetriols **48**, **54**, and **60** were used to synthesize the photolabile spacers **50**, **56**, and **62** differing in the position of the diazirino group. These compounds are not commercially available and were prepared from the different starting materials using specific reactions shown in Figure 19.

A key step in these syntheses is the oxidation of the secondary hydroxy group with hypochlorous acid giving the corresponding ketone. A three-step sequence yields in each case the diazirine according to the procedure by Church and Weiss (*43*).

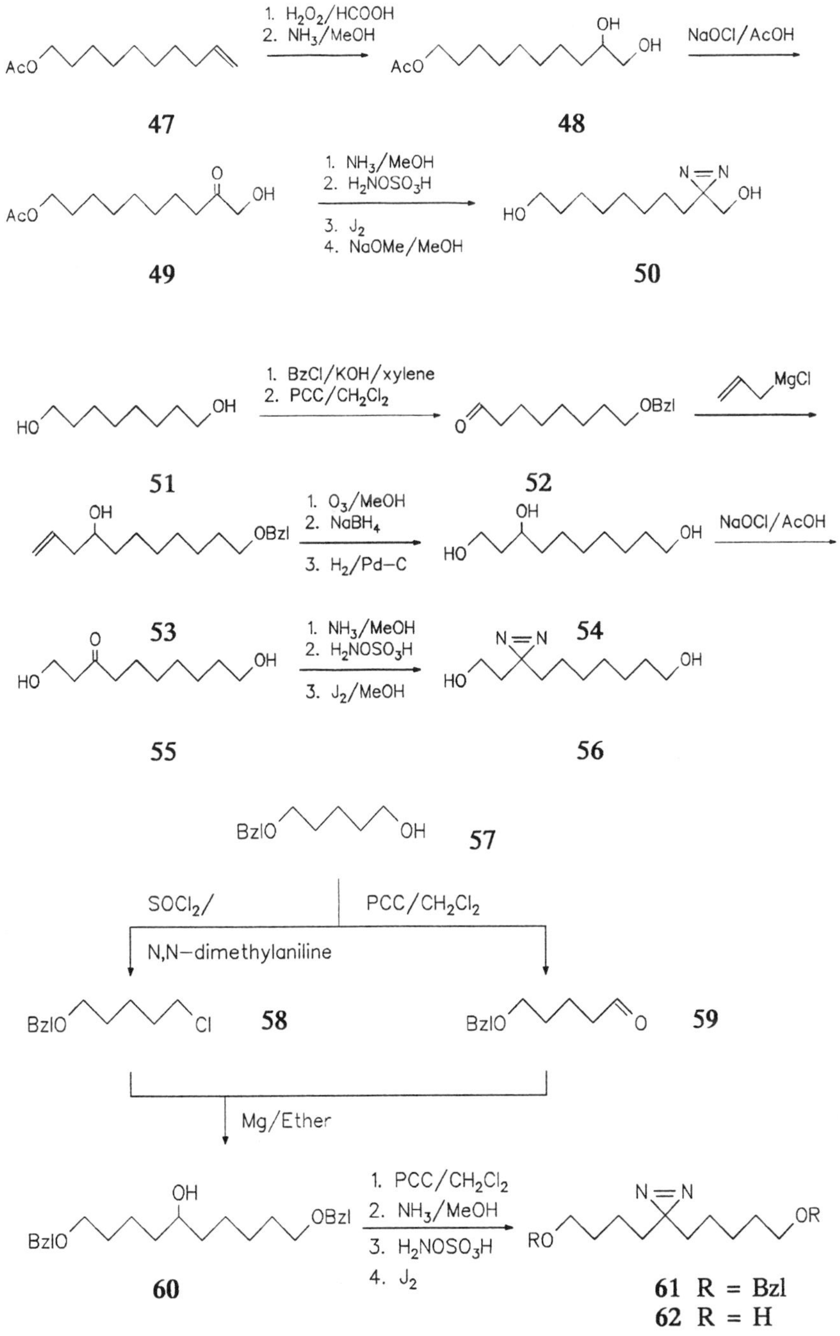

Figure 19   Synthesis of three isomeric, photolabile spacers.

The diazirino group is stable under the conditions of glycosylation with $SnCl_4$ in methylene chloride and therefore the photolabile diantennary ligands **63**, **64**, and **65** are prepared in the same way as the SMOs with unmodified hydrocarbon spacers.

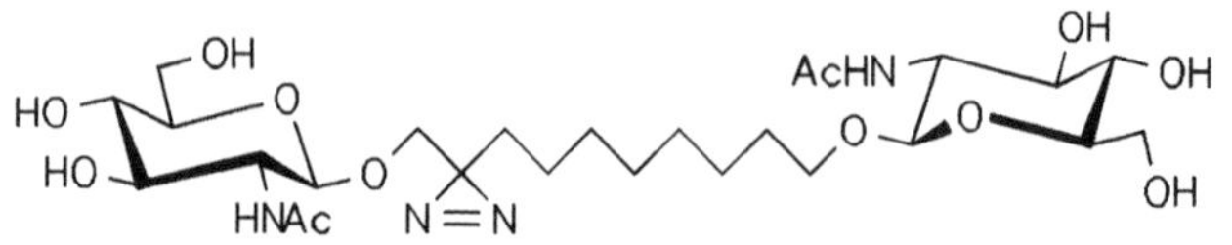

63

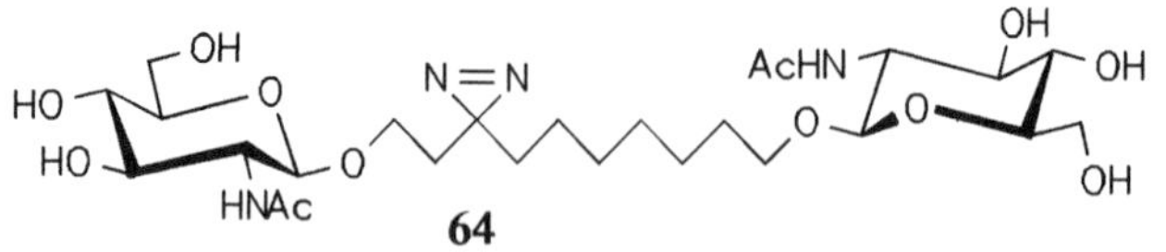

64

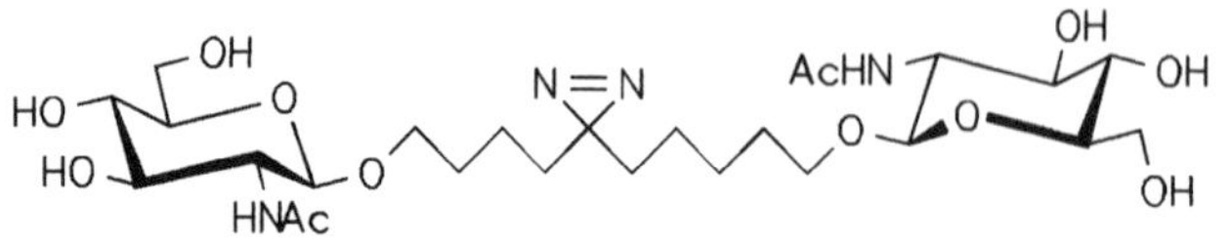

65

Radiolabelling

Removal of the N-phthaloyl protecting group with sodium borohydride in aqueous isopropanol (*44*) yielding the free GlcN terminal residues of the SMOs: the amino groups can then be radiolabelled selectively with [³H]acetic anhydride in methanol at low temperature. The yields are about 50% and the specific radioactivity of commercial [³H]acetic anhydride (6.7 Ci/mmol) is sufficient for a significant radio-photoaffinity labelling.

Hydrophobic Glycosides of GlcNAc

The SMOs consisting only of the two terminal GlcNAc residues and the aliphatic spacer are rather hydrophobic. It is known that hydrophobic interaction of structural elements in the ligand can enhance affinities for the receptor binding site significantly. Specific binding can even be converted to unspecific interaction if the hydrophobic effect is stronger than bonding of the natural carbohydrate moiety to its binding site. In order to differentiate between binding due to the biantennary GlcNAc terminals in SMOs **43-46** and **63-65** on one hand, and the strongly hydrophobic element introduced by the spacer, on the other, two simple glycosides were also synthesised for comparison.

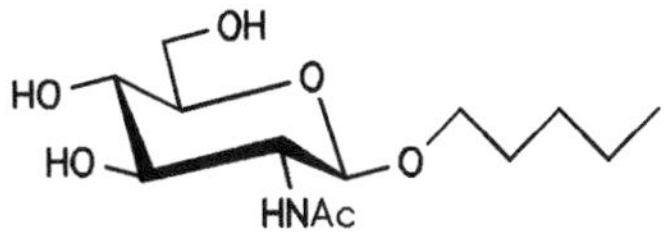

66 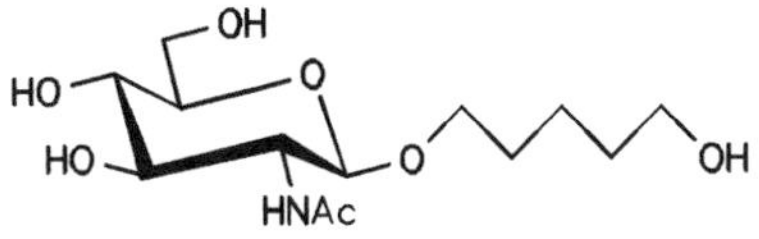 **67**

Compound **66** is the SMO **44** cut in half. The other, **67**, carries a hydroxy group at the end of the aglyconic carbon chain. Syntheses of these compounds were carried out in a similar way as described for the SMOs **43-46**.

Introducing Rigid Structures into the Spacer Decreases Conformational Flexibility

All the SMOs so far described are structurally flexible due to the long, linear spacer. This should give the GlcNAc end groups enough freedom to bind unhindered to either of the two potential binding subsites. As structural hybrids between the flexible SMOs and the much more rigid structure of the pentasaccharide **35** a partial structure of **35**, the spacer-modified trisaccharide glycoside **72** was synthesised.

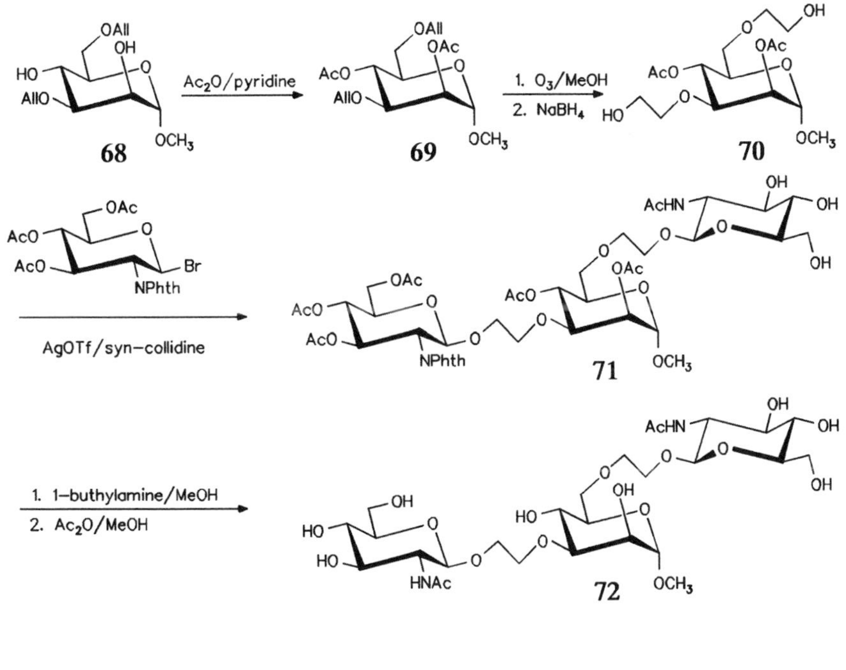

Figure 20   Synthesis of a relatively rigid spacer-modified trisaccharide.

The compound consists of the center mannosyl unit as a potentially directing element for the branch specificity of Gal-T. The flanking (1→3)- and (1→6)-linked mannosylresidues are replaced by two ethylene spacers. The synthesis begins with the selective allylation of α-methyl mannoside in the presence of tri-n-butyl-tin oxide (45) and follows the sequence outlined in Figure 20 (46).The spacer-modified trisaccharide 72 was prepared primarily to determine which structural elements can account for the observed branch specificity. The monogalactosylated products 73a,b were synthesised enzymatically and investigated by n.m.r-spectroscopy, which demonstrates the absence of branch specificity in the case of compound 72 (47).

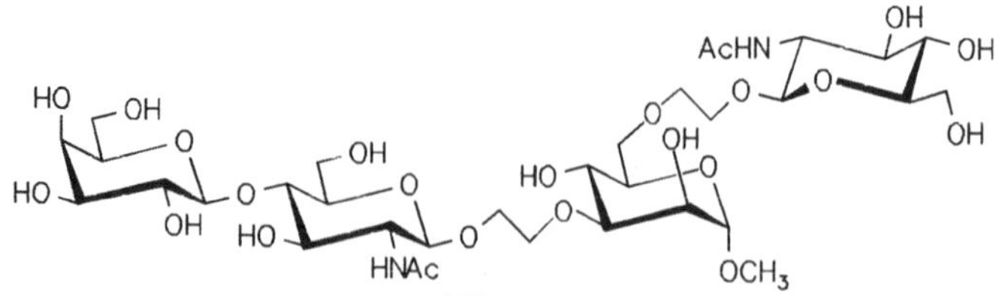

73a

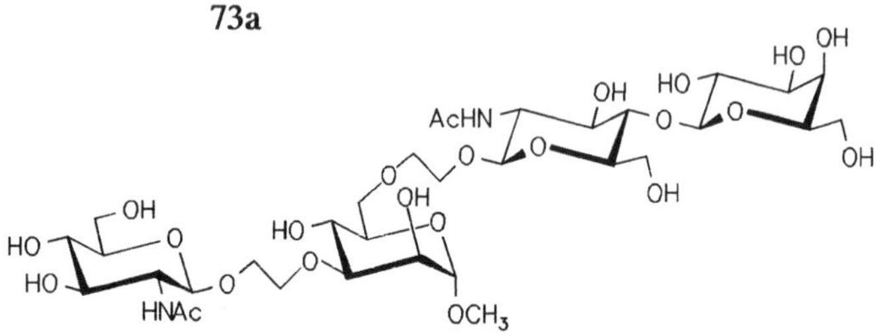

73b

Apparently the center mannose only has a directing effect in the intact pentasaccharide **AEDCB** (Figure 17), where it determines the conformational arrangement of the two rigid antennae **EA** and **CB** (Figure 17). There are indications from photoaffinity labelling experiments that the center part of the spacer-modified disaccharide **65**, corresponding to the center mannose **D** in compound **35**, probably has less contact with the Gal-T binding site than the parts closer to the terminal GlcNAc groups.

In compound **79**, the center part in the flexible SMOs is replaced by a rigid but hydrophilic cyclohexenediol unit, prepared as described in Figure 21. The compound is a mixture of diastereomers, which were not separated prior to transgalactosylation.

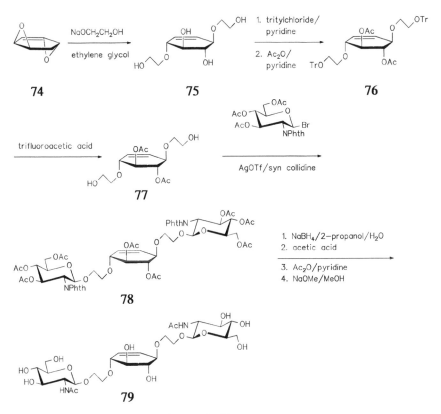

Figure 21   Synthesis of a relatively unflexible SMO.

The monogalactosylated SMOs derived from compounds **44, 72** and **79** were all prepared by preparative enzymic transglycosylation with UDP-Gal and Gal-T and purified by chromatography on Biogel.

Acceptors Competing for the Binding Site. Relative Rates of Galactosylation are an Indication of Binding Affinities

Glycosylation by Gal-T occurs by a so-called **ordered bi-bi-mechanism** (*48*), with GlcNAc as a standard acceptor substrate. All other acceptors were compared with the standard by competitive binding. Since the donor substrate is applied under conditions of saturation, the comparative initial rates between galactosylation of GlcNAc-R and GlcNAc at given concentrations (relative rates) are an indication for their relative affinities to the receptor. Although assays to measure rates of galactosylation by Gal-T do exist, they are complicated, expensive and not applicable to the determination of rates for two competing acceptor substrates.

The essence of determining Gal-T-activity with the radioactive donor substrate UDP-[^{14}C]Gal is the separation of donor from product or products, which is usually done by ionophoreses or ion-exchange chromatography of a denatured aliquot. Here we use thin layer chromatography to separate all the radioactive components and determine their relative amounts by scanning. The method is quick and relatively inexpensive. Three different solvent mixtures (Table 5) were used to separate the different products of galactosylation from the donor substrate. Since relative rates of galactosylation of certain acceptors in comparison with the standard GlcNAc can be over 100, this difference has to be compensated by increasing the concentration of GlcNAc in order to obtain comparable amounts of galactosylation products (Table 5).

Table 5   Conditions for Measuring Relative Rates of Galactosylation.

Acceptor-substrate	[GlcNAc]	solvent mixture
43-46, 63, 64, 65	0.5 M	6:4:3 1-butanol/pyridine/water
66, 67	0.1 M	4:2:1 ethyl acetate/methanol/water
72, 79	0.05 M	3:2:1 ethyl acetate/methanol/water

The method is quite accurate and consistent. Consistency can be checked by letting acceptors other than GlcNAc, such as compounds **63** and **72**, compete with each other and compare the results with the standard relative rates. Some measured relative rates are listed in Table 6.

There is no doubt that hydrophobic structural elements in SMOs **43-46** and **63-65** and the simple glycosides **66** and **67** increase the affinity of acceptors to Gal-T in comparison with GlcNAc. This well-known phenomenon cannot, however, explain the very high affinities of SMOs **44** and **63**. It is clear that two free GlcNAc residues in the biantennary structures **44** and **63** are essential for optimal binding. Omitting one GlcNAc either by taking only half of the molecule as in compounds **66** and **67** or blocking one of the GlcNAc residues by galactosylation, drastically decreases the affinity of such acceptor substrates.

The low affinities of the structures **72** and especially **79** are not as easy to comprehend. Apparently the central mannose unit **D**, on its own, does not contribute much to the general affinity of the natural pentasaccharide or heptasaccharide. The decreased conformational flexibility in **72** and of course also in **79**, when compared with the highly flexible structures **44** and **63**, may be responsible for this rather strange phenomenon.

All these results support the assumption of two separate GlcNAc binding subsites in Gal-T, yet the most convincing evidence is supplied by the kinetic comparison of the acceptors **43-46**.

Table 6   SMOs and Glycosides Arranged According to their Relative Rate of Galactosylation.

Relative rate of galactosylation	Acceptor-substrate
145	**44**
145	**63**
62	**45**
20	**66**
18	**67**
16	**79**
10	**72**
1	

All four compounds have similar polarities and carry two terminal GlcNAc groups. Only the length of the spacer differs decreasing by one methylene group at a time. The optimal ligand is compound **44**, with the "right" spacer length. In comparison both shorter homologues show highly decreased affinities, whereas the longer one does not differ significantly (Figure 22).

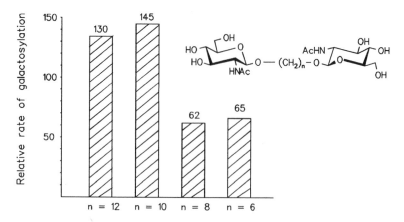

Figure 22   Rate of galactosylation for a homologoues series of SMOs in comparison to GlcNAc.

The "mechanical" model of three different Gal-T ligand complexes (Figure 23) explains why the two GlcNAc end groups have to be kept at a great enough or the "right" distance for optimal binding to occur.

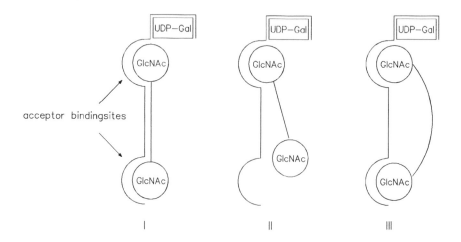

Figure 23   Mechanical model of three different Gal-T ligand complexes.

Irreversible Inhibition of Galactosyltransferase by three Photolabile SMOs

A spacer replacing the three monosaccharides C, D, E in the natural ligand 35 of Gal-T can be used to covalently modify those parts in the enzymes binding site which normally interact with the corresponding carbohydrate structures. Compounds 63, 64 and 65 carrying one photolabile diazirine group each at a different position along the hydrocarbon chain should react, when photolysed, with the complementary area of the receptor. Because the two ends of the SMOs are not distinguishable by the enzyme, theoretically six different locations (Figure 18) may be covalently modified depending on which side of the active site the SMO attaches. This should give an overall impression of the binding area, provided that modification actually takes place and if the modified segments can be isolated and their structures determined. Such experiments should give results similar to those described for alpha-amylase in the preceding chapter.

Gal-T, although acting by an ordered bi-bi mechanism, binds acceptors even in the absence of UDP-Gal (*49*). In the presence of UDP-Gal, the SMOs would rapidly be turned into much weaker binding acceptors - one ligand being blocked by galactosylation - whereby the efficiency of photoaffinity labelling is greatly reduced.

The decay rate for all three compounds is approximately 3 minutes under the previously described conditions. This means that after 10 minutes irradiation, 80 - 90% of the diazirines are photolysed. In a control experiment without the photolabile ligand present, Gal-T retains 95% of its activity after 20 minutes of irradiation in buffer.

Since determination of enzyme activity is based on a transfer assay with UDP-[^{14}C]Gal, the acceptor properties of the photolysed SMOs were determined and proved to be the same as those of the precursor. Also using t.l.c., the galactosylated, intact SMO and the galactosylated products of its photolysis were indistinguishable in the solvent mixture used. After incubation of Gal-T with photolabile acceptor and irradiation, remaining activity could be determined by adding UDP-[^{14}C]Gal to the mixture and carrying out the t.l.c. assay. The presence of UDP-Gal during irradiation protects the enzyme, proving the specific blocking of the active site in the absence of the donor substrate. The results of photodeactivation are reproducible and the differences in efficiency of the three SMOs as photoaffinity labels are significant (Figure 24).

The most effective compound is 64 with the diazirino group in position 3 (Figure 24). The SMO 65, carrying the diazirine in the center part, where it corresponds to the potential binding area for the mannose unit D, has only a weak effect. It may be speculated that here there is only loose contact between ligand (SMO or pentasaccharide) and receptor. This would also provide an explanation for the missing branch-specificity when the spacer-modified trisaccharide 72 is galactosylated.

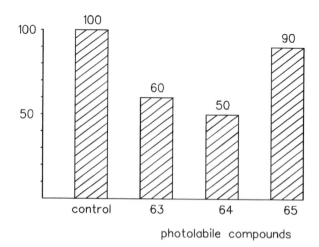

Figure 24   Photodeactivation of Gal-T by three different SMOs.

        One of the SMOs, **63**, was radioactively labelled by introducing a N-[³H]acetyl group. Irradiation with Gal-T, which had to be stabilised with BSA against aggregation, was carried out under the aforementioned conditions. In a control experiment, GlcNAc in high concentrations was added to the mixture to be irradiated. After thorough dialysis, the protein mixture was submitted to SDS polyacrylamide gel electrophoresis. The results are shown in Figure 25 (*50*).

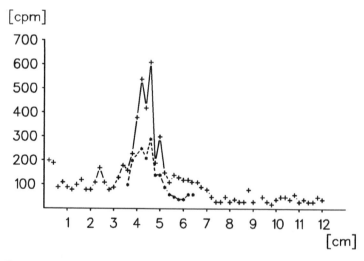

Figure 25   SDS-PAGE of radioactively labelled Gal-T.

Commercial Gal-T shows a relatively broad, diffuse double band upon staining. This is due to differences in glycosylation. Radioactivity is almost exclusively associated with Gal-T. BSA, present in 5-fold concentration, becomes only insignificantly and unspecifically labelled. Radioactivity is greatly reduced when photoaffinity labelling was carried out in the presence of large concentrations of GlcNAc.

Conclusions

Oligosaccharides, as natural ligands of receptors in general, can be "falsified" by positioning acyclic spacers to correspond to one or more monosaccharides in the native structure, without loss of affinity. The spacers, equipped with photolabile groups, are sterically unobtrusive and, if placed in different parts of the oligosaccharide structure, allow the regioselective labelling of the receptors binding site. Since these syntheses are comparatively simple, such spacer-modified oligosaccharides are excellent probes for structurally analyzing binding areas. The principle of reactive, spacer-modified oligosaccharides is generally useful for the irreversible inhibition of carbohydrate receptors.

Literature Cited

1. Blithe, D. L.; *Trends in Glycoscience and Glycotechnology*, **1993**, *5*, 81-98.
2. Drickamer, K. *Nature* **1992**, *360*, 183-186.
3. Jacobs, J. W.; Schultz, P. G.; Sugasawara, R.; Powell, M. *J. Am. Chem. Soc.*, **1987**, *109*, 2174; Shokat, K. M.; Ko, M. K.; Scanlan, T. S.; Kochersperger L.; Yonkovich, S.; Thaisriwongs, S.; Schultz, P. G. *Angew. Chem.* **1990**, *102*, 1339-1346.
4. Reymond, J.-L.; Janda, K. D.; Lerner, R. A. *Angew. Chem.* **1991**, *103*, 1690-1692.
5. Plückthun, A. *Ch. i. u. Z.* **1990**, *24*, 182-198; *Bio-Technology* **1991**, *9*, 545-551.
6. Kuriki, T. *Trends in Glykoscience and Glycotechnology* **1992**, *4*, 567-572.
7. Quiocho, F. A. *Ann. Rev. Biochem.* **1986**, *55*, 287-315.
8. Spurlino, J. C.; Lu, G.-Y.; Quiocho, F. A. *J. Biol. Chem.* **1991**, *266*, 5202-5219.
9. Bayley, H; Knowles, J. R. *Methods Enzymol.* **1977**, *46*, 69-114; Chowdhry, V.; Westheimer, F. W. *Ann. Rev. Biochem.* **1979**, *48*, 293-325.
10. Ziser, L. Dissertation Freiburg **1991**.
11. Midgley, P. J. W.; Parkar, B. A.; Holman, G. D.; Thieme, R.; Lehmann, J. *Biochim. Biophys. Acta* **1985**, *812*, 27-32.
12. Singh, A.; Thornton, E. R.; Westheimer, F. W. *J. Biol. Chem.* **1962**, 237, 3006-3008.
13. Bayley, H. in Chemistry of Diazirines, M. T. H. Liu ed. Vol. 2, p. 75-99, CRC Press, Boca Raton, Florida.
14. Petry, St.; Lehmann, J. *Carbohydr. Res.* **1993**, *239*, 133-142.
15. Wallenfels, K.; Malhotra, O. P. *Adv. Carbohydrate Chem.* **1961**, *16*, 239-298.
16. C.-St.; Lehmann, J.; Steck, J. *Tetrahedron* **1990**, *46*, 3129-3134.
17. Kuhn, C.-St.; Lehmann, J. *Carbohydr. Res.* **1987**, *160*, C6-C8.
18. Jegge, S.; Lehmann, J. *Carbohydr. Res.* **1984**, *133*, 247-254.
19. Connolly, D. T.; Townsend, R. R.; Kawaguchi, K.; Bell, W. R.; Lee, Y. C. *J. Biol. Chem.* **1982**, *257*, 939-945.
20. Blanc-Muesser, M.; Vigne, L.; Driguez, H.; Lehmann, J.; Steck, J.; Urbahns, K. *Carbohydr. Res.* **1992**, *224*, 59-71.

21. Pasero, L.; Mazzei, Y.; Abadie, B; Moinier, D.; Fougereau, M.; Marchis-Mouren, G. *Biochem, Biophys. Res. Commun.* **1983**, *110*, 726-732.
22. Payan, F.; Haser, R.; Pierrot, M.; Frey, M.; Astier, J. P.; Duée, E.; Buisson, G. *Acta Crystallogr.* **1980**, *B 36*, 416-421; Buisson, G.; Duée, E.; Payan, F. *EMBO J.* **1987**, *6*, 3909-3916; Buisson, G.; Duée, E.; Payan, F. *Food Hydrocoll.* **1987**, *1*, 399-406.
23. Takeo, K.; Nakagen, M.; Teramoto, Y.; Nitta, Y. *Carbohydr. Res.* **1990**, *201*, 261-275.
24. Wallenfels, K.; Földi, P.; Niermann, H.; Bender, H.; Linder, D. *Carbohydr. Res.* **1978**, *61*, 359-368.
25. Lehmann, J.; Brockhaus, M. *Methods Carbohydr. Chem.* **1980**, *8*, 301-304.
26. Lehmann, J.; Ziser, L. *Carbohydr. Res.* **1990**, *205*, 93-103.
27. Lehmann, J.; Ziser, L. *Carbohydr. Res.* **1992**, *225*, 83-97.
28. Bergmeyer, H. U.; Bergmeyer, J.; Graßl, M. (eds.), Methods of Enzymatic Analysis, 3. edition, Vol. 4, Verlag Chemie, Weinheim, Deerfield Beach (Florida), Basel, 1984, pp 146-177.
29. Seigner, C.; Prodanov, E.; Marchis-Mouren, G. *Biochim Biophys. Acta* **1987**, *913*, 200-209.
30. Dixon, M. *Biochem. J.* **1953**, *55*, 170-171.
31. Smith, C. A.; Brew, K. *J. Biol. Chem.* **1977**, *252*, 7294-7299.
32. Strous, G. J. A. M.; Berger, E. G. *J. Biol. Chem.* **1982**, *257*, 7623-7628.
33. Lee, T. K.; Wong, L.-J. C.; Wong, S. S. *J. Biol. Chem.* **1983**, *258*, 13166-13171.
34. Wong, C.-H.; Ichikawa, Y.; Krach, T.; Gautheron-Le Narvor, C.; Dumas, D. P.; Look, G. C. *J. Am. Chem. Soc.* **1991**, *113*, 8137-8145.
35. Rao, A. K.; Mendecinio, J. *Biochemistry* **1978**, *17*, 5632-5638.
36. Paquet, M. R.; Narasimhan, S.; Schachter, H.; Moscarello, M. A. *J. Biol. Chem.* **1984**, *259*, 4716-4721.
37. Paulsen, H.; Lebhuhn, R. *Carbohydr. Res.* **1984**, *125*, 21-45.
38. Lehmann, J.; Petry, St. *Carbohydr. Res.* **1990**, *204*, 141-144.
39. Tahir, S. H.; Hindsgaul, O. *Can. J. Chem.* **1986**, *64*, 1771-1780.
40. Lee, Y. C. in "Carbohydrate Recognition in Cellular Function", Ciba Foundation Symposium 145, eds G. Bock and S. Harnet, John Willey & Sons: Chichester - New York - Brisbane - Toronto - Singapore - **1989**; pp 80-95.
41. Hanessian, S.; Banoub, J. *Carbohydr. Res.* **1977**, *59*, 261-267.
42. Ats, C. S.; Lehmann, J.; Petry, St. *Carbohydr. Res.*, in press.
43. Church, R. F. R.; Weiss, M. J. *J. Org. Chem.* **1970**, *35*, 2465-2471.
44. Dasgupta, F.; Garegg, P. J. *J. Carbohydr. Chem.* **1988**, *7*, 701-707.
45. Ogawa, T.; Katano, K.; Sasajirna, K.; Matsui, M. *Tetrahedron*, **1981**, *37*, 2779-2786.
46. Ats, S.-C.; Lehmann, J.; Petry, St. *Carbohydr. Res.* **1992**, *233*, 141-150.
47. Ats, S.-C.; Hunkler, D.; Lehmann, J. *Carbohydr. Res.*, in press.
48. Khatra, B. S.; Herries, D. G.; Brew, K. *Eur. J. Biochem.* **1974**, *44*, 537-560.
49. Ram, B. P.; Munjal, D. D. *CRC Critical Rev. Biochem.* **1984**, *17*, 257-311.
50. Lehmann, J.; Petry, St. *Liebigs Ann. Chem.* **1993**, 1111-1116.

RECEIVED March 30, 1994

Chapter 13

Ganglioside Lactams as Analogs of Ganglioside Lactones

Goran Magnusson, Michael Wilstermann, Asim K. Ray, and Ulf Nilsson

Organic Chemistry 2, Chemical Center, The Lund Institute
of Technology, University of Lund, P.O. Box 124, S–221 00 Lund,
Sweden

Ganglioside lactams are hydrolytically stable analogs of unstable
ganglioside lactones. The conformations of GM_3-lactam and -lactone are
very similar. GM_3-lactam-BSA conjugate was a very potent immunogen,
which gave rise to monoclonal antibodies that cross-reacted with the natural
GM_3-lactone but not with GM_3-ganglioside.
The antibodies were used for staining of melanoma cells containing GM_3-
lactone. The synthesis of GM_3-, GM_4-, GM_2-, and GM_1-lactams is briefly
discussed. The conformation of the NeuAcα2-3Gal part was very similar in
GM_3- and GM_4-lactam, whereas it was different for GM_2-lactam.

Gangliosides are glycolipids that have one or more sialic acid unit attached to the
backbone of neutral saccharides. Gangliosides are often present in increased amounts on
tumor cells and they have been found to function as adhesion molecules that define
metastasis sites (1). A ganglioside (sialyl-Le^x) present on leucocytes is recognized by
selectin proteins on the surface of vascular endothelial cells, thus causing leucocytes to
roll on the endothelium and finally stop and invade the tissue. This phenomenon is
important in connection with inflammatory processes (2).
The sialic acid residue(s) of gangliosides is normally attached to the 3- or 6-position of a
galactose residue or, in disialic acid-containing gangliosides, to the 8-position of another
sialic acid unit. Since gangliosides are hydroxy acids, they can form lactones (internal
esters) in a number of ways as exemplified in Fig. 1 for the gangliosides GM_3 and
GD_3.
   GM_3-ganglioside (1, Fig. 1) is present in increased amounts on mouse melanoma
cells (3). Antibodies raised against melanoma cells recognize GM_3-ganglioside and also
the GM_3-lactone 2. A comparative immunization of mice with GM_3-lactone 2 and GM_3-
ganglioside 1, showed 2 to be a stronger immunogen than 1 (3).
   Ganglioside lactones are formed reversibly at low pH (4-8). Since gangliosides are
acidic glycolipids, dense packing in a cell membrane might cause a local drop in pH,
which might induce lactonization. Thus, the immune system might react to the high
amount of GM_3-ganglioside (1) on melanoma cells by forming antibodies (9) and T
cells (10) against the strongly immunogenic GM_3-lactone (2), present in low amount in
equilibrium with 1. It is well known that saccharides that contain sialic acid act as "self"
and thus are only weakly immunogenic.
   Gangliosides with more than one sialic acid unit can form one or more lactone rings
(Fig. 1), which might be important for fine tuning of the ganglioside structure *in vivo*.
The rigidity of ganglioside lactones as compared with the normal open form makes the
lactones conformationally more defined, which might be one reason for their greater
immunogenicity.

0097–6156/94/0560–0233$08.00/0

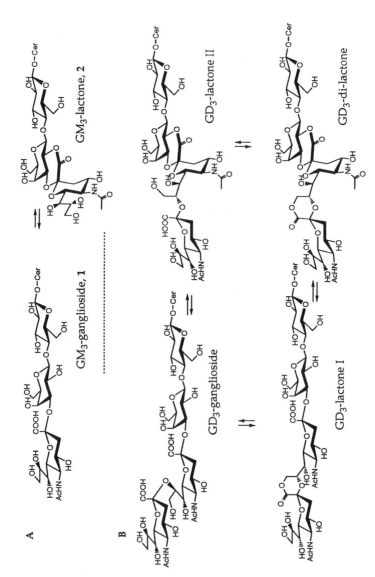

Figure 1. Equilibrium between gangliosides and their lactones, as exemplified by GM$_3$-ganglioside (A), and GD$_3$-ganglioside (B).

Ganglioside lactones are however in practice poor immunogens due to their ease of hydrolysis and thereby low equilibrium concentration at neutral pH. In contrast, the corresponding lactams should be hydrolytically more stable and for this reason should be useful lactone substitutes for immunization. A proviso is of course that the over-all shapes are similar and that the antibodies and cells obtained by immunization will cross-react with the natural ganglioside lactones. We have synthesized GM$_3$-lactam (Scheme 1) as the soluble glycoside 6 and the BSA conjugate 11 *(11)*. Ganglioside lactams have not been reported earlier despite the fact that they consist of naturally occurring monosaccharides. Thus, they might be natural products yet to be discovered.

The conformation of GM$_3$-lactone (2) has been determined by NMR *(12)*. A highly rigid structure was proposed, with the lactone ring in a chair-like conformation. However, based on NMR analysis of GM$_3$-lactone and GM$_3$-lactam, we suggested that a boat-like conformation is preferred instead *(11)*. Only the boat-like conformation of GM$_3$-lactone and -lactam is compatible with the interatomic distances required by the NMR data (chemical shifts and NOE effects). Molecular mechanics using the MM2(91) force field *(13, 14)* resulted in boat-like conformations for both the lactones and lactams even when the starting conformation was chair-like *(11)*.

The low-energy boat conformations of GM$_3$-lactone and GM$_3$-lactam were superimposed and RMS-fitted *(14)* to give an RMS-value of ~0.1 Å. The superimposed structures are depicted in Fig. 2. The overall van der Waals surface has been removed where the individual surfaces are closer together than the indicated distance setting. Practically no surface part is farther away than 0.5 Å. This shows that the two compounds have very similar shapes, which is an important condition for using ganglioside lactams as substitutes for lactones.

Mice were immunized with GM$_3$-lactam-BSA (11) and a large number (>300) of antigen-specific hybridomas were established *(15)*. The majority of the monoclonal antibodies (Mabs) from these hybridomas recognized the sialyl-lactam-galactose portion of the antigen. Mabs that recognized BSA itself or a glucose-BSA conjugate (Glc-BSA) were not processed further.

Eight anti-GM$_3$-lactam Mab's were chosen at random. They were all of the IgG-type and bound to the immunogen (11) attached to microtiter plates but did not bind to GM$_3$-ganglioside (1). Two of the antibodies (P5-1 and P5-3) bound also to GM$_3$-ganglioside lactone (2). The binding of antibodies P5-1 and P5-3 was inhibited by the GM$_3$- and GM$_4$-lactam glycosides 6 and 14 whereas GM$_3$-TMSEt glycoside, the GM$_2$-lactam 20, asialo GM$_2$-TMSEt glycoside, and asialo GM$_1$-TMSEt glycoside were inefficient as inhibitors *(15)*. That GM$_2$-lactam 20 was inactive indicated that the antibodies recognized the side of GM$_3$-lactam and -lactone that in 20 carries a sterically hindering GalNAc residue. Alternatively, the conformation of GM$_2$-lactam 20 (see below) is sufficiently different from the conformations of the GM$_3$- and GM$_4$-lactams 6 and 14, for 20 to loose its inhibitory activity.

Binding of a third antibody (P3) to GM$_3$-lactam-BSA (11) was inhibited by the TMSEt glycosides of GM$_3$-, GM$_4$-, and GM$_2$-lactam (6, 14, and 20). Obviously, the GalNAc residue of 14 did not hinder the binding. Thus, antibodies P3, P5-1, and P5-3 seem to recognize different saccharide epitopes.

It is principally important that the antibodies raised against GM$_3$-lactam recognize natural GM$_3$-lactone and not the open form GM$_3$. The antibodies are potentially valuable reagents for biomedical investigations of ganglioside lactones *in vivo*. Other ganglioside lactams may be generally useful substitutes for unstable ganglioside lactones both for immunization and as inhibitors of lactone-protein binding.

Staining of Melanoma Cells with Anti-GM$_3$-Lactam Antibodies.

The existence of ganglioside lactones on cell surfaces *in vivo* has been debated for a long time. In spite of the fact that lactones have been detected in ganglioside isolates,

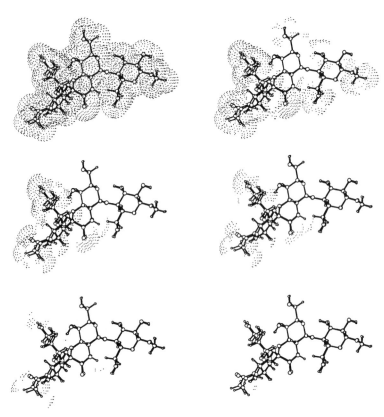

Figure 2. Least-squares-fitting of energy minimized (MM2-91) GM₃-lactone and GM₃-lactam. The parts of the individual van der Waals (dot)surfaces, that are closer together than the noted values, were removed.

their existense as natural membrane components has not been demonstrated unequivocally, since lactones may be formed during the isolation process. Several attempts have been made to produce GM₃-lactone-specific antibodies for use as reagents. However, the antibodies obtained reacted with both GM₃-lactone and the open form GM₃-ganglioside (*16*). A lactone-specific antibody was recently reported that recognized the lactone ring formed between two terminal sialic acid residues (in GD₃) but not the lactone ring formed between sialic acid and galactose (*17*).

We have used the GM₃-lactam-specific antibodies P5-1, P5-3, and P3 for staining of cultured mouse melanoma (B16) cells. The DH2 antibody (*16*), which recognizes both GM₃ and GM₃-lactone gave, as expected, a more intense staining of the melanoma cells. Since our antibodies do not recognize the open-form GM₃ structure, the staining clearly indicates that GM₃-lactone is present on the B16 cells (*18*). There remains the possibility that the antibodies induce lactonization of GM₃ on the cell surface. This is however, not very probable since they did not recognize GM₃ ganglioside (**1**) coated on microtiter plates (*15*).

Synthesis of Ganglioside Lactams

The syntheses of ganglioside lactams are based on the standard protection and glycosylation methods of carbohydrate chemistry. It was anticipated that the lactam-forming step would occur spontaneously once a suitable amino ester was formed. This was indeed the case, as illustrated by the successful synthesis of GM$_3$-lactam 6 (11). Here, the amino ester, obtained from nickel boride reduction of the corresponding azide 5, cyclized readily in pyridine (step i, Scheme 1).

Synthesis of GM$_3$-lactam; Scheme 1 (11). The lactose derivative 3, used as the acceptor in the sialylation reaction, was obtained by glycosylation (19) of 2-(trimethylsilyl)ethyl 2,3,6-tri-O-benzyl-β-D-glucopyranoside (20) with 3,4,6-tri-O-acetyl-2-azido-2-deoxy-α-D-galactopyranosyl bromide (21), followed by protection/deprotection to give the desired acceptor lactoside 3 (step a-e, Scheme 1). Sialylation (step f) was performed using the sialic acid xanthate 4 with silver triflate/methylsulfenyl bromide as promoter (22). The desired α-linked GM$_3$ trisaccharide 5 was obtained in 71% yield after chromatographic removal of 4% of the isomeric β-glycoside. This corresponds to an α/β-ratio of ~20:1 and 75% yield in the sialylation step. No material was observed that emanated from glycosylation of HO-4' of the lactose derivative. It should be noted that both the yield, regio- and stereoselectivity was unusually high in this step.

Reduction of the azide group of the trisaccharide derivative 6 was originally done by nickel boride (11). We later found that hydrogen sulfide reduction gives the desired amine in a cleaner reaction. The amino ester lactamizes spontaneously but the reaction is accelerated by pyridine. The complete sequence of deacetylation, reduction of the azide, lactamization, and hydrogenolysis (steps g-j) was performed in an over-all yield of 54%, thus furnishing the GM$_3$-lactam TMSEt glycoside 6, suitable as inhibitor in the specificity determinations of monoclonal antibodies (15) as described above. Reduction of the azide group of 5 before deacetylation gave an amine that on lactamization gave the GM$_3$-lactam 15 (Scheme 3) having only one unprotected hydroxyl group (HO-4') and therefore suitable for glycosylation in the synthesis of GM$_2$ and GM$_1$ lactams (see below).

Compound 6 was acetylated and then treated with dichloromethyl methyl ether/zinc chloride (23), which gave the corresponding α-chloro saccharide 7 in quantitative yield, (step k, l). The chloro sugar was used for glycosylation of 2-bromoethanol, thus giving (step m) the corresponding 2-bromoethyl glycoside 8 in 54% yield as an inseparable β/α mixture (85:15). Treatment with methyl 3-mercaptopropionate gave 9 (step n) and deacetylation (step o) gave the GM$_3$-lactam 10, carrying a spacer-arm suitable for coupling to protein. The sequence of reactions leading to the spacer-arm saccharide via a 2-bromoethyl glycoside is analogous to our earlier syntheses of carbohydrate antigens (24, 25).

Coupling of the spacer-arm saccharide to bovine serum albumin (BSA) (11) gave the glycoconjugate 11 having ~18 mol of saccharide per mol of BSA, as determined by sulfur combustion analysis. This compound was used as immunogen in the production of monoclonal anti-GM$_3$-lactam antibodies as described (15).

Synthesis of GM$_4$-lactam; Scheme 2. Sialylation of 2-(trimethylsilyl)ethyl 2-azido-6-O-benzoyl-2-deoxy-β-D-galactopyranoside (12) with the sialic acid xanthate 4, followed by acetylation of HO-4, gave the desired GM$_4$-azide derivative 13 in 58% over-all yield (step a, b, Scheme 2). No β-glycoside was obtained. Nickel boride reduction, deacylation, and treatment with 4-dimethylaminopyridine and pyridine (step c-e), gave in 61% overall yield, the GM$_4$-lactam glycoside 14, suitable for investigation of antibody specificity (15).

Synthesis of GM$_2$-lactam; Scheme 3 and 4. We have used both GM$_3$-lactam and the open form GM$_3$ azide as acceptors for glycosylation with galactosamine derivatives, thus leading to GM$_2$-type compounds (Scheme 3 and 4). The lactam acceptor 15 was glycosylated (26, step a) with four different methylthio galactoside donors (16-19) to

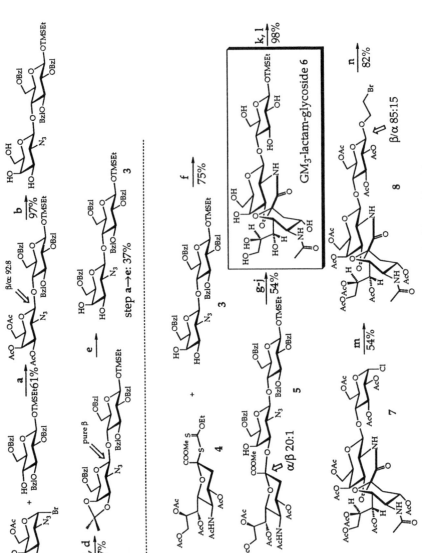

Scheme 1. Synthesis of GM₃-lactam-TMSEt saccharide and its BSA conjugate.

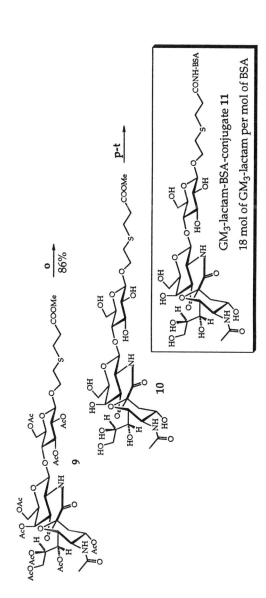

GM$_3$-lactam-BSA-conjugate **11**

18 mol of GM$_3$-lactam per mol of BSA

Scheme 1. *Continued*

a: Ag-silicate/CH$_2$Cl$_2$, MS 4Å/18 h/r.t. b: MeONa/MeOH/r.t. c: Me$_2$C(OMe)$_2$, camphor-SO$_3$H, 48h/r.t.; 78%; MeOH/H$_2$O 10:1. d: BzlBr/NaH/DMF/12 h/23°C, 86%. e: AcOH/H$_2$O, 1.5h/85° f: CH$_3$CN/CH$_2$Cl$_2$, MS 3Å/1.5 h/r.t.; AgOTf/MeSBr, 2 h/−78°. g: MeONa/MeOH. h: NiCl$_2$·6H$_2$O/H$_3$BO$_3$/NaBH$_4$, EtOH. i: Pyridine/12 h/r.t. j: H$_2$/Pd–C/AcOH. k: Ac$_2$O/pyridine/24 h/r.t.; 98%. l: Cl$_3$CHOMe/ZnCl$_2$/CHCl$_3$, 12 h/r.t.; 100%. m: HOCH$_2$CH$_2$Br/AgOTf/MS 3 Å/CH$_2$Cl$_2$/16 h/−28° → r.t.; 54%. n: HSCH$_2$CH$_2$COOMe/Cs$_2$CO$_3$/DMF/2.5 h/r.t.; 82%. o: MeONa/MeOH/4 h/r.t.; 86%. p: H$_2$NNH$_2$·H$_2$O/EtOH/12 h/r.t. q: HCl/dioxane/DMSO/DMSO/tBuNO$_2$. r: H$_3$NSO$_3$H/DMSO. s: BSA/Buffer/12 h/r.t.; freeze-drying.

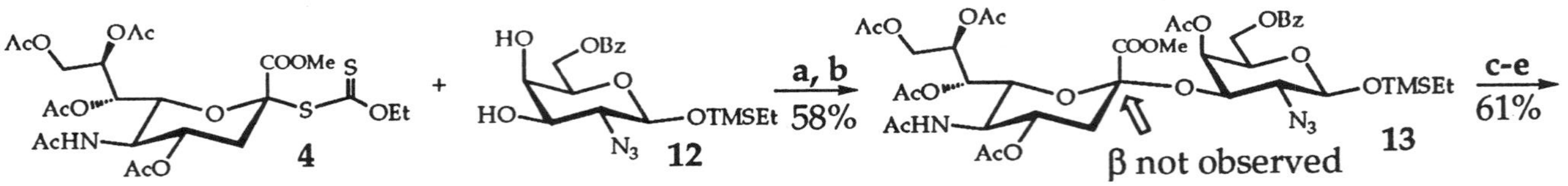

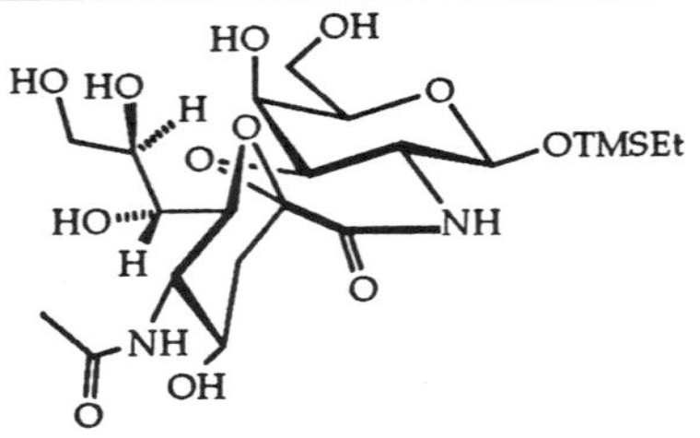

a: AgOTf/MeSBr/Mol. sieves 3Å, CH$_3$CN/CH$_2$Cl$_2$ 3:2, 3.5 h, -78°C. b: Ac$_2$O/pyridine, ~12 h. c: NiCl$_2$·6H$_2$O/ H$_3$BO$_3$/NaBH$_4$, EtOH. d: MeONa/MeOH e: DMAP/pyridine/Et$_3$N, room temp., 48 h.

GM$_4$-lactam-saccharide 14. Potential building block for other ganglioside lactams.

Scheme 2. Synthesis of GM$_4$-lactam-TMSEt saccharide.

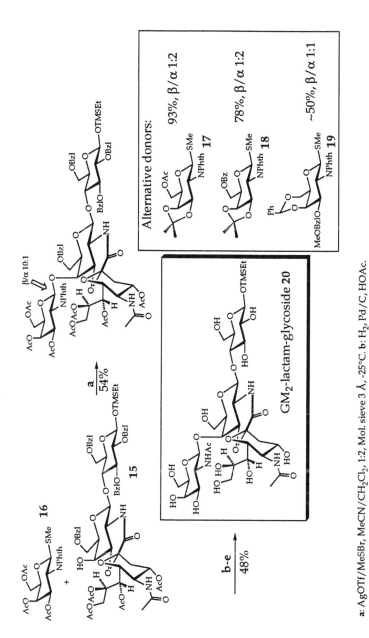

Scheme 3. Synthesis of GM$_2$-lactam-TMSEt saccharide.

a: AgOTf/MeSBr, MeCN/CH$_2$Cl$_2$, 1:2, Mol. sieve 3 Å, -25°C. b: H$_2$, Pd/C, HOAc.
c: H$_2$NNH$_2$·H$_2$O. d: Ac$_2$O, pyridine. e: NaOMe/MeOH.

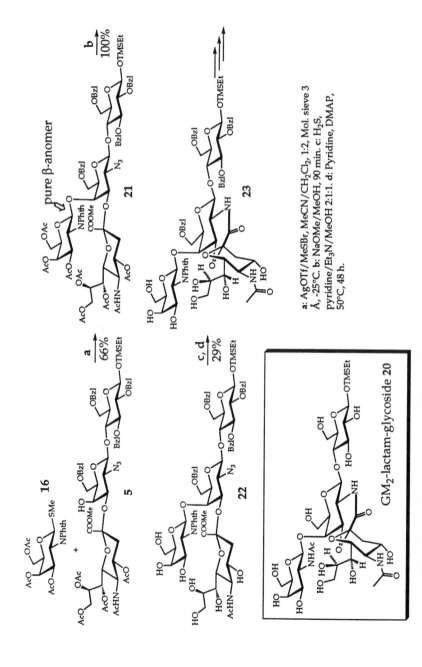

a: AgOTf/MeSBr, MeCN/CH$_2$Cl$_2$, 1:2, Mol. sieve 3
Å, -25°C. b: NaOMe/MeOH, 90 min. c: H$_2$S,
pyridine/Et$_3$N/MeOH 2:1:1. d: Pyridine, DMAP,
50°C, 48 h.

give the desired GM$_2$-lactam derivative. Methyl 3,4,6-tri-O-acetyl-2-deoxy-2-phthalimido-1-thio-β-D-galactopyranoside **16** (*27*) gave moderate yield with good stereoselectivity (β/α 10:1), whereas the acetal variants gave the desired β glycoside as a minor component, which is unexpected for donors having a 2-phthalimido substituent. Deprotection of the primary glycosylation product (step b-e, Scheme 3) gave GM$_2$-lactam glycoside **20** in 48% overall yield.

Glycosylation of the GM$_3$ azide **5** with methyl 3,4,6-tri-O-acetyl-2-deoxy-2-phthalimido-1-thio-β-D-galactopyranoside **16** (step a, Scheme 4) gave the desired tetrasaccharide **21** in 66% yield as the pure β anomer. This is in contrast to the results obtained in the synthesis of the GM$_2$-lactam derivative (Scheme 3), which gave ~10% of the α-anomer, probably due to more steric crowding in the transition state in the latter case. Deacetylation (step b) gave **22** in quantitative yield. The ensuing azide reduction and lactamization (step c, d) proceeded in low yield (29%), the product **23** being the same as one intermediate *en route* to GM$_2$-lactam **20** as depicted in Scheme 3.

Synthesis of GM$_1$-lactam; Scheme 5. The GM$_3$-lactam **15**, having a free HO-4' group, was glycosylated (*29*) in yields varying between 0 and 45%, using the Galβ1-3GalNPhth thioglycoside donor **24** (step a). The reason for the varying yields seems to be the lability of the product **25**. It splits into the starting GM$_3$-lactam and the hemiacetal of the donor (probably formed on the tlc plate by hydrolysis of the corresponding triflate) on prolonged treatments with the reaction medium or silica gel. Similarily, attempted dephthaloylation of the GM$_1$-lactam saccharide **25** led to decomposition. The reason for the lability is probably the strained nature of the compound. This is indicated by the large deshielding of the anomeric hydrogen atom in the Gal-NPhth residue as compared to that of the corresponding GM$_2$-lactam saccharide. The deshielding might well have its cause in forced positioning of a carbonyl oxygen of the phthaloyl ring close to the anomeric hydrogen. The unsuccessful removal of the phthalic acid residue is yet another indication of the need for more efficient β-directing protecting groups for amino sugars.

The synthesis of the Galβ1-3GalNPhth thioglycoside donor deserves some comment (Scheme 6). Glycosylation (step a) of 2-(trimethylsilyl)ethanol with 3,4,6-tri-O-acetyl-2-azido-2-deoxy-α-D-galactosyl bromide (*21*) with silver silicate as the promoter (*28*) gave **26** in 72% yield and a high β/α ratio (23:1). Deacetylation and chromatography (step b) gave the crystalline triol **27**, devoid of the α-glycosidic contaminant that was present after the glycosylation step. Benzylidenation (step c) gave 2-(trimethylsilyl)ethyl 2-azido-4,6-benzylidene-2-deoxy-β-D-galactopyranoside **28**, suitable as glycosyl acceptor. Glycosylation (step d) with phenyl 2,3,4,6-tetra-O-acetyl-1-thio-β-D-galactopyranoside (*28*), using N-iodosuccinimide-triflic acid (*29*) as promoter, gave the desired disaccharide **29** in 87% yield, provided that acid-washed molecular sieves (AW 300, Aldrich) were used. With non-acid-washed sieves, the corresponding orthoester was the main product. Treatment of the orthoester under the conditions shown in step d caused a smooth transformation to the desired disaccharide **29**. Similar glycosylations leading to the Galβ1-3GalNAc-residue have been reported although yields vary and the reproducibility of some methods has been questioned (*30-34*). Removal of the benzylidene group, acetylation of the exposed hydroxyl groups, reduction of the azide group and phthaloylation of the resulting amine (step e-h) gave the TMSEt glycoside **31** suitable for anomeric activation. In the present case, the TMSEt group was replaced by acetyl followed by treatment with trimethylsilylmercaptan-trimethylsilyltriflate (*35*), thus furnishing the desired thioglycoside donor **24** (step i, j) in 89% overall yield. Transformation of TMSEt glycosides into 1-O-acyl saccharides proceeds with retention of anomeric configuration (*20*). Other useful reactions of TMSEt glycosides (*35*) are transformation into hemiacetals (*20*), 1-chloro saccharides (*23, 37, 38*), and methoxymethyl glycosides (*37*).

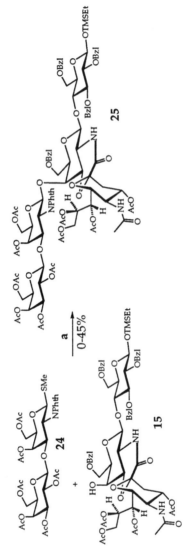

Scheme 5. Synthesis of GM$_1$-lactam-TMSEt saccharide.

a: NIS/TfOH, MeCN/CH$_2$Cl$_2$ 3:1, Mol. sieves 3 Å, -45°C.

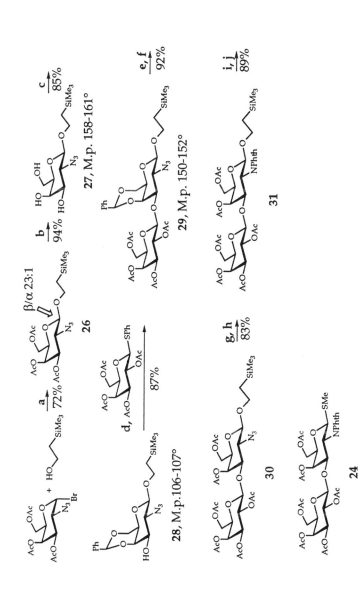

Scheme 6. Synthesis of a Galβ1-3GalNAc donor.

a: Ag-silicate, mol. sieves, 4 Å, CH₂Cl₂, room temp., 1 h. b: NaOMe/MeOH, room temp., 1 h. c: α,α-dimethoxytoluene, MePhSO₃H, MeCN, room temp., 24 h. d: NIS/TfOH, MeCN/CH₂Cl₂ 3:1, mol. sieves 3 Å, acid-washed, –45°, 2 h. e: AcOH/H₂O 4:1, 90°, 2 h. f: Ac₂O, pyridine. g: NiCl₂·6H₂O/H₃BO₃/NaBH₄, EtOH. h: phthalic anhydride, pyridine, Ac₂O. i: BF₃Et₂O, Ac₂O. j: Me₃SiSMe, TMSOTf.

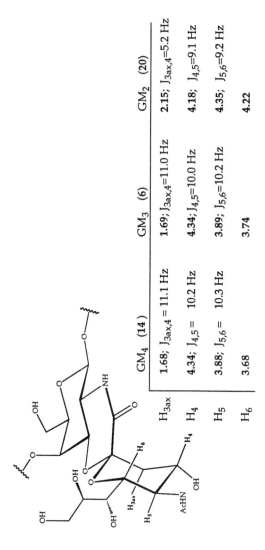

	GM$_4$ (14)	GM$_3$ (6)	GM$_2$ (20)
H$_{3ax}$	1.68; J$_{3ax,4}$ = 11.1 Hz	1.69; J$_{3ax,4}$=11.0 Hz	2.15; J$_{3ax,4}$=5.2 Hz
H$_4$	4.34; J$_{4,5}$ = 10.2 Hz	4.34; J$_{4,5}$=10.0 Hz	4.18; J$_{4,5}$=9.1 Hz
H$_5$	3.88; J$_{5,6}$ = 10.3 Hz	3.89; J$_{5,6}$=10.2 Hz	4.35; J$_{5,6}$=9.2 Hz
H$_6$	3.68	3.74	4.22

Figure 3. Chemical shifts and coupling constants in the sialic acid residues of GM$_4$-, GM$_3$-, and GM$_2$-lactams reveal large conformational differences.

The conformation of GM_3-, GM_4-, and GM_2-lactams.

The conformation of GM_3-lactam is very similar to that of GM_3-lactone (*11, 18*) as shown in Figure 2. This is a very important prerequisite for the use of the hydrolytically stable GM_3-lactam as a substitute for the unstable GM_3-lactone in immunizations leading to anti-GM_3-lactam/lactone antibodies (*15*). Strong deshielding (δ 4.34 ppm) of H_4 of the sialic acid residue is caused by the close proximity of the lactam (and lactone) carbonyl oxygen and can be used as a conformational probe in these compounds. The chemical shift of H_4 in analogous open form GM_3-saccharides is ~3.6 ppm. The close proximity between the carbonyl oxygen and H_4 requires that the lactam ring is present in a boat-like conformation, contrary to the chair conformation suggested by others (*12*).

The protons H_3, H_4, H_5, and H_6 of the sialic acid residues in GM_3-lactam (**6**) and GM_4-lactam (**14**) have very similar chemical shifts and coupling constants (Fig. 3), thus suggesting very similar conformations. In GM_2-lactam (**20**) however, the chemical shifts of the corresponding protons are quite different. H_{3ax} and H_5 are deshielded ~0.45 ppm compared to the case in GM_3-lactam and GM_4-lactam. Furthermore, the $J_{3ax,4}$ coupling constant is only 5.2 Hz in GM_2-lactam **20**, as compared to ~11 Hz in both GM_3- and GM_4-lactams **6** and **14**, indicating a conformational change in the sialic acid residue (Fig. 3). In addition, the proton H_6 in GM_2-lactam **20** is deshielded ~0.5 ppm and H_4 is shielded 0.16 ppm (compared with GM_3-lactam and GM_4-lactam). These deshielding, shielding, and coupling constant effects indicate that the conformation of GM_2-lactam (and probably GM_1-lactam) is quite different from the conformations of GM_3-lactam and GM_4-lactam. The reason is probably that in GM_2-lactam there is a GalNAc-induced steric crowding that either forces the lactam ring into a chair-like conformation or the sialic acid ring into a boat-like conformation. The differences in preferred conformations between GM_3/GM_4- and GM_2-lactam saccharides may well have large consequences for the different compound's ability to function in biological receptor systems. Similar preferred conformations are expected to exist for the natural ganglioside lactones.

Acknowledgments. This work was supported by grants from the Swedish Natural Science Research Council, the Swedish Research Council for Engineering Sciences, and Symbicom AB, Sweden.

Literature cited
1.   Hakomori, S.-i. *Curr. Opin. Immun.* **1991**, *3*, 646-653.
2.   Aruffo, A. *Trends Glycosci. Glycotechn.* **1992**, *4*, 146-151.
3.   Nores, G. A., Dohi, T., Taniguchi, M., Hakomori, S.-i. *J. Immun.* **1987**, *139*, 3171-3176.
4.   Wiegandt, H. *Physiol. Biol. Chem. Exp. Pharmacol.* **1966**, *57*, 190-222.
5.   Gross, S. K., Williams, M. A., McCluer, R. H. *J. Neurochem.* **1980**, *34*, 1351-1361.
6.   Riboni, L., Sonnino, S., Acquotti, D., Malesci, A., Ghidoni, R., Egge, H., Mingrino, S., Tettamanti, G. *J. Biol. Chem.* **1986**, *261*, 8514-8519.
7.   Bassi, R., Riboni, L., Sonnino, S., Tettamanti, G. *Carbohydr. Res.* **1989**, *193*, 141-146.
8.   Maggio, B., Ariga, T., Yu, R. K. *Biochemistry* **1990**, *29*, 8729-8734.
9.   Dohi, T., Nores, G., Hakomori, S.-i. *Cancer Res.* **1988**, *48*, 5680-5685.
10. Harada, Y., Sakatsume, M., Taniguchi, M. *Jpn. J. Cancer Res.* **1990**, *81*, 383-387.
11. Ray, A. K., Nilsson, U., Magnusson, G. *J. Am. Chem. Soc.* **1992**, *114*, 2256-2257.
12. Yu, R. K., Koerner, T. A. W., Ando, S., Yohe, H. C., Prestegaard, J. H. *J. Biochem.* **1985**, *98*, 1367-1373.

13. Burkert, U. & Allinger, N. L., Molecular Mechanics, American Chemical Society, Washington D. C. 1982.
14. Molecular construction and energy minimization were made with the MacMimic/MM2(91) package: InStar Software, Ideon Research Park, S-223 70 Lund, Sweden.
15. Ding, K., Rosén, A., Ray, A. K., Magnusson, G. *Glycoconj. J.* **1993**, *9*, 303-306.
16. Dohi, T., Nores, G., Hakomori, S.-i. *Cancer Res.* **1988**, *48*, 5680-5685.
17. Bouchon, B., Levery, S. B., Clausen, H., Hakomori, S.-i. *Glycoconj. J.* **1992**, *9*, 27-38.
18. Magnusson, G., Ding, K., Nilsson, U., Ray, A. K., Rosén, A., Sjögren, H.-O. in "Complex Carbohydrates in Drug Research", K. Bock, H. Clausen, P. Krogsgaard-Larsen, H. Kofod, Eds., Proceedings of the 36th Alfred Benzon Symposium, Copenhagen June 6-10, 1993.
19. van Boeckel, C. A. A., Beetz, T. *Recl. Trav. Chim. Pays-Bas*, **1987**, *106*, 596-598.
20. Jansson, K., Ahlfors, S., Frejd, T., Kihlberg, J., Magnusson, G., Dahmén, J., Noori, G., Stenvall, K. *J. Org. Chem.* **1988**, *53*, 5629-5647.
21. Lemieux, R. U., Ratcliffe, R. M. *Can. J. Chem.* **1979**, *57*, 1244-1251.
22. Lönn, H., Stenvall, K. *Tetrahedron Lett.* **1992**, *33*, 115-116.
23. Jansson, K., Noori, G., Magnusson, G. *J. Org. Chem.* **1990**, *55*, 3181-3185. Kovác, P., Whittaker, N. F., Glaudemans, C. P. J. *J. Carbohydr. Chem.* **1985**, *4*, 243-254. Gross, H., Farkas, I., Bognár, R. *Z. Chem.* **1978**, *18*, 201-210.
24. Dahmén, J., Frejd, T., Magnusson, G., Noori, G., Carlström, A.-S. *Carbohydr. Res.* **1984**, *129*, 63-71, and references cited therein.
25. Magnusson, G., Nilsson, U., Ray, A. K., Taylor, K. G. in "Carbohydrate Antigens", P. J. Garegg and A. A. Lindberg, Eds. ACS Symposium Series 519, Am. Chem. Soc. Washington, DC, 1993.
26. Dasgupta, F., Garegg, P. J. *Carbohydr. Res.* **1988**, *177*, c13-c17.
27. Hasegawa, A., Nagahama, T., Ohki, H., Kiso, M. *J. Carbohydr. Chem.* **1992**, *11*, 699-714.
28. Ferreier, R. J., Furneux, R. H. *Carbohydr. Res.* **1976**, *52*, 63-68.
29. (a) Konradsson, P., Mootoo, D. R., McDevitt, R. E., Fraser-Reid, B. *J. Chem. Soc. Chem. Commun.* **1990**, 270-272. (b) Veeneman, G. H., van Leeuwen, S. H., van Boom, J. H. *Tetrahedron Lett.* **1990**, *31*, 1331-1334.
30. Lemieux, R. U., Sabesan, S. *Can. J. Chem.* **1984**, *62*, 644-654.
31. Ito, Y., Nunomura, S., Shibayama, S., Ogawa, T. *J. Org. Chem.* **1992**, *57*, 1821-1831.
32. Catelani, G., Marra, A., Paquet, F., Sinaÿ, P. *Carbohydr. Res.* **1986**, *115*, 131-140.
33. Lüning, B., Norberg, T., Tejbrant, J. *Glycoconj. J.* **1989**, *6*, 5-19.
34. Paulsen, H., Paal, M. *Carbohydr. Res.* **1984**, *135*, 53-69.
35. Pozsgay, V., Jennings, H. J. *Tetrhedron Lett.* **1987**, *28*, 1375-1376.
36. Magnusson, G. *Trends Glycosci. Glycotechn.* **1992**, *4*, 358-367.
37. Jansson, K., Magnusson, G. *Tetrahedron*, **1990**, *46*, 59-64.
38. Kartha, K. P. R., Jennings, H. J. *Tetrahedron Lett.* **1990**, *31*, 2537-2540.

RECEIVED February 22, 1994

Chapter 14

Stereocontrolled Approaches to *O*-Glycopeptide Synthesis

Yoshiaki Nakahara[1], Hiroyuki Iijima[1], and Tomoya Ogawa[1,2]

[1]The Institute of Physical and Chemical Research (RIKEN),
Wako-shi, Saitama 351–01, Japan
[2]Faculty of Agriculture, University of Tokyo, Yayoi, Bunkyo-ku,
Tokyo 113, Japan

Synthetic studies on the poly O-glycosylated
fragments of a sialoglycoprotein, glycophorin A,
are described with respect to the following
subjects: 1. Glycosylation of L-Ser/Thr
derivatives, 2. Synthesis of sialyl glycosides,
3. Synthesis of glycopeptides.

The chemical synthesis of complex oligosaccharides and their
conjugates is a promising technology not only to probe the
biological mechanisms in which these glycoconjugates participate
but also to create new bio-regulating agents. The N- and O-linked
glycans present in various glycoproteins*(1,2)* play biologically
important roles through the process of carbohydrate recognition
by specific receptors. Such biological phenomena as cell-cell and
receptor-ligand interactions, viral infections, and cancer metastasis
have often been interpreted by the significant involvement of
those glycans.

Glycophorin A is a major sialoglycoprotein found in human
erythrocyte membranes. The protein structure and carbohydrate
branching have been well-characterized*(3,4)*, although its
biological function within the membrane is uncertain. The N-
terminal region of the protein is highly O-glycosylated and this
forms a carbohydrate cluster structure. The polymorphic N-
terminal pentapeptide sequences are responsible for MN blood
group antigenic determinants.

Several laboratories have reported syntheses of glycophorin
A fragments of various size, in which the mono- [α-D-Gal*p*NAc-(1→,

0097–6156/94/0560–0249$08.00/0

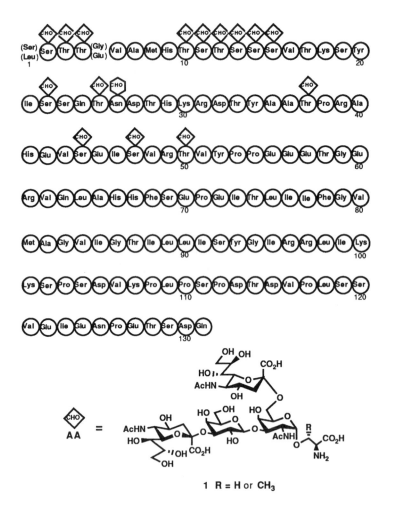

1 R = H or CH₃

Tn epitope] and di-saccharide [β-D-Galp-(1→3)-α-D-GalpNAc-(1→, T epitope] glycosides have been studied in depth. In contrast, there has been less investigation into the synthesis of O-glycoprotein glycans containing sialic acid residues(5-8).

Previously, we reported the syntheses of sialic acid-containing di- **2**(5), tri- **3**(7), and tetrasaccharides **4**(8) linked to L-serine. The sialic acid residue was introduced with moderate efficiency at O-6 of GalpNAc and O-3 of Galp residues using sialyl glycosyl chloride and AgOTf as a promoter. These glycans were then coupled with a L-serine derivative in the later stages using the trichloroacetimidate method. In the glycosylation of the tri- and tetrasaccharides, the reaction was not stereoselective with the undesired β-glycoside being produced preferentially(**3**; α/β=15/32, **4**; α/β=3/8).

Paulsen et al.*(9)* also reported a remarkable β-selectivity (α/β = 9/16) in a similar glycosylative process using a L-serine derivative with the trisaccharide thioglycoside **5** when activated with DMTST[dimethyl(methylthio)sulfonium trifluoromethane-sulfonate].

Recently, several new methodologies have been developed for the stereoselective synthesis of sialyl α-glycosides by taking advantage of stereo-directing anchimeric assistance (**6**)*(10)* or a β-oriented acetonitrilium cation intermediate (**7**)*(11)*.

These advances in glycosylation techniques prompted us to search for a new synthetic route to fragments of O-glycoproteins which contain sialic acid.

Here we describe a novel strategy for the syntheses of α-D-Neu$p$5Ac-(2→6)-α-D-Gal$p$NAc-(1→3)-Ser(Thr) (**2**), α-D-Neu$p$5Ac-(2→3)-β-D-Gal$p$-(1→3)-[α-D-Neu$p$5Ac-(2→6)-]-α-D-Gal$p$NAc-(1→3)-Ser(Thr) (**1**) blocks, and oligopeptides containing them. The former glycan **2** is known as a tumor associated antigen sialosyl Tn epitope and would be a suitable prototype for synthetic sialoglycopeptides.

The synthetic strategy was designed so that the necessary α glycosidic linkage between the Gal$p$NAc and serine (threonine) residues would be established at an early stage of the synthesis as shown below.

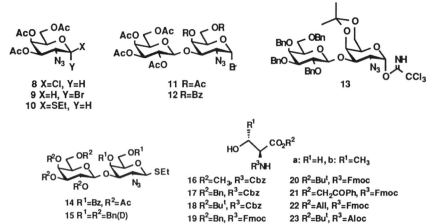

Glycosylation of L-serine (L-threonine) derivatives

Since a 2-azido-2-deoxy-D-galactose derivative was first demonstrated by Paulsen et al.*(12)* to be an indispensable precursor of α-D-Gal$p$NAc glycosides, many derivatives of α-linked Gal$p$NAc-Ser(Thr) have been successfully synthesized through a variety of glycosylation reactions, as summarized in Table 1.

8 X=Cl, Y=H
9 X=H, Y=Br
10 X=SEt, Y=H

11 R=Ac
12 R=Bz

13

14 R^1=Bz, R^2=Ac
15 R^1=R^2=Bn(D)

16 R^2=CH_3, R^3=Cbz
17 R^2=Bn, R^3=Cbz
18 R^2=Bu^t, R^3=Cbz
19 R^2=Bn, R^3=Fmoc

20 R^2=Bu^t, R^3=Fmoc
21 R^2=CH_2COPh, R^3=Fmoc
22 R^2=All, R^3=Fmoc
23 R^2=Bu^t, R^3=Aloc

a: R^1=H, b: R^1=CH_3

Table 1.   Parameters of a Variety of Glycosylation Reactions

	donor	acceptor	promoter	solvent	yield (α,β)	ref.
1	8	16a	Hg(CN)$_2$, HgCl$_2$	CH$_3$NO$_2$	65 (α)	13
2	8	16b	Hg(CN)$_2$, HgCl$_2$	CH$_3$NO$_2$	45 (α)	13
3	8	19a	Ag$_2$CO$_3$, AgClO$_4$	CH$_2$Cl$_2$-toluene	65 (α), 9 (β)	14
4	8	19b	Ag$_2$CO$_3$, AgClO$_4$	CH$_2$Cl$_2$-toluene	60 (α), 4 (β)	14
5	8	20a	Ag$_2$CO$_3$, AgClO$_4$	CH$_2$Cl$_2$-toluene	70 (α / β = 8 / 1)	15
6	8	20b	Ag$_2$CO$_3$, AgClO$_4$	CH$_2$Cl$_2$-toluene	78 (α / β = 9 / 1)	15
7	8	23a	Ag$_2$CO$_3$, AgClO$_4$	CH$_2$Cl$_2$-toluene	91 (α / β = 18 / 1)	16
8	8	23b	Ag$_2$CO$_3$, AgClO$_4$	CH$_2$Cl$_2$-toluene	95 (α / β = 20 / 1)	16
9	9	23b	Ag$_2$CO$_3$, AgClO$_4$	CH$_2$Cl$_2$-toluene	70 (α / β = 20 / 1)	16
10	9	22a	Ag$_2$CO$_3$, AgClO$_4$	CH$_2$Cl$_2$-toluene	68 (α)	17
11	9	22b	Ag$_2$CO$_3$, AgClO$_4$	CH$_2$Cl$_2$-toluene	60 (α), 3 (β)	17
12	10	17a	DMTST	CH$_2$Cl$_2$-toluene	94 (α / β = 7 / 2)	9
13	11	18a	AgOTf	CH$_2$Cl$_2$	61 (α), 11 (β)	18
14	11	18b	AgOTf	CH$_2$Cl$_2$	54 (α)	18
15	11	21a	AgOTf	toluene	52 (α), 25 (β)	19
16	11	21b	AgOTf	toluene	39 (α), 31 (β)	19
17	12	17a	Ag$_2$CO$_3$, AgClO$_4$	CH$_2$Cl$_2$-toluene	81 (α)	20
18	12	17b	Ag$_2$CO$_3$, AgClO$_4$	CH$_2$Cl$_2$-toluene	85 (α)	20
19	12	19a	AgClO$_4$	CH$_2$Cl$_2$-toluene	71 (α / β = 7 / 1)	21
20	12	19b	AgClO$_4$	CH$_2$Cl$_2$-toluene	61 (α)	22
21	12	20a	AgClO$_4$	CH$_2$Cl$_2$-toluene	54 (α / β = 3 / 1)	15
22	12	20b	AgClO$_4$	CH$_2$Cl$_2$-toluene	56 (α / β = 3 / 1)	15
23	12	23a	AgClO$_4$	CH$_2$Cl$_2$-toluene	76 (α / β = 8 / 1)	16
24	12	23b	AgClO$_4$	CH$_2$Cl$_2$-toluene	71 (α / β = 9 / 1)	16
25	12	22b	Ag$_2$CO$_3$, AgClO$_4$	CH$_2$Cl$_2$-toluene	57 (α), 22 (β)	17
26	13	17a	TMSOTf	CH$_2$Cl$_2$	53 (α), 34 (β)	23
27	14	17a	DMTST	CH$_2$Cl$_2$-toluene	87 (α / β = 5 / 1)	9
28	15	17a	MeOTf	ether	91 (α / β = 4 / 1)	9

We have recently shown that glycosyl fluorides [30(24), 31(24), 38-41(25)] are useful alternatives to those glycosyl donors(8-15). The properly protected glycosyl fluorides were synthesized as follows.

The hemiacetal 24(26), prepared in 4 steps from D-galactose, was protected with a bulky silyl group (t-BuPh$_2$SiCl, imidazole, DMF, 60°, 83%) to give 25. After deacetylation with NaOMe-MeOH, selective benzylidenation (1,1-dimethoxytoluene, p-TsOH, CH$_3$CN) afforded **27**(91%). For preparation of the fluorides **30** and **31**, **27** was first benzylated to give **28** (BnBr, NaH, THF, reflux, overnight, 85%). Desilylation of **28** with Bu$_4$NF-AcOH(27) in THF gave **29** which, when treated with DAST (diethylaminosulfur trifluoride)(28) in THF, produced the readily separable isomers **30** (78%) and **31**(13%).

For preparation of the disaccharide fluorides, **27** was glycosylated with 2,3,4,6-tetra-O-acetyl-α-D-galactopyranosyl bromide in the presence of AgOTf and 2,6-di-t-butylpyridine in dichloroethane to give **32** (65%), α-isomer (5%), and two stereoisomeric orthoesters (14%). The orthoesters were readily converted into **32** by treatment with TMSOTf in CH$_2$Cl$_2$.

Debenzylidenation of **32** (80%aq.AcOH, 60°, 100min, 98%) was followed by deacetylation (NaOMe-MeOH-toluene, quant.) to afford the hexaol **33**, which after isopropylidenation in 2 steps (1. acetone, CuSO$_4$, p-TsOH, 3h; 2. 2,2-dimethoxypropane, p-TsOH, CH$_3$CN, 15min) gave **34** (73%) and a regioisomeric 4,6:4',6'-di-O-isopropylidene derivative (19%). Benzylation of **34** (BnBr, NaH, THF, 60°, 18h, 82%) afforded **35**, which was converted in 2 steps (1. Bu$_4$NF-AcOH, THF, 1day, 89%; 2. DAST, THF, quant.) into the fluoride **37** in a 4:1 α:β ratio.

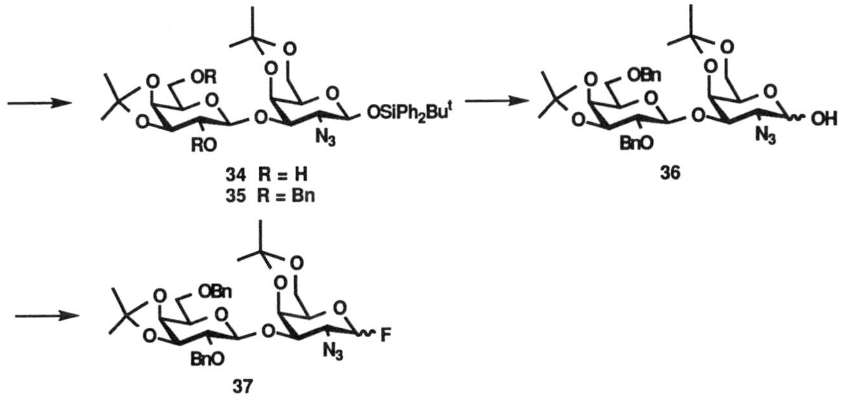

34 R = H
35 R = Bn

36

37

Glycosylation reactions of the Fmoc-protected serine and threonine esters with the glycosyl fluorides were successfully completed using the promoters developed by Suzuki et al(*29*). With zirconocene-perchlorate, the reaction (Cp_2ZrCl_2, $AgClO_4$, CH_2Cl_2, -20° ~ -10°) of **21a** and **30** gave a mixture of glycosides, from which the α-**38** and β-isomers **39** were isolated in 68 and 18% yield respectively. The hafnocene promoter (Cp_2HfCl_2-$AgClO_4$) was equally useful for this reaction. The stereochemical outcome of these glycosylations were not directly attributable to the configuration of the donor fluorides, since reaction of the β-fluoride **31** gave rise to approximately the same anomeric ratio as the α-fluoride.

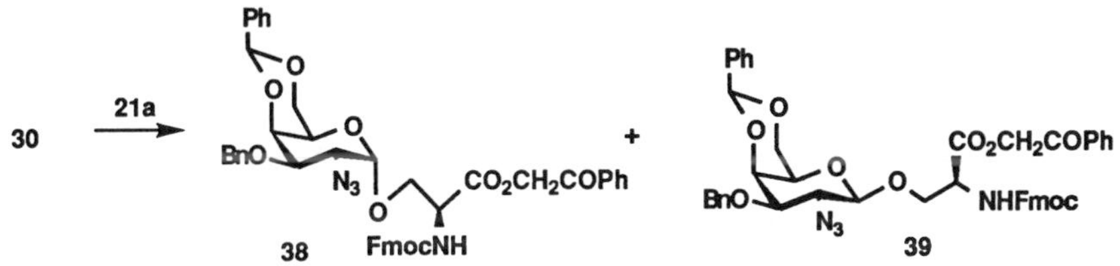

Glycosylation of the Fmoc serine and threonine allyl esters (**22a,b**) with **30** or **31** displayed better α-selectivity (α:β = 7 ~ 10:1) yielding **40a** (68%) and **40b** (71%).

However, coupling of the disaccharide donor **37** with **22a** resulted in a low yield of the products (α-glycoside **43**, 31% and β-glycoside **44**, 4%) due to the lability of the isopropylidene groups in the acidic reaction media. The major by-products that formed were from transisopropylidenation (→**45**) and self-condensation of donor molecules (→tetrasaccharides).

The corresponding dibenzylidene derivatives **46 - 49**, [prepared in 4 steps from **35** (1. 80%aq.CF$_3$CO$_2$H, 0°, 1h, 2. 1,1,-dimethoxytoluene, p-TsOH, CH$_3$CN, 30min, → exo 33%, endo 44%, 3. Bu$_4$NF, AcOH, THF, 99%, 4. DAST, CH$_2$Cl$_2$, 94% α:β = 1:1)], under the same glycosylation conditions afforded the desired disaccharides in good yields.

The glycosyl fluoride **46** was reacted with **22a** or **22b** in the presence of Cp$_2$ZrCl$_2$-AgClO$_4$ in CH$_2$Cl$_2$ to give α-glycoside **50a** (70%) or **50b** (72%) and β-glycoside **51a** (10%) or **51b** (7%). The other diastereomers **47 - 49** were eqally reactive with **22a** or **22b**, approximately the same α,β-product ratio being given.   During the course of the reaction the exo/endo configuration of 3',4'-benzylidene group was equilibrated, therefore the products **50(a,b)** and **51(a,b)** were isolated respectively as nearly 1:1 exo/endo mixture.   After debenzylidenation, the monosaccharide diols **42a,b** and disaccharide tetraols **52a,b** were used for further elongation of the glycan chain by sialyl glycosylation.

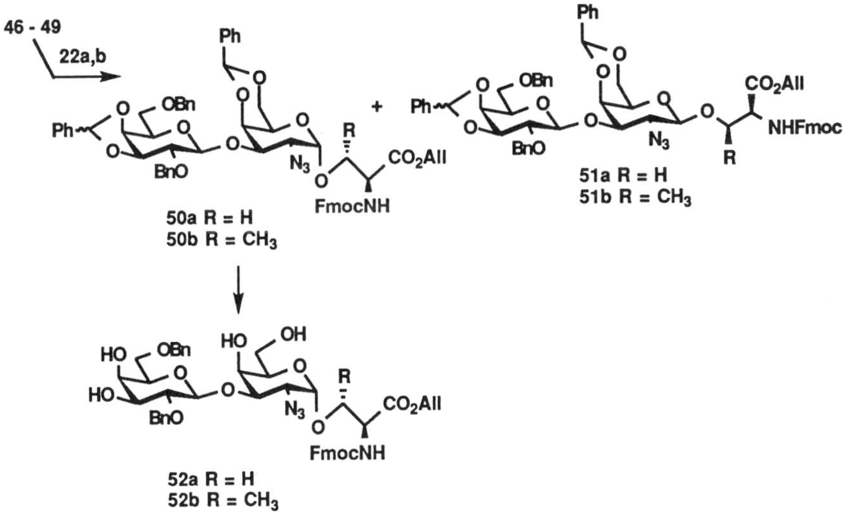

50a R = H
50b R = CH₃

51a R = H
51b R = CH₃

52a R = H
52b R = CH₃

Synthesis of Sialyl Glycosides

In a previous report from this laboratory(*10*), we demonstrated that the C-3β phenylthio-substituted sialic acid derivative **53** can be utilized to produce sialyl α-glycosides. This compound was successfully utilized for the synthesis of complex sialoglycans in the first total synthesis of ganglioside GD₃(*30*).

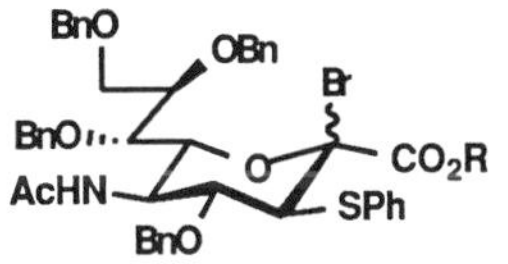

53 R = CH₃
54 R = Bn

This sialylation method would also be utilized for the stereocontrolled synthesis of sialoglycopeptides.   To simplify deprotection of the product, the benzyl ester derivative **54** was chosen as the glycosyl donor.   Helferich glycosylation of **42a** and **42b** with 1.5 equivalents of **54** in CCl₄   afforded the coupling products **55a** and **55b** exclusively in 66 and 77% yields, respectively.   The stereochemistry of the newly formed glycosidic linkage was assumed on the basis of the reaction mechanism and would be assignable after full deprotection of the synthetic glyco-oligopeptides.

The regiochemistry of the coupling reaction was evidenced by the ^{1}H-nmr spectra of the corresponding acetates **56a** and **56b** showing the deshielded signals for GalN$_3$ H-4 protons at δ 5.45 and 5.44 ppm, respectively.   The disaccharides were treated with AcSH to convert the azide group to the acetamide, giving **57a** (92%) and **57b** (68%).   These compounds were then deallylated with Pd(Ph$_3$P)$_4$ and N-methylaniline(17) in THF to give **58a** and **58b** in quantitative yields.   The sulfide group, which had been used to direct the stereochemistry for the sialyl glycosylation was then removed(10).   The desulfurized products were purified through a combination of normal and reverse phase column chromatographies to give pure **59a** and **59b** in 93 and 80% yields, respectively, the desired building blocks for glycopeptide synthesis using the Fmoc strategy.

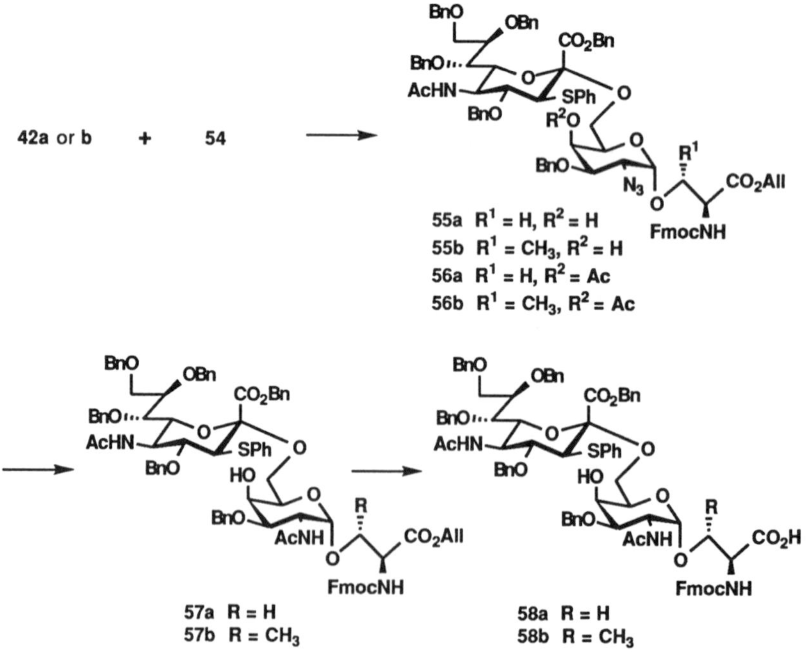

42a or b   +   **54**   ⟶

55a R^1 = H, R^2 = H
55b R^1 = CH$_3$, R^2 = H
56a R^1 = H, R^2 = Ac
56b R^1 = CH$_3$, R^2 = Ac

57a R = H
57b R = CH$_3$

58a R = H
58b R = CH$_3$

59a R = H
59b R = CH₃

Based upon the results obtained from the disaccharide synthesis, the natural glycan units corresponding to **1** were next synthesized.

Glycosylation of **52a** with an excess amount (4 eq) of **54** gave a mixture of the disialo-derivative **60a** (8%) and monosialo-products **61** (7%), **62** (6%). Given the low solubility of the tetraol **52a** in the solvent (CCl_4-CH_2Cl_2), it is not surprising that the sialylation proceeded in low yield even with the excess glycosyl donor present.

Efficient introduction of the sialic acid residues was realized in the separate steps via more soluble intermediates.   Reaction of **52(a,b)** with a slight excess (1.3 eq) of t-BuMe₂SiCl and imidazole in DMF at room temp. for 1-2 h selectively produced the mono-silylated derivatives **63(a,b)** in 91-95% yield, which were soluble in CCl₄.

Reaction of **63(a,b)** with **54** (1.8 eq) in the presence of Hg salts, molecular sieves, and 2-methyl-2-butene (10-15 eq) in CCl₄ gave the coupling products **64a** and **64b**, which were isolated by gel permeation chromatography followed by silica gel chromatography in 75 and 82% yield, respectively.   The 2-methyl-2-butene was used as a scavenger of PhSBr which was generated by elimination from **54** and tended to add to the double bond of allyl esters.

The 2→3 linkage in the glycosylation was proven by ¹H-nmr spectra of the corresponding carbamates, which were produced by the rapid derivatization of **64(a,b)** with trichloroacetyl isocyanate done in the nmr tube *(31)*.   The downfield signals of the H-4 protons for Gal*p* and Gal*p*N₃ residues were observed at δ 5.13 and 5.50 ppm for **64a** and δ 5.09 and 5.52 ppm for **64b**.   Treatment of **64(a,b)** with 80% aq.CF₃CO₂H afforded the triols **62(a,b)** quantitatively.

Sialylation using **62(a,b)** with **54**(2.5 eq) and the same reaction conditions produced **60(a,b)** as the sole coupling products both in 69% yield.

Treatment of **60(a,b)** with a mixture of AcSH and pyridine, gave the acetamide (**65a,b**, 93-94%) without any side products caused by the addition of AcSH to the allyl ester double bond.

Deallylation of **65(a,b)**[Pd(Ph₃P)₄, N-methylaniline, THF, →**66a,b**, 94-97%] followed by desulfurization (Ph₃SnH, AIBN, benzene, reflux, 6-7h) unexpectedly provided the lactone derivatives (**67a,b**, 57-69%) instead of the dibenzyl ester **68(a,b)**.

When the desulfurization reaction of **66b** was run at a lower temperature (60° C) for 24h, a mixture of **67b** (37%) and **68b** (19%) was produced. However, further decreases in temperature (< 50°

C) did not give improved yields of **68b**.   Although the exclusive formation of **68(a,b)** was not attained in this synthetic route, the model studies with the α-D-Neup5Ac-(2→3)-Galp disaccharide implied that cleavage of 1',4-lactone would be facilitated by treatment with aqueous weak base in the final stage of synthesis.

Therefore, having the necessary glycosyl serine and threonine derivatives in hand, attention was next focused on the synthesis of the oligopeptides.

	R¹	R²
65a	H	All
65b	CH₃	All
66a	H	H
66b	CH₃	H

Synthesis of Glycopeptides

Syntheses of MN type pentapeptides of glycophorin A were pursued using **59(a,b)** and the Fmoc strategy.

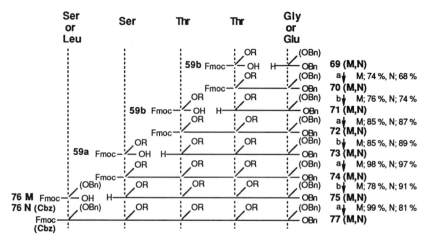

R = protected oligosaccharide, Fmoc = 9-fluorenylmethoxycarbonyl, Cbz = benzyloxycarbonyl, Bn = benzyl,   a : EEDQ, CH₂Cl₂, room temp., 2 - 6 days,   b : morpholine, room temp., 1 - 3 h

The peptide coupling reactions proceeded smoothly using EEDQ (2-ethoxy-1-ethoxycarbonyl-1,2-dihydroquinoline) as the condensing agent in CH_2Cl_2 to give oligopeptide derivatives in 68-99% yield.   The N-terminal Fmoc group was selectively removed with morpholine in 74-91% yield.   Finally, the pentapeptide derivative **77M** was deprotected in 2 steps (1. morpholine, 2. H_2, 20%Pd(OH)₂-C, aq.MeOH), while **77N** was hydrogenated.

78 R^1 = CH₂OH, R^2 = H
79 R^1 = CH₂CH(CH₃)₂, R^2 = CH₂CH₂CO₂H

The synthetic sialoglycopeptides were purified by gel permeation (Sephadex LH-20, water) and ion-exchange chromatography (Mono Q, 0.05-0.3M aq.NH_4HCO_3) to give the M type **78**(91%) and the N type **79**(88%).

Similarly, the tetrasaccharide-linked serine and threonine building blocks were used for the synthesis of M type heptapeptide fragment of natural glycophorin A. The procedure is outlined in the figure shown below.

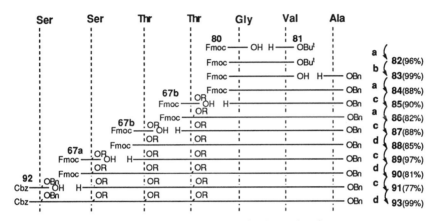

R = protected oligosaccharide, Fmoc = 9-fluorenylmethoxycarbonyl,
Bn = benzyl, Cbz = benzyloxycarbonyl
a : EEDQ, CH_2Cl_2, room temp., 4-7days, b : 80%CF_3CO_2H, CH_2Cl_2, room temp.,
20h, c : morpholine, room temp., 1-1.5h, d : IIDQ, CH_2Cl_2, room temp., 1-5days

IIDQ (2-isobutoxy-1-isobutoxycarbonyl-1,2-dihydroquinoline) was the activating agent used when the amino components had carbohydrate substituents. All the steps of the condensation and subsequent deblocking of N-terminal Fmoc group were successful and the structures of the synthetic substances were determined by FAB-Mass and ^{1}H-NMR spectroscopies.

The protected heptapeptide-dodecasaccharide **93** thus obtained was hydrogenated with 20%Pd(OH)$_2$-C in aqueous THF to afford **94** quantitatively. The conserved tri-lactone structure was proved by the signal for the three H-4 protons of the Gal*p* residues at δ 5.29 ppm. The lactone persisted in D_2O solution even after several days at room temperature (~24° C). An addition of $NaHCO_3$ to the solution (0.03M, pH 7.5-8) provided slow hydrolysis of the lactone ring: after 11 days, the signal at 5.29 was no longer detectable. The target molecule **95** was isolated via gel permeation chromatography.

The ^{1}H-NMR and FAB-MS data of the synthetic sialoglyco-peptides **78**, **79**, and **95** were all in good agreement with the proposed structures. The configuration of the Neup5Ac glycosides was established as α by comparison of the H-3 signals with those reported for the related, natural glycoprotein fragments *(32-35)*.

93

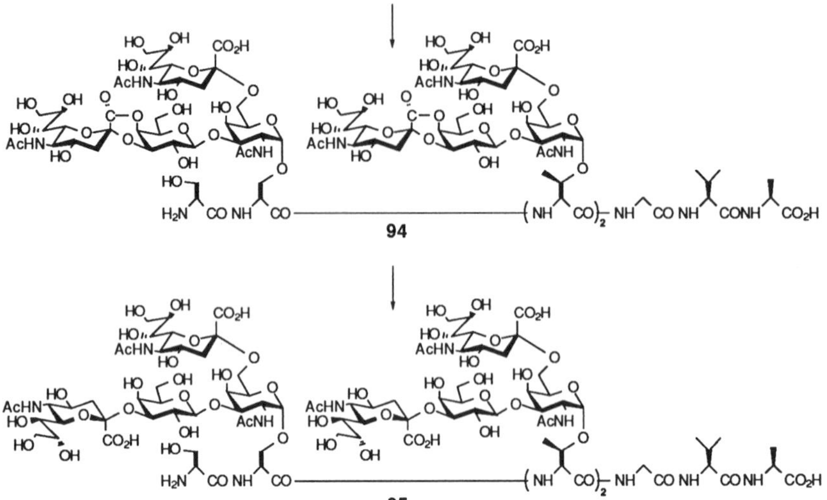

94

95

In conclusion, syntheses of N-terminal fragments of glycophorin A carrying trimers of α-D-Neup5Ac-(2→6)-α-D-GalpNAc-(1→ or α-D-Neup5Ac-(2→3)-β-D-Galp-(1→3)-[α-D-Neup5Ac-(2→6)]-α-D-GalpNAc-(1→ were achieved through stereocontrolled glycosylations with the designed glycosyl donors and acceptors.   Peptide condensations were efficiently carried out in solution utilizing the Fmoc methodology.

Acknowledgements.   A part of this work was financially supported by a Grant-in-Aid for Scientific Research from the Ministry of Education, Science and Culture, and also by the Special Coordination Funds of the Science and Technology Agency of the Japanese Government.   We thank Dr. J. Uzawa and Mrs. T. Chijimatsu for NMR, Mr. Y. Esumi for FAB-MS, Ms. M. Yoshida and her staff for elemental analyses, and Ms. A. Takahashi for technical assistance.

References
 1. Montreuil, J. *Adv. Carbohydr. Chem. Biochem.* **1980,** *37* ,157
 2. Montreuil, J. *Comprehensive Biochemistry* **1982** *19 BII,* 1

3. Tomita, M.; Furthmyer, H.; Marchesi, V. T. *Biochemistry*
1978, *17*, 4756
4. Siebert, P. D.; Fukuda, M. *Proc. Natl. Acad. Sci. USA* **1986**, *83*,
1665
5. Iijima, H.; Ogawa, T. *Carbohydr. Res.,* **1988**, *172*, 183
6. Paulsen, H.; von Deessen, U. *ibid.* **1988**, *175*, 283
7. Iijima, H.; Ogawa, T. *ibid.* **1989**, *186*, 95
8. Iijima, H.; Ogawa, T. *ibid.* **1989**, *186*, 107
9. Paulsen, H.; Rauwald, W.; Weichert, U. *Liebigs Ann. Chem.*
1988, 75
10. Ito, Y.; Ogawa, T. *Tetrahedron Lett.,* **1988**, *29*, 3987;
Tetrahedron **1990**, *46*, 89
11. Kanie, O.; Kiso, M.; Hasegawa, A. *J. Carbohydr. Chem.* **1988**, *7*,
501
12. Paulsen, H.; Kolar, C.; Stenzel, W. *Chem. Ber.* **1978**, *111*, 2358
13. Ferrari, B.; Pavia, A. A. *Carbohydr. Res.,* **1980**, *79*, C1
14. Kunz, H., Birnbach, S. *Angew. Chem. Int. Ed. Engl.* **1986**, *25*,
360
15. Paulsen, H.; Adermann, K. *Liebigs Ann. Chem.* **1989**, 751
16. Paulsen, H.; Merz, G.; Brockhausen, I. *ibid.* **1990**, 719
17. Ciomer, M.; Kunz, H. *SYNLETT.* **1991**, 593
18. Bencomo, V. V.; Jacquinet, J.-C.; Sinaÿ, P. *Carbohydr. Res.*
1982, *110*, C9
19. Lüning, B.; Norberg, T.; Tejbrant, J. *Glycoconj. J.* **1989**, *6*, 5
20. Paulsen, H., Paal, M.; Schultz, M. *Tetrahedron Lett.* **1983**, *24*,
1759
21. Paulsen, H.; Schultz, M. *Carbohydr. Res.* **1987**, *159*, 37
22. Paulsen, H.; Schultz, M. *Liebigs Ann. Chem.* **1986**, 1435
23. Kinzy, W.; Schmidt, R. R. *Carbohydr. Res.* **1987**, *164*, 265
24. Nakahara, Y.; Iijima, H.; Shibayama, S.; Ogawa, T., *Tetrahedron
Lett.,* **1990**, *31*, 6897; *Carbohydr. Res.,* **1991**, *216*, 211
25. Nakahara, Y.; Iijima, H.; Ogawa, T. in preparation
26. Grundler, G.; Schmidt, R. R. *Liebigs Ann. Chem.* **1984**, 1826
27. Kinzy, W.; Schmidt, R. R. *ibid.* **1985**, 1537
28. Rosenbrook,Jr. W.; Riley, D. A.; Larty, P. A. *Tetrahedron Lett.*
1985, *26*, 3;   Posner, G. H.; Haines, S. R. *ibid* **1985**, *26*, 5
29. Suzuki, K.; Maeta, H.; Matsumoto, T.; Tsuchihashi, G. *ibid* **1988**,
29, 3567, 3571;   Suzuki, K.; Maeta, H.; Matsumoto, T. *ibid.*
1988, *30*, 4853
30. Ito, Y.; Numata, M.; Sugimoto, M.; Ogawa, T. *J. Am. Chem. Soc.,*
1989, *111*, 8508
31. Kinzy, W.; Schmidt, R. R. *Carbohydr. Res.* **1989**, *193*, 33

32. Vliegenthart, J. F. G.; Dorland, L.; van Halbeek, H.; Haverkamp, J. in " *Sialic Acids* " Schaur, R., Ed; Springer-Verlag, Wien, **1982**, pp 127-172

33. Gerken, T. A. Arch. Biochem. Biophys., **1986,** *247,* 239

34. Dill, K., Hu, S.; Huang, L. J. *Carbohydr. Chem.*, **1990,** *9,* 863

35. Linden, H.-U.; Klein, R. A.; Egge, H.; Peter-Katalinic, J.; Dabrowski, J.; Schindler, D.   Biol. Chem. Hoppe-Seyler, **1989,** *370,* 661

RECEIVED January 24, 1994

Chapter 15

Synthetic Probe for Study of Molecules Related to Selectin Family

N. E. Nifant'ev[1], A. S. Shashkov[1], Y. E. Tsvetkov[1], A. B. Tuzikov[2], I. V. Abramenko[3], D. F. Gluzman[3], and N. V. Bovin[2]

[1]N. D. Zelinsky Institute of Organic Chemistry, Russian Academy of Sciences, 117913 Moscow, Russia
[2]M. M. Shemyakin Institute of Bioorganic Chemistry, Russian Academy of Sciences, 117871 Moscow, Russia
[3]R. E. Kavetsky Institute for Oncology Problems, Ukrainian Academy of Sciences, 252022 Kiev, Ukraine

Increased expression of cancer-associated antigens is a characteristic feature of cancer cell membranes. SiaLe^a tetrasaccharide being a cancer associated antigen and ligand for selectins is one of the most interesting structures in this respect. In order to reveal and study receptors for SiaLe^a, we have synthesized the probe which contains SiaLe^a tetrasaccharide and biotin moieties covalently linked to the polyacrylamide carrier *via* spacers. The preparation of the probe and its application for the study of endogenous lectins of cancer and leukemia cells are described in this paper.

The carbohydrate antigen Neu5Acα(2→3)Galβ(1→3)[Fucα(1→4)]GlcNAc (SiaLe^a, known also as cancer antigen CA19-9) is one of the most significant tumor markers (*1*) in practical cancer diagnostics. Glycoproteins and glycolipids terminated by SiaLe^a tetrasaccharides were found in the cytoplasmatic membrane of cancer cells and in blood as a product of tumor cells shedding. Though SiaLe^a serves as a useful tumor marker, the role of this cancer-associated antigen in processes of malignancy, tumor survival following attack by the immune system, invasion, metastasis and in other processes is still unclear. Taking these factors into consideration, the discovery of selectins, one of the receptors of which is SiaLe^a (*2-7*) is rather intriguing. The presence of selectins on both tumor and normal cells surrounding them (i.g. endothelial cells, lymphocytes, neutrophils, monocytes, eosinophils, platelets) compels us to examine the functional significance of intercellular interactions between SiaLe^a and selectins.

The aim of the described research was to prepare a convenient tool for direct examination of SiaLe^a-binding molecules on cell membranes. In designing the probe, we took into account that naturally-occurring glycoproteins display heterogeneity in glycosylation pattern and can exhibit undesired binding properties. That is why the target synthetic probe should contain only SiaLe^a tetrasaccharidic moieties. We also expected that the affinity of cell surface oligosaccharide-binding molecules could be low at the monovalent ligand level (8), and that is why the probe should be polyvalent in order to produce the maximal binding with selectins. Finally, the probe had to contain the label groups. Taking these considerations into account the target probe **1** contains the SiaLe^a tetrasaccharide and biotin moieties covalently linked to the polyacrylamide chains *via* a spacer. The polyacrylamide chain was chosen as polymeric carrier because it is inert and does not bind with cell-surface molecules. The random coiling of the polyacrylamide matrix is capable of arranging the ligands to the complementary protein molecule. Recently it was shown, that the probes of this type, but with another oligosaccharide determinants are convenient tools for the study of endogenous human cell lectins binding with blood group antigens, and for revealing the endogenous lectins on the plasma membrane, organelles, and in the cytoplasm of human cells, including transformed ones (9-12).

1

Preparation of the Probe 1

The initial step in the construction of the probe **1** was the synthesis of spacer-armed SiaLe^a tetrasaccharide **14** (Scheme 2) (Nifant'ev, N. E.; Shashkov, A. S.; Tsvetkov, Y. E.; Tuzikov, A. B.; Maslennikov, I. V.; Abramenko, I. V.; Gluzman, D. F.; Bovin, N. V. *Bioorg. Khim.* **1993**, in press.) For the preparation of this compound the triol **9** (Scheme 1) and ethyl 2-thio-sialoside **10** were selected as sialyl acceptor and sialyl donor, respectively, with NIS-TfOH as promoter. Hasegawa *et al.* showed (*13*) the advantages of the combination of these reagents for the synthesis of oligosaccharides containing the Neu5Acα(2→3)Gal fragment.

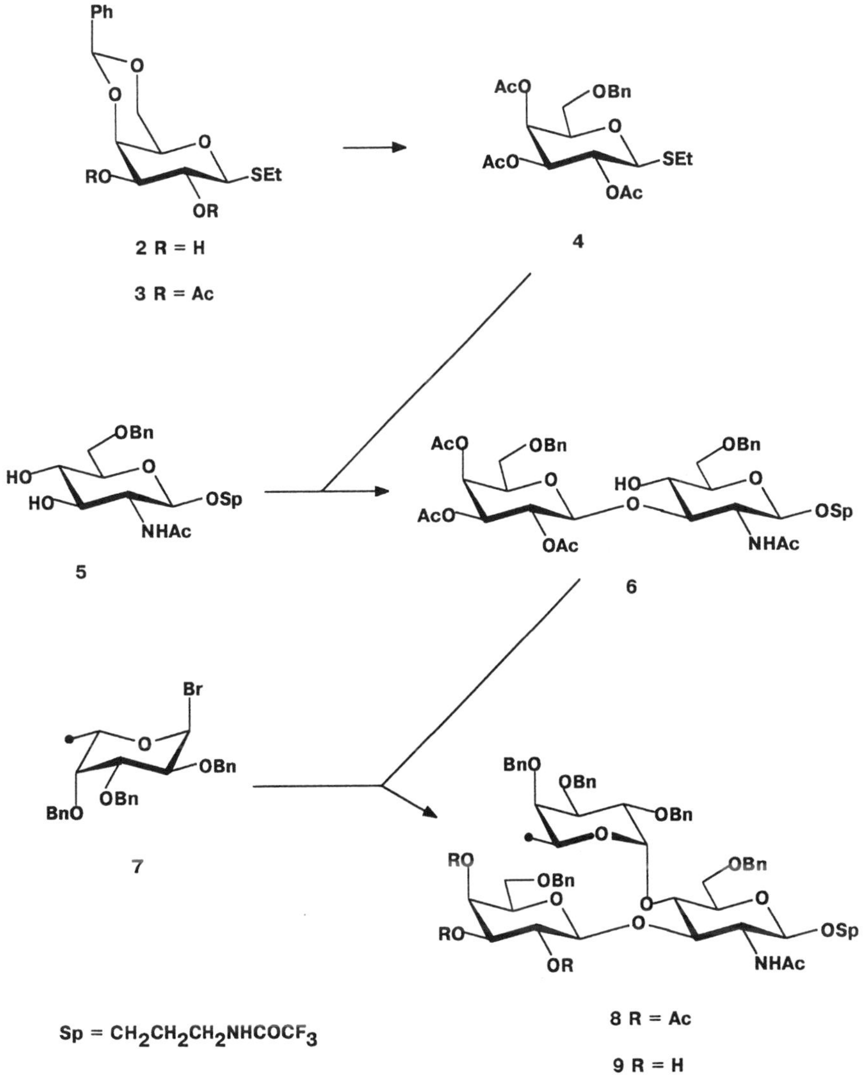
Ph
O
O
O
RO
SEt
OR
2 R = H
3 R = Ac
AcO
OBn
O
AcO
SEt
OAc
4
OBn
HO
O
HO
OSp
NHAc
5
AcO
OBn
O
AcO
OAc
HO
O
OBn
O
OSp
NHAc
6
Br
O
OBn
OBn
BnO
7
BnO
OBn
OBn
O
RO
OBn
O
RO
OR
O
O
OBn
O
OSp
NHAc
8 R = Ac
9 R = H
Sp = CH2CH2CH2NHCOCF3

Scheme 1

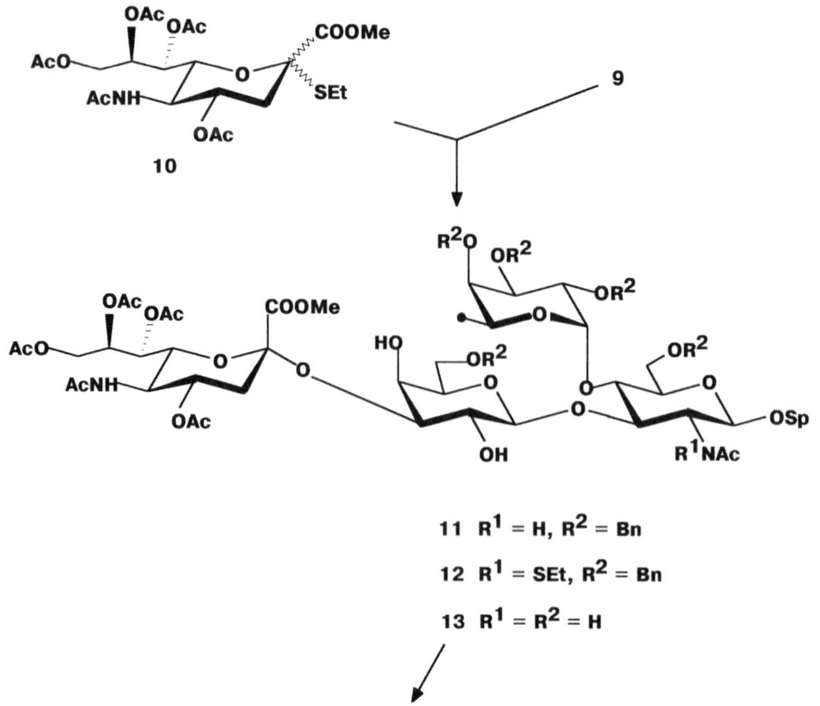

11 R^1 = H, R^2 = Bn

12 R^1 = SEt, R^2 = Bn

13 R^1 = R^2 = H

Neu5Acα(2-3)Galβ(1-3)GlcNAcβ-O(CH$_2$)$_3$NH$_2$

4

|

Fucα1   **14** (K$^+$-salt)

Sp = CH$_2$CH$_2$CH$_2$NHCOCF$_3$

Scheme 2

The synthetic parways for the preparation of the triol **9** are presented in Scheme 1. Thus, the treatment of ethyl 4,6-O-benzylidene-1-thio-β-D-galactopyranoside **2** (*14*) or its 2,3-diacetate **3** with NaCNBH$_3$ in acidic tetrahydrofuran (*15*) and subsequent acetylation gave ethyl 2,3,4-tri-O-acetyl-6-O-benzyl-1-thio-β-D-galactopyranoside **4** in the yields of 48 and 91%, respectively. Glycosylation of the glucosaminide **5** (*16*) containing the spacer group by **4** was performed in dichloromethane using CuBr$_2$-HgBr$_2$-Bu$_4$NBr mixture (*17*) as a promoter. This reaction, which is analogous to one reported previously (*14*), was regio- and stereoselective and gave (1→3)-linked disaccharide **6** (87.5%, [α]$_D$ -7.4° (c 3, CHCl$_3$)). The presence of the free OH-group in **6** at position 4 of the GlcNAc-residue was proved by highfield location of H-4GlcNAc signal (δ 3.45 ppm) in ^{1}H NMR spectrum (in CDCl$_3$) and by the fact that this signal was simplified by deuterium exchange upon addition of CD$_3$OD. The β configuration of the Gal unit in **6** was indicated by the $J_{1,2}$ value of 7.9 Hz. Further halide-ion catalyzed (*18*) glycosylation of **6** by 2,3,4-tri-*O*-benzyl-α-L-fucopyranosyl bromide afforded the substituted Le^a trisaccharide **8** (71%, [α]$_D$ -62.6^0 (c 2, CHCl$_3$)). The α configuration of Fuc residue in **8** was proved by the data of its ^{1}H NMR spectrum which contain a doublet for H-1Fuc at δ 5.12 ppm with $J_{1,2}$ of 3.8 Hz. *O*-Deacetylation of which gave the triol **9** in 90% yield, [α]$_D$ -37.8^0 (c 2, CHCl$_3$).

In continuation of the synthesis of tetrasaccharide **14**, we performed the NIS-TfOH promoted glycosylation of triol **9** by 2-thiosialoside **10** (*Syntesome*, Munchen) in acetonitrile at -40°C which afforded the target substituted tetrasaccharide **11** in 19% yield, [α]$_D$ -8.2^0 (c 1, CHCl$_3$). Compound **11** was then subjected to catalytic hydrogenolysis over Pd/C followed by saponification by aqueous KOH to give target spacer-armed SiaLe^a tetrasaccharide **14** as the K+-salt (76%, [α]$_D$ -57.4^0 (c 1, H$_2$O)).

Unexpectedly, the reaction of **9** and **10** gave two tetrasaccharidic products with the SiaLe^a backbone. In addition to the target product of **11**, its N$_{GlcNAc}$-thioethylated derivative **12** was also formed in the yield of 29%, [α]$_D$ -26.3^0 (c 1, CHCl$_3$). ^{1}H NMR spectrum of **12** contain signals for S-Et group (at δ 1.20 and 2.90 ppm for CH_3 and CH_2 protons) whereas a doublet for NH-group of GlcNAc residue was absent. The N-SEt group can be easily removed upon treatment with Pd/C, so catalytic hydrogenolysis of **12** as well as the tetrasaccharide **11** gave, in almost quantitative yield, the same product **13** in both cases. Therefore, in spite of the side N-thioethylation, the glycosylation of **9** was rather effective from the point of view of the total yield (48%) of the desired products of α(2→3)-sialylation.

Another interesting peculiarity of the reaction of compounds **9** and **10** under the conditions used was the formation of a group of isomeric tetrasaccharide lactones in a total yield of 30%. One of the components which contained a β-(2→4)-glycosidic and (1→3)-intramolecular ester linkages, was isolated by HPLC as compound **15**, after catalytic hydrogenolysis of the mixture. We also found out that no loss of the tetrasaccharide **11** due to side lactonisation took place. This was proved by the fact that after saponification of debenzylated lactonic fraction mainly two isomeric tetrasaccharides **16** and **17** were obtained in yields of 45 and 38% respectively, while no traces of tetrasaccharide **14** were indicated.

Sp = CH$_2$CH$_2$CH$_2$NHCOCF$_3$

15

The structures of spacer-armed tetrasaccharides **14-16**, particularly the directions of glycosidic linkages and anomeric configurations were determined by ^{1}H- and ^{13}C-

NMR spectroscopy. The substitution of the Gal residue at position 3 in **14** and **16** and at position 4 in **17** were confirmed by the low-field location of C-3$_{Gal}$ signal in ^{13}C-NMR spectra of **14** and **16** and of C-4$_{Gal}$ signal in the spectrum of **17**. The α-configuration of the Neu5Ac in **14** and the β-configuration in **16** and **17** were proved by characteristic chemical shift values of H-3eq. (δ 2.69, 2,46, and 2,39 ppm for **14**, **16**, and **17**, respectively).

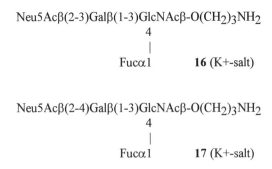

Neu5Acβ(2-3)Galβ(1-3)GlcNAcβ-O(CH$_2$)$_3$NH$_2$
4
|
Fucα1   **16** (K+-salt)

Neu5Acβ(2-4)Galβ(1-3)GlcNAcβ-O(CH$_2$)$_3$NH$_2$
4
|
Fucα1   **17** (K+-salt)

The general procedure (*9,19*) for the preparation of the glycoconjugates of this type was used for the construction of the probe **1**. The sequential treatment of poly(4-nitrophenylacrylate) (100 μg-equiv., degree of polymerisation 1000; *Syntesome*) with spacer-armed tetrasaccharide **14** (20 μmol), d-biotin 6-aminohexylamide (5 μmol, *Syntesome*), and excess of ethanolamine were performed in dimethylformamide in the presence of triethylamine. Probe **1** was isolated (95% yield) by chromatography on a Sephadex LH-20 column. Its structure was proved by ^{1}H-NMR spectral data. In particular, the ratio of SiaLe^a, biotin, and ethanolamine moieties in probe **1** was found to be 20:5:75. The molecular weight (M$_r$) was estimated by the gel-chromatography on a TSK HW-55 column using proteins and dextrans as standards. The M$_r$ was about 40kDa using a protein for comparison or about 20 kDa compared to dextran standards. In the same manner, the probes with the ratios of SiaLe^a tetrasaccharide, biotin, and ethanolamine moieties of 10:5:85 and 30:5:65 were prepared.

Application of the Probe 1 and Related Probes for the Study of Endogenous Lectins

The probe **1** was used for the study of endogenous lectins of the following cells: 1) blood mononucleocytes from healthy donors and 2) from patients with chronic lymphocytic leukemia, hairy cell leukemia, and acute lymphoblastic leukemia (content of transformed cells was not less than 90%); 3) cancer embrionic antigen(CEA)-positive (stained with anti-CEA monoclonal antibodies Dakopatts, Denmark) and Egp34-positive (stained with monoclonal antibodies DAKO-Ber-EP4) human cancer cells from pleural exudate. To determine the probe's binding specificity in more detail, we also used a series of analogous probes bearing other determinants representing Lewis system oligosaccharides: Le^a, Le^x, Le^y, and their fragments Fucα1→2Gal, Fucα1→3GlcNAc, Fucα1→4GlcNAc, and Neu5Acα residues. Their syntheses were described previously (*9,10,12,19-23*). The staining experiments were performed as follows: cells were fixed for 3 min in 10% formalin vapors, incubated for 1 h at 20°C with probe solution (0.1 mg/ml) in phosphate buffered saline, and washed. Probe binding with the cells was observed as cell preparation staining after streptavidine-enzyme conjugate addition with the following development of the substrate by peroxidase or phosphatase (*10,11*). The reaction was assumed to be positive when intensive staining of 90% of cells was observed; the negative reaction was taken when a staining of only few cells or staining of low intensity took place. For example, no staining was indicated upon treatment of all cell preparations by Glcβ-probe. Reaction on plasma membrane reflects cell-surface lectin binding, and the diffuse reaction on cytoplasm reflects "soluble" lectin binding. In some cases organelle specific binding was also observed, as well as simultaneous staining of three or two mentioned types. In particular, single cancer cells were stained by probe 1 on membrane, but cancer cells emboli - as diffuse emboli encirclement and at cell contacts.

One of the most interesting aspects of malignancy is the participation of cell surface glycoconjugates and endogenous lectins in the processes of cell-cell interactions. Increased expression of cancer-associated carbohydrate antigens (CAA) is a characteristic feature of transformed cell membranes (*24-29*). It is reasonable to assume that the cancer cells might express also the molecules complementary to their CAA. Gabius and coworkers (*30*) were the first to have found that specific binding sites for the synthetic probe biotin-BSA-Galβ(1→3)GalNAcβ as well as for the plant lectin recognizing the disaccharide ligand Galβ(1→3)GalNAc were co-expressed on the same cells (cell lines of human breast carcinoma). In the previous work (*11*), we found that Galβ(1→3)GalNAcα (Thomsen-Friedenreich) and GalNAcα(1→3)GalNAcβ (Forssman) specific probes bound intensely cancer cells. In the present work, we used the synthetic tetrasaccharide SiaLe[a]-containing probe 1 to reveal corresponding receptors on cancer cells isolated from the serous membranes of the pleural cavity in oncologic patients. The carbohydrate antigen SiaLe[a] (CA19-9) is a serum specific marker of pancreatic and stomach carcinomas (*1*). In contrast to normal cells which contain the isomeric tetrasaccharide SiaLe[x] (Neu5Acα(2→3)Galβ(1→4)[Fucα(1→3)]GlcNAc) sialylated Lewis A occurs only on cancer cells or as a minor component on a limited number of normal and benign cells. The synthetic probe with SiaLe[a] determinant may mimic naturally occurring antigens in the respect of its interaction with complementary lectins as selectins, which recognise sialylated Lewis structures, or related cancer lectins. Therefore, such probes may be useful for effecting the carcinogenesis process. As a first step, we have tried to reveal SiaLe[a]-recognizing molecules on the surface of the above mentioned cells.

Peripheral blood mononucleocytes of healthy persons, benign tumor patients and cells from patients with chronic lymphocytic leukemia, hairy cell leukemia, and acute lymphoblastic leukemia did not stain with SiaLe[a] probe, or Le[a] and Le[x] ones. In contrast, cells of serous effusions (lung and breast cancer) bound the SiaLe[a] probe intensely. Binding was specific because probes Le[a], Le[x], and the fucodisaccharides did not stain the same cancer cells. For example the cells of only 3 from 26 cancer patients (1 case uterus cancer, 1 - stomach, and 1 - unknown origin) stained by Le[a]. It is interesting that a significant correlation between probes SiaLe[a] and Le[y] binding to the cancer cells studied was found. Le[y] (Fucα(1→2)Galβ(1→4)[Fucα(1→3)]GlcNAc) is a cancer-associated carbohydrate antigen but typical for other than SiaLe[a] tumor localizations. It is doubtful that both probes bind the same receptor; most likely coexpression of the different carbohydrate-binding molecules was observed. Probe with the Neu5Acα residue as a ligand bound practically all cancer cells tested (14 from 15) but by another manner: both membrane and cytoplasm as well as some organelles were stained, indicating another type lectin binding with this probe.

It is well documented that selectins of normal and malignant cells recognize SiaLe^x and SiaLe^a (2-5), moreover SiaLe^a has a greater affinity for the selectin than does SiaLe^x (31), which was discovered first as a selectin receptor. Apparently, we have revealed selectins or closely related cancer-associated molecules. This may also be confirmed by two additional facts. Firstly, cytokine stimulation increases staining of the cancer cells by probe 1 (unpublished results by authors) that is typical for L-selectin, and secondly, the binding is Ca^{2+}-dependent.

It should be noted that the SiaLe^a probe 1 binds cancer cells but is inert to leukemia cells. In a recent paper (7), it was shown that leukemia cell lines express mainly SiaLe^x in contrast to cancer cell lines which express both SiaLe^x and SiaLe^a, but only latter mediates adhesion of the cancer to the endothelial cells. The conclusion (7) is that the *SiaLe /selectin* cell adhesion system plays an important role in the adhesion of epithelial cancer cells to endothelial cells, while its role in the adhesion of leukemia cells seems to be minor. Our results obtained on the cells from cancer and leukemia *patients* with the help of the synthetic SiaLe^a probe agree with this conclusion. Moreover, one can speculate that not only the SiaLe^a tetrasaccharide but also complementary molecules (selectin or selectin-like) play a role in hetero- and homotype cancer cell interaction during invasion and metastasis.

Do the revealed lectins *in vivo* come in contact with SiaLe^a on the same cell or neighbor cells in emboli? How is lectin expression controlled relative to expression of the complementary carbohydrate ligand? Is SiaLe^a expression essential for "cancer type" contacts, tumor survival and metastasis? These and other questions remain to be solved using different approaches and technics, including synthetic probes.

Literature Cited.

1. Magnani, J. L.; Nilsson, B.; Brockhaus, M.; Zopf, D.; Steplewski, Z.; Koprowski, H.; Ginsburg, V. *J. Biol. Chem.*, **1982**, *57*, 14365.
2. Varki, A. *Curr. Opinion Cell Biol.* **1992**, *4*, 257.
3. Feizi, T. *Trends in Biochem. Sci.* **1991**, *16*, 84.
4. Walz, G.; Aruffo, A.; Kolanus, W.; Bevilacqua, B.; Seed, B. *Science* **1990**, *225*, 1132.
5. Larkin, M.; Ahern, T. J.; Stoll, M. S.; Shaffer, M.; Sako, D.; Brien, J. O.; Yuen, C.-T.; Lawson, A. M.; Childs, R. A.; Barone, K. M.; Langer-Safer, P. R.; Hasegawa, A.; Kiso, M.; Larsen, G. R.; Feizi, T. *J. Biol. Chem.* **1992**, *267*, 13661.
6. Takada, A.; Ohmori, K.; Yoneda, T.; Tsuyuoka, K.; Hasegawa, A.; Kiso, M.; Kannagi, R. *Cancer Res.* **1993**, *53*, 354.
7. Larsen, G. R.; Sako, D.; Ahern, T. J.; Shaffer, M.; Erban, J.; Sajer, S. A.; Gibson, R. M.; Wagner, D. D.; Furie, B. C.; Furie, B. *J. Biol. Chem.* **1992**, *267*, 11104.
8. Korchagina, E. Y.; Bovin, N. V. *Bioorgan. Khim.* **1992**, *18*, 283.
9. Bovin, N. V. In: *Lectins and Glycobiology* (H.-J. Gabius and S. Gabius Eds.), Springer-Verlag, Berlin, **1993**, P. 23-28.
10. Abramenko, I. V.; Gluzman, D. F.; Korchagina, E. Y.; Zemlyanukhina, T. V.; Bovin, N. V. *FEBS Lett.* **1992**, *307*, 283.

11. Abramenko, I. V.; Gluzman, D. F.; Bovin, N. V. *Experim. Oncol.* **1993**, *15 (3)*, 40.
12. Bovin, N. V.; Korchagina, E. Y.; Zemlyanukhina, T. V.; Gluzman, D. F.; Abramenko, I. V. *Experim. Oncol.* **1993**, *15 (2)*, 41.
13. Hasegawa, A.; Nagahama, T.; Ohki, H.; Hotta, K.; Ishida, H.; Kiso, M. *J. Carbohydr. Chem.* **1991**, *10*, 493.
14. Sato, S.; Ito, Y.; Ogawa, T. *Carbohydr. Res.* **1986**, *155*, C1.
15. Garegg, P. -J.; Hultberg, H.; Wallin, S. *Carbohydr. Res.* **1982**, *108*, 97.
16. Bovin, N. V.; Zemlyanukhina, T. V.; Chagiashvili, C. N.; Khorlin, A. Y. *Khim. Prirodn. Soed.* **1988**, 777.
17. Sato, S.; Mori, M.; Ito, Y.; Ogawa, T. *Carbohydr. Res.* **1986**, *155*, C6.
18. Lemieux, R. U.; Hendriks, K. B.; Stick, R. V.; James, K. *J. Amer. Chem. Soc.* **1975**, *97*, 4056.
19. Bovin, N. V.; Korchagina, E. Y.; .Zemlyanukhina, T. V.; Byramova, N. E.; Galanina, O. E.; Zemlyakov, A. E.; Ivanov, A. E.; Zubov, V. P.; Mochalova, L. V. *Glycoconj. J.* **1993**, *10*, 142.
20. Byramova, N. E.; Tuzikov, A. B.; Bovin, N. V. *XVI Int. Carbohydrate Symp.* **(1992**, Paris), Abstract book, p. 208.
21. Gabius, H.-J.; Gabius, S.; Zemlyanukhina, T. V.; Bovin, N. V.; Brinck, U.; A.Danguy, Josho, S. S.; Kayser, K.; Schottelius, J.; Sinowatz, F.; Tietze, L. F.; Vidal-Vanaclocha, F.; Zanetta, J.-P. *Histol. Histopath.* **1993**, *8*, 369.
22. Matrosovich, M. N.; Mochalova, L. V.; Marinina, V. P.; Byramova, N. E.; Bovin, N. V. *FEBS Lett.* **1990**, *272*, 209.
23. Bovin, N. V.; Khorlin, A. Y. *Bioorgan. Khim.* **1987**, *13*, 1405.
24. Gluzman, D. F.; Bovin, N. V.; Abramenko, I. V.; Sclyarenko, L. M. *Experim. Oncol.* **1992**, *14 (1)*, 3.
25. Gabius, H.-J.; Bardosi, A. *Progress in histochemistry and cytochemistry* **1991**, *22*, 1.
26. Monsigny, M.; Roche, A.-C; Kieda, C.; Midoux, P.; Obrenovitch, A. *Biochimie* **1988**, *70*, 1633.
27. Raz, A.; Lotan, R. *Cancer and Metastasis Rev.* **1987**, *6*, 433.
28. Lotan, R.; Raz, A. *J. Cell. Biochem.*, **1988**, *37*, 107.
29. Hakomori, S. *Adv.Cancer. Res.* **1989**, *52*, 257.
30. Gabius, H.-J.; Schroter, C.; Gabius, S.; Brinck, U.; Tietze, L.-F. *J. Histochem. Cytochem.* **1990**, *38*, 1625.
31. Nelson, R. M.; Dolich, S.; Aruffo, A.; Cecconi, O.; Bevilacqua, M. P. *J. Clin. Invest.*, **1993**, *91*, 1157.

RECEIVED December 13, 1993

Chapter 16

Chemical Synthesis of Sialylated Glycoconjugates

Richard R. Schmidt

Fakultät Chemie, Universität Konstanz, Postfach 5560 M 725
D–78434 Konstanz, Germany

The biological importance of glycoconjugates has recently evoked great endeavours in the chemical syntheses of glycosphingolipids and especially gangliosides (sialylated glycosphingolipids). Sialylation results using glycosyl halides and thioglycosides and the appropriate promoters are discussed. The utilization of the "nitrile effect" led to very efficient syntheses for Lewis X (*lactoneo-*) and Lewis A antigens (*lacto*-glycosphingolipid series) as well as improvements in selective α-sialylation. The demand for catalytic activation of sialyl donors could be fulfilled with sialyl phosphites. Their ready formation and activation with catalytic amounts of acid provides, in nitrile solvents, a very convenient highly α-selective sialylation method. Of particular note is the direct phosphite/phosphate exchange affording biologically important sialyl phosphates by simple means.

In the chemical synthesis of gangliosides (sialylated glycosphingolipids), sialylation, that is attachment of N-acylated neuraminic acid, has been carried out by various methodologies (*1-3*). When halogenoses of O,N-acylated neuraminic acid esters are used as donors (L = Hal in Scheme 1) which are activated by silver or mercury salts as promoter (= **P**) only modest yields of the desired α-products are obtained (*1-10*), especially when secondary hydroxyl groups are used as the acceptor. Thus, 3b-O-unprotected lactose and lactosamine derivatives afford the desired building blocks for gangliosides of the *ganglio* (GM_1 etc.) or the *lactoneo* series (sialyl Lewis X), respectively (Scheme 1), in low yields. Neighboring group assistance from the 3β-position of the neuraminic acid molecule was employed (*11,12*) to increase the yield. However, the introduction and removal of the auxiliary groups requires additional steps which lower the efficiency of this approach. Recently, thioglycosides of neuraminic acid derivatives have been proposed as sialyl group donors (L = SR in Scheme 1) (*13,14*). The requirement for at least equimolar amounts of thiophilic reagents [N-Iodosuccinimide (NIS) (*14,15*), DMTST (*14*), methylsulfenyl bromide (*16*) or silver triflate (*4-8,16*)] as promoters **P** constitutes a major disadvantage in this approach; this is even more so because often up to two- to fourfold excess of the promoter is required for reaction. Therefore, improvements in the sialylation step in terms of yield of α-product and in ease of performance of the reaction were urgently required.

0097–6156/94/0560–0276$08.18/0

Scheme 1

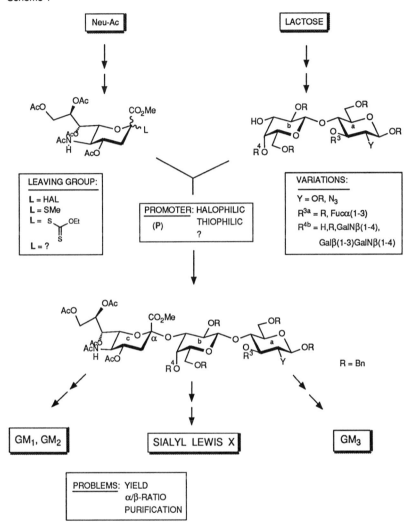

The Nitrile Effect

(i) General Aspects

The influence of solvents on the anomeric ratio of glycosylation reactions is well known (*17-19*). For instance, the participation of ether as a solvent results in S_N1-type conditions in the generation of equatorial oxonium ions [inverse anomeric effect (*20*)] which favor the formation of more thermodynamically stable axial products (Scheme 2). The dramatic effect of nitriles as participating solvents has only been observed in a few cases (*21-29*).

Several years back, we reported a highly α-selective glycosylation with halogenoses of uronic acid derivatives by their treatment with silver perchlorate in acetonitrile at -15°C, followed by the acceptor (21). It was found that the reaction is controlled by formation of intermediate β-nitrilium-nitrile conjugates, which are generated *in situ* under S_N1 conditions (21,22). Further results with different glycosyl donors and catalysts exhibited the formation, transformation, and reactivity of these highly reactive nitrilium-nitrile conjugates (21,22,27). From these results, a general principle could be derived as shown in Scheme 2 (21,27,30): provided a very good leaving group **L**, a strong catalyst **C** (preferentially TMSOTf), and low temperatures are employed, then the glycosyl donor undergoes heterolytic cleavage of **L⁻** in the presence of a nitrile solvent to give a LC⁻-complex. Thus, a highly reactive carbenium ion intermediate is formed which will be attacked by nitriles (for stereoelectronic reasons) preferentially to afford axial intermediates (α-face attack) resulting in kinetically controlled α-nitrilium-nitrile conjugate formation. These intermediates are rapidly converted in the presence of all types of acceptors into the equatorial products, the β-products in Scheme 2, in high yields and diastereoselectivities. The high reaction rate at low temperatures (up to -80°C) would indicate that intramolecular acceptor transfer via an eight-membered, cyclic transition state takes place. On the other hand, equatorial (β-)nitrilium-nitrile conjugates are thermodynamically more stable (inverse anomeric effect), thus favoring via equilibration subsequent formation of the axial product, the α-product in Scheme 2. Presumably, the equilibration rate between the α- and β-onium ions is much faster for ethers than for nitriles; therefore, ethers afford preferentially axial products, α-products in Scheme 2. From these considerations, it follows that upon nitrile participation the α/β-ratio in the products is practically independent of the configuration of the glycosyl donor leaving group, provided it is a good leaving group. Otherwise, S_N2-type reactions with generally lower reaction rates will be competitive.

(ii) Application to Lewis X (Le^X) Synthesis

Glycosphingolipids which have been found to be "tumor associated antigens" (31-35) have recently become the subject of great interest. Their presence in tissues may reveal tumorigenesis, support diagnosis, and hopefully lead to immunotherapy for cancer (33,34). A prominent tumor-associated antigen is the Lewis X (Le^X) determinant Galβ(1-4)[Fucα(1-3)]GlcNAc which was found in monomeric, dimeric, and trimeric form (Scheme 3, n = 0,1,2) (33). These glycosphingolipids are members of the *lactoneo* family for which an efficient total synthesis is outlined in this paper based on the retrosynthetic strategy depicted in Scheme 3 (15,32).

Previous syntheses of Le^X antigens reported by other groups have concentrated on different strategies, different building blocks and different protective group patterns (35-37). We based two strategies on azido lactose for which convenient protective group patterns could be developed (31; Windmüller R.: Ph.D. thesis, Universität Konstanz, to be submitted). However, the finding that trichloroacetimidates of azidoglucose, independent of their anomeric configuration, are ideal glycosyl donors for β-glycosylation in nitrile solvents (15,27) led to a most efficient synthesis of the *lactoneo* and *lacto* families of glycosphingolipids (15,32). This approach starts from readily available 1-O-silyl protected (TBS, TDS etc) 2-azido-4,6-O-benzylidene-2-deoxy-β-D-glucopyranosisde (38), O-(2,3,4-tri-O-benzyl-α-L-fucopyranosyl)trichloroacetimidate as fucosyl donor (39), and O-(2,3,4,6-tetra-O-acetyl-α-D-galactopyranosyl)trichloroacetimidate as galactosyl donor (19). Representative results of this strategy are exhibited in Scheme 4.

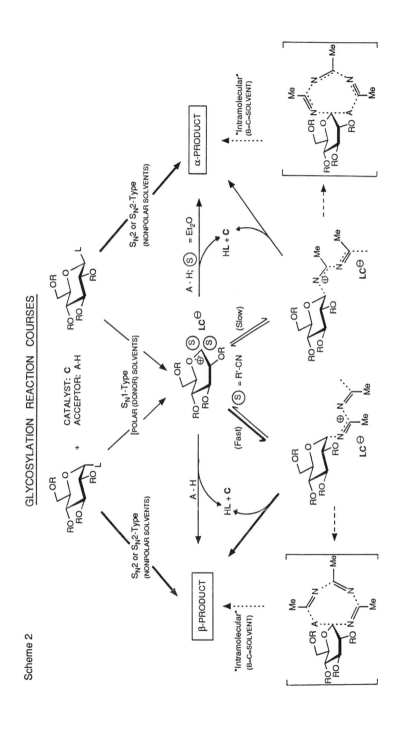

Scheme 2

Scheme 3

SYNTHESIS OF THE MONOMERIC, DIMERIC, TRIMERIC, AND TETRAMERIC Le^x ANTIGEN

RETROSYNTHETIC ANALYSIS

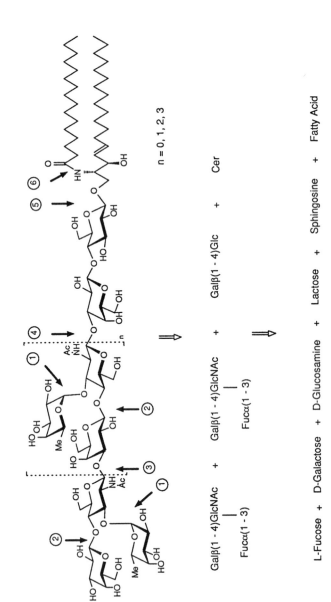

n = 0, 1, 2, 3

Galβ(1 - 4)GlcNAc   +   Galβ(1 - 4)GlcNAc   +   Galβ(1 - 4)Glc   +   Cer
       | |
Fucα(1 - 3)   Fucα(1 - 3)

⟹

L-Fucose + D-Galactose + D-Glucosamine + Lactose + Sphingosine + Fatty Acid

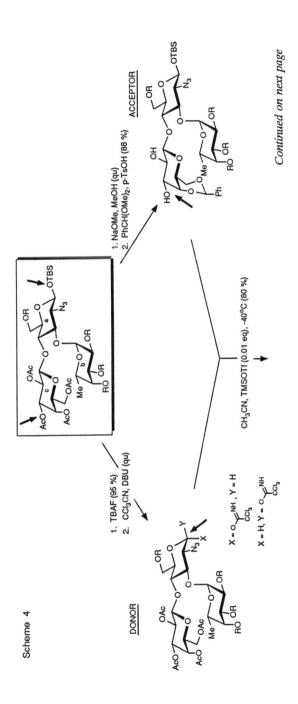

Scheme 4

Continued on next page

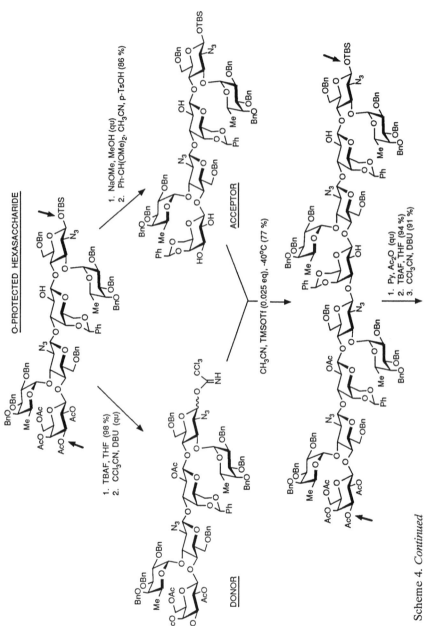

Scheme 4. *Continued*

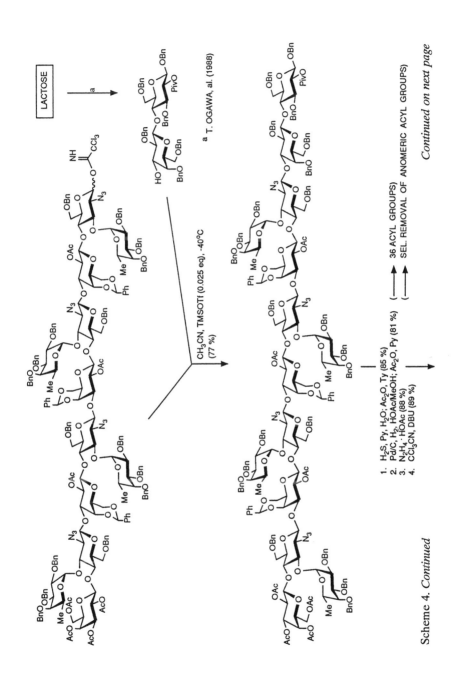

Scheme 4. Continued

Continued on next page

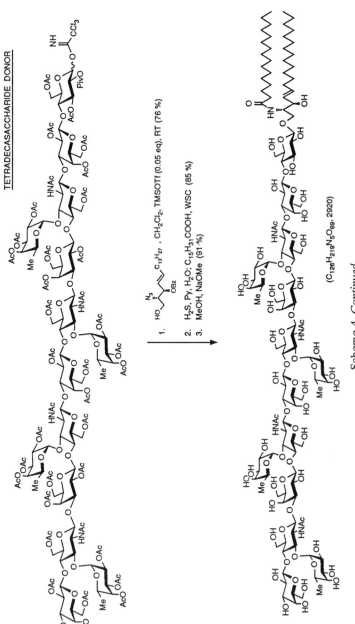

Scheme 4. *Continued*

Improvements in α-fucosylation (for instance of the 3a-O-position of azidoglucose, as shown above) with the O-benzyl protected fucosyl donor by applying the inverse procedure (I.P.: i.e., addition of the donor to an acceptor-catalyst solution) (*40*), ensuing regioselective reductive ring opening of the O-benzylidene moiety with $NaCNBH_3/HCl$ in the derived disaccharide, and then selective β–galactosylation at the 4a-O-position afforded the decisive trisaccharide shown in the frame of Scheme 4, in high overall yield. The versatility of this monomer Le^X building block is demonstrated by its ready transformation into a donor and an acceptor, respectively, thus providing in acetonitrile at -40°C and the presence of catalytic amounts of TMSOTf, the dimer Le^X hexasaccharide in high yield. By the same means, this hexasaccharide building block can be again transformed into a donor and an acceptor, respectively. When these are reacted, the tetramer Le^X dodecasaccharide was obtained, which again exhibits the same structural features. Transformation into a donor and then reaction with a known lactose acceptor (*41*) gave - again with the help of the "nitrile effect" - exclusively the β-configuration, thus affording a tetradecasaccharide in high yield. Then azido group reduction, followed by N-acetylation of the derived amino group, hydrogenolytic O-debenzylation and debenzylidenation, per-O-acetylation, and selective removal of the 1-O-acetyl group were carried out. The selective removal of the 1-O-acetyl group, out of 36 acyl groups present, is quite remarkable and proceeded in astonishing 88 % yield. Transformation of the 1-O-deprotected tetradecasaccharide into a trichloroacetimidate provided the required donor for the "azidosphingosine glycosylation procedure" (*42*) which gave tetrameric Le^X in high overall yield. The ^{1}H-NMR spectra (Scheme 5) demonstrates the purity of this material.

(iii) Application to Sialylation

The results obtained for the sialyl donors thus far seem to preclude S_N2-type sialylation of O-nucleophiles due to steric hindrance at the anomeric center and due to low nucleophilicity of oxygen acceptors (*15,43,44*). Obviously, under S_N1-type conditions a promoter generates from the sialyl donors a (solvent separated) ion pair which in, nitrile solvents, should again be intercepted under kinetically controlled conditions from the β-face (axial attack), thus leading to β-nitrilium-nitrile conjugates (*27,30*). This process is independent of the anomeric configuration of the donor. Reaction of the β-nitrilium-nitrile conjugates with the acceptor, prior to slow transformation into α-nitrilium-nitrile species, either directly or via intramolecular acceptor transposition (*30,44*), should lead to the equatorial glycoside which is the desired α-glycoside in neuraminic acid. This could be clearly shown by us (*15*) (see Scheme 6) and others (*45*) for thio group activated neuraminic acid derivatives which led to improved α-selectivities.

For instance, the monomer Le^X trisaccharide of Scheme 4 was readily transformed into a 2c,3c-O-unprotected acceptor (Scheme 6) which gave upon sialylation in acetonitrile, a versatile sialyl Le^X building block, albeit in rather low yield (29% isolated yield; 46% based on consumed acceptor; the excess sialyl donor could not be recovered). Although this yield can be further improved by increased steric accessibility of the 3c-OH group, as found by Hasegawa and collaborators (*45*), the desired catalytic activation of readily accessible sialyl donors is still not available.

Scheme 5

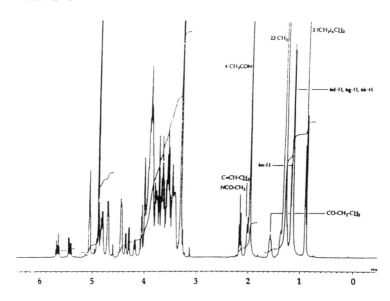

Tetrameric Le^x : ^{1}H-NMR (400 MHz, CD$_3$OD)

Scheme 6

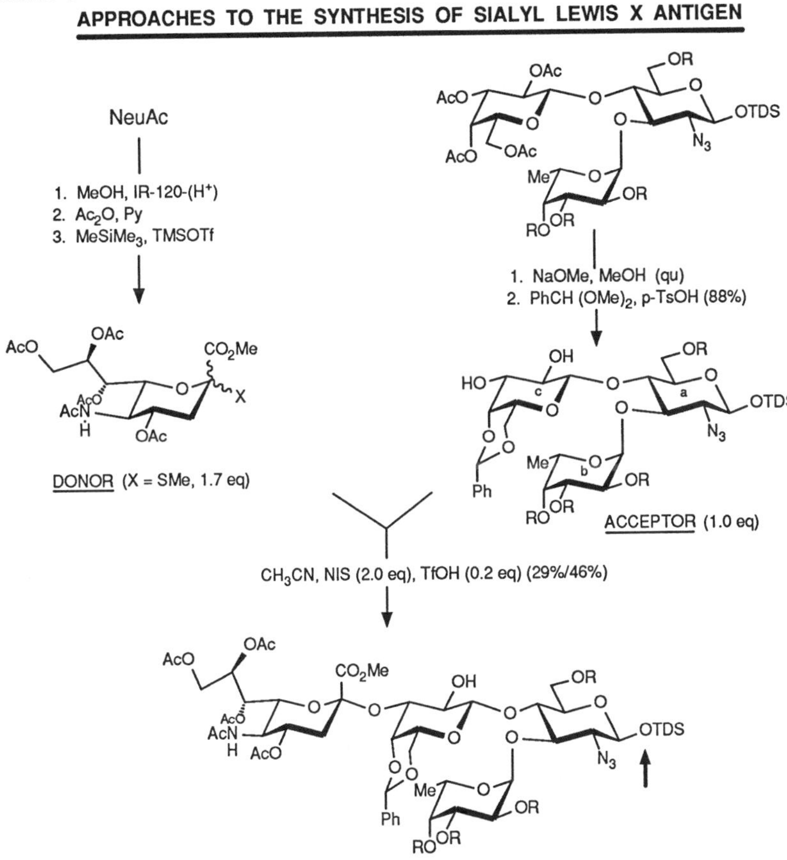

APPROACHES TO THE SYNTHESIS OF SIALYL LEWIS X ANTIGEN

NeuAc

1. MeOH, IR-120-(H$^+$)
2. Ac$_2$O, Py
3. MeSiMe$_3$, TMSOTf

DONOR (X = SMe, 1.7 eq)

1. NaOMe, MeOH (qu)
2. PhCH (OMe)$_2$, p-TsOH (88%)

ACCEPTOR (1.0 eq)

CH$_3$CN, NIS (2.0 eq), TfOH (0.2 eq) (29%/46%)

Sialyl Donors Accessible to Catalytic Activation

(i) The Concept

Provided the assumptions regarding the "nitrile effect", as discussed above, are also applicable to sialyl donors, then the generation of the decisive (solvent separated) ion pair requires in nitriles only a simple, acid sensitive leaving group; it will be removed from the anomeric center by catalytic amounts of a promoter (a true catalyst! for instance, TMSOTf, TfOH etc) under the condition that the leaving group, because of its low basicity, does not consume the catalyst. Thus, under kinetic control equatorial glycosides (α-products) should be accessible by simple means in a catalytic procedure. Hence, neither an acidity increase supporting any side

Scheme 7

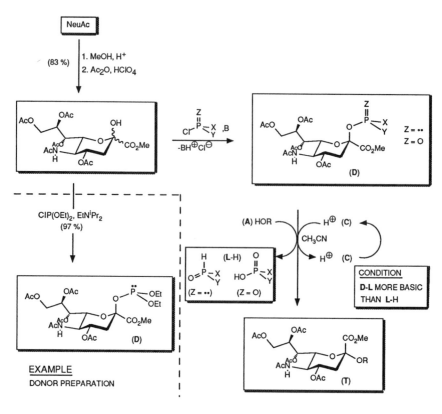

reactions will occur, nor any expensive halo- or thiophilic promoters (see introduction) will be used up in at least equimolar amounts during the reaction course.

Consideration of various leaving groups led us to phosphite (and phosphate) moieties and their derivatives (43,44,46) (Scheme 7). Thus, a readily available O-acetylated neuraminic acid derivative will be transformed into a sialyl donor (D) which with alcohols as acceptors (AH) and in nitrile solvents will provide the target molecules (T). The catalyst (C) due to the basicity of the leaving group on the sialyl donor (D) will preferentially attack D. However, the cleavage product thus generated from the leaving group (= LH) will release the catalyst C because of its low basicity, with the help of the proton available from the acceptor hydroxy group. Obviously, the phosphite/phosphonate system (Z = ··) seems to be especially suitable because basicity and strong leaving group character in the phosphite species (= L) are combined with relatively low acidity and basicity in the released phosphonate species (= LH). Additionally, phosphitylation at the anomeric position can be performed in high yields as indicated in Scheme 7 for an ethyl ester (43); other esters could be also readily obtained (44).

(ii) Sialylation - A Comparison of Methods

Comparison of the sialyl donor properties of diethyl sialyl phosphite [Scheme 7; X = OP(OEt)$_2$ in Scheme 8] with ethyl xanthate and methylthio, respectively, as leaving groups [Scheme 8, X = SCS(OEt), SMe] and differently protected lactose derivatives as acceptors clearly exhibited the relative ease of carrying out the phosphite activation reaction and the requirement for only catalytic amounts of TMSOTf for the activation process. As discussed above, the steric accessibility of the 3b-OH group of lactose is a decisive factor in regards to the yield; however, sialyl phosphites also give higher yields with less reactive acceptors. Therefore, sialyl phosphites became important sialyl donors (*43,44,46*) due to the ease of their formation and their convenient activation by addition of catalytic amounts of TMSOTf.

Meanwhile, phosphites were also successfully employed as leaving groups with other sugars (Müller, T.: to be published). Additionally, highly stereoselective ring closure reactions for the generation of α-glycosides of neuraminic acid derivatives could be carried out (*47*).

(iii) Application to Ganglioside Syntheses in the *ganglio* and *lactoneo* Series

Scheme 8 shows the successful transformation of sialyllactose into GM$_3$ in part by making use of a published protocol (*9,10*). Sialyllactose derived from 3b,4b-O-unprotected lactose (Scheme 8) also seems to be an ideal starting material for the synthesis of other important gangliosides of the *ganglio* series (for instance GM$_2$, GM$_1$, GD$_{1a}$, BGM$_1$, etc) (*48*). To this end, a Galβ(1-3)GalN-donor was prepared, which was then successfully employed in the GM$_1$-synthesis, as outlined in Scheme 9 (Brescello, R. al. unpublished results). The yield of 46% in the glycosylation reaction with the very unreactive 4b-OH group of sialyllactose was the primary reason to design an alternative route to this molecule and other desired compounds in this series (Stauch, T., unpublished results). Application of sialyl phosphite donors to the preparation of building blocks for sialyl Le^x synthesis, which are interesting cell-cell adhesion epitopes, was also very successful. The importance of steric demand in sialylation led us to synthesize a modified monomer Le^x building block, as depicted in Scheme 10, by making use of the "nitrile effect".

A dimer Le^x hexasaccharide was obtained in very high yield; its treatment with sodium methanolate/methanol afforded a 2c-, 2f-, 3f-, 4f-O-unprotected acceptor. Sialylation under the conditions as described above gave selective 3f-O-attack with formation of the desired sialyl dimer Le^x heptasaccharide building block in 60% yield. This compound can be transformed into the unprotected target molecule

Sialylation of Phosphates

CMP-Neu5Ac is an especially important compound which is required for the enzymatic sialylation of glycoconjugates (biosynthesis of gangliosides and sialylated glycoproteins) (*1*). To this end, we have investigated the direct chemical reaction of sialyl phosphites with phosphorous acid derivatives as acceptors (*49*). Due to the sensitivity of sialyl phosphites to mild acid catalysis as well as the operation of the thermodynamic anomeric effect, a direct β-selective reaction with phosphorous acids as acceptors is expected (thermodynamic control of product formation due to acid sensitivity of the product). This would result in sialyl phosphates having phosphonate as the leaving group which, because of its neutrality, should not interfer in the reaction. Similar reaction behavior has been already observed for various O-

Scheme 8

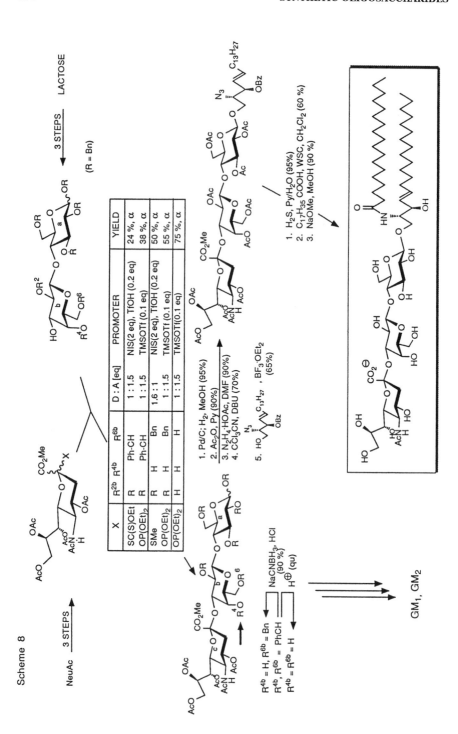

X	R^{2b}	R^{4b}	R^{6b}	D : A [eq]	PROMOTER	YIELD
SC(S)OEt	R	Ph-CH		1 : 1.5	NIS(2 eq), TfOH (0.2 eq)	24 %, α
OP(OEt)$_2$	R	Ph-CH		1 : 1.5	TMSOTf (0.1 eq)	38 %, α
SMe	R	H	Bn	1,6 : 1	NIS(2 eq), TfOH (0.2 eq)	50 %, α
OP(OEt)$_2$	R	H	Bn	1 : 1.5	TMSOTf (0.1 eq)	55 %, α
OP(OEt)$_2$	H	H	H	1 : 1.5	TMSOTf (0.1 eq)	75 %, α

1. Pd/C; H$_2$ MeOH (95%)
2. Ac$_2$O, Py (90%)
3. N$_2$H$_4$·HOAc, DMF (90%)
4. CCl$_3$CN, DBU (70%)
5. (65%)

NaCNBH$_3$, HCl (90 %)
H⊕ (qu)

R^{4b} = H, R^{6b} = Bn
R^{4b}, R^{6b} = PhCH
R^{4b} = R^{6b} = H

GM$_1$, GM$_2$

1. H$_2$S, Py/H$_2$O (95%)
2. C$_{17}$H$_{35}$ COOH, WSC, CH$_2$Cl$_2$ (60 %)
3. NaOMe, MeOH (90 %)

Scheme 9

SYNTHESIS OF GM₁ VIA A GM₃ INTERMEDIATE

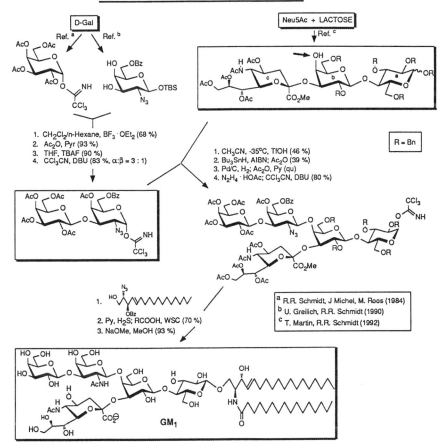

1. CH₂Cl₂/n-Hexane, BF₃·OEt₂ (68 %)
2. Ac₂O, Pyr (93 %)
3. THF, TBAF (90 %)
4. CCl₃CN, DBU (83 %, α:β = 3 : 1)

1. CH₃CN, -35°C, TfOH (46 %)
2. Bu₃SnH, AIBN; Ac₂O (39 %)
3. Pd/C, H₂; Ac₂O, Py (qu)
4. N₂H₄·HOAc; CCl₃CN, DBU (80 %)

R = Bn

1.
2. Py, H₂S; RCOOH, WSC (70 %)
3. NaOMe, MeOH (93 %)

[a] R.R. Schmidt, J Michel, M. Roos (1984)
[b] U. Greilich, R.R. Schmidt (1990)
[c] T. Martin, R.R. Schmidt (1992)

GM₁

Scheme 10

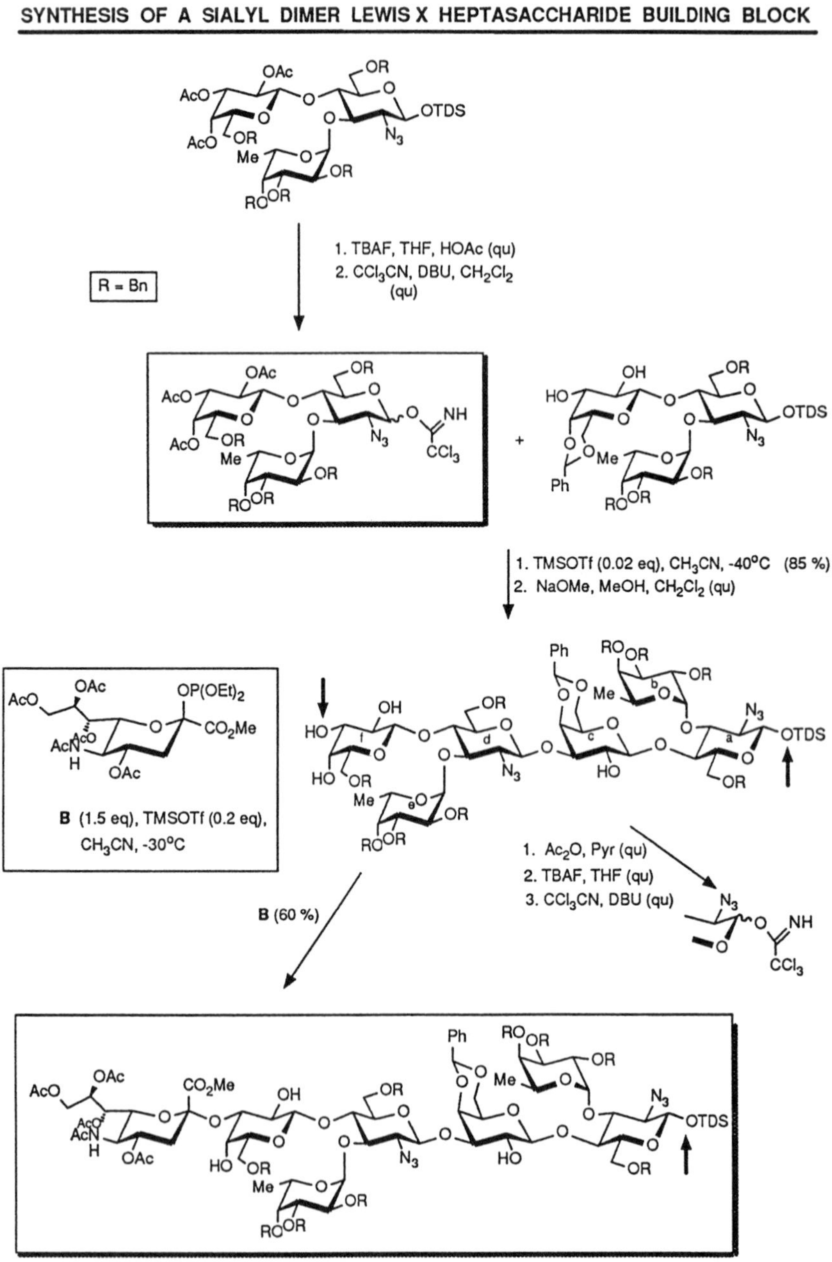
SYNTHESIS OF A SIALYL DIMER LEWIS X HEPTASACCHARIDE BUILDING BLOCK
R = Bn
1. TBAF, THF, HOAc (qu)
2. CCl3CN, DBU, CH2Cl2 (qu)
1. TMSOTf (0.02 eq), CH3CN, -40°C (85 %)
2. NaOMe, MeOH, CH2Cl2 (qu)
B (1.5 eq), TMSOTf (0.2 eq), CH3CN, -30°C
1. Ac2O, Pyr (qu)
2. TBAF, THF (qu)
3. CCl3CN, DBU (qu)
B (60 %)
OTDS
OP(OEt)2
CO2Me

glycosyl-trichloroacetimidates (*19*), but CMP-Neu5Ac was not accessible via this approach.

Reaction of diethyl sialyl phosphite with various simple phosphorous acid mono- and diesters indeed resulted in a phosphite/phosphate exchange providing sialyl phosphates in good yield (*49*). However, the direct formation of CMP-Neu5Ac with CMP failed because of its insolubility in solvent systems favorable for the reaction. Therefore, CMP was first transformed into the 2',3',N^4-triacetyl derivative (Scheme 11 = Tri-Ac-CMP) which exhibited a reasonable solubility in an CH$_3$CN/THF/DMF mixture. Reaction with diethyl sialyl phosphite at -15°C and slowly raising the temperature to room temperature afforded the desired acetylated CMP-Neu5Ac as solid material in 50% yield. Base catalyzed deacetylation furnished CMP-Neu5Ac as crystalline material in 68% yield (*49*).

Scheme 11 **SYNTHESIS OF NATURAL CMP-Neu5Ac AND UMP-Neu5Ac**

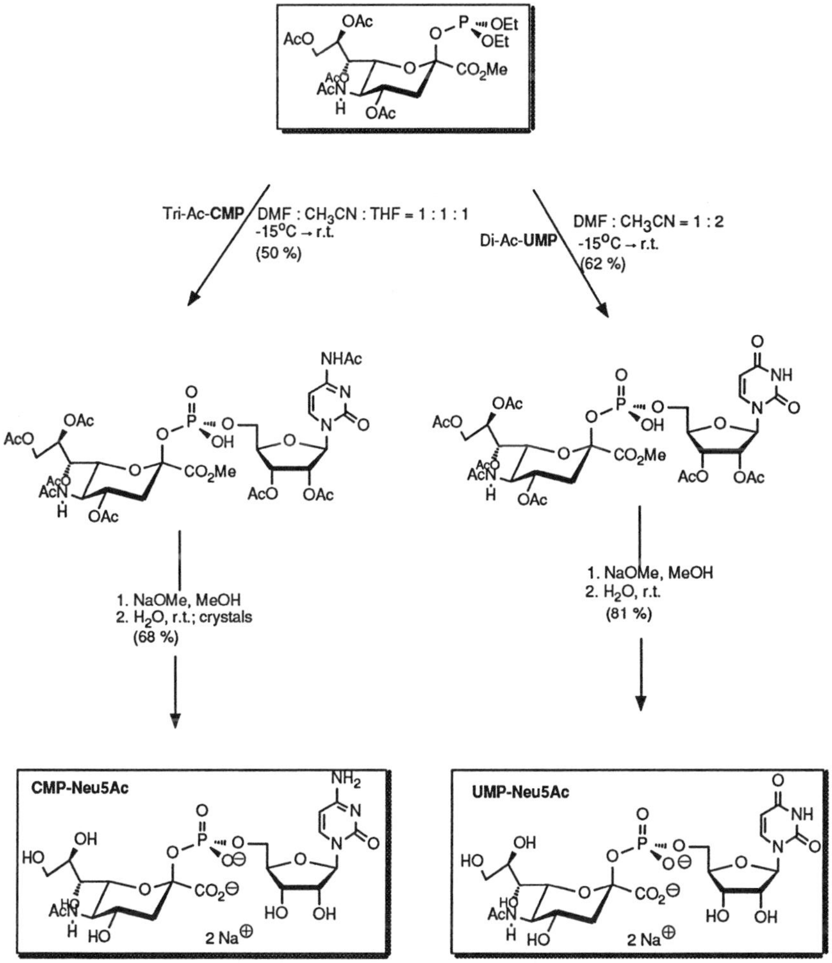

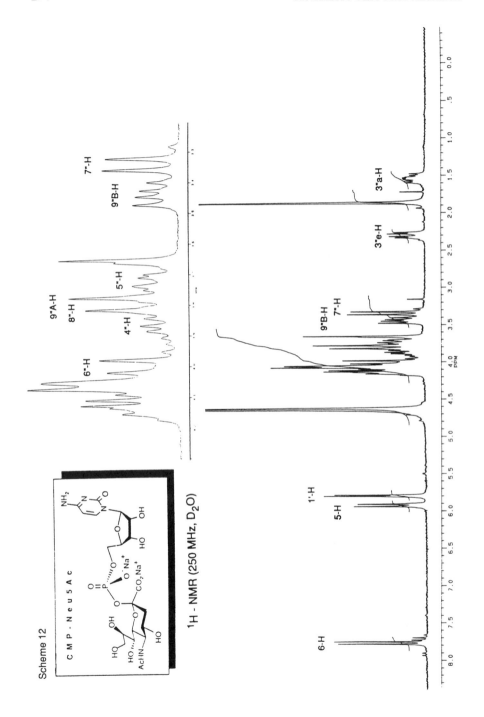

Scheme 12

C M P - N e u 5 A c

^{1}H - NMR (250 MHz, D$_2$O)

The ^{1}H-NMR spectra of the material obtained is depicted in Scheme 12, as proof of its high purity. This method could be also successfully applied to other types of biologically important phosphorous acid ester (for instance O-acetylated UMP, as shown in Scheme 11; (Martin, T.J.; al. to be published), thus exhibiting the efficiency and versatility of this process.

Conclusion

The importance and requirements for catalytic activation of sialyl donors led to the acid sensitive phosphite moiety as leaving group. In combination with the "nitrile effect" a simple, highly α-selective, chemical sialylation method could be developed which makes these important gangliosides and sialyl phosphates much more readily available.

Acknowledgements

This work was supported by the *Deutsche Forschungsgemeinschaft,* the *Fonds der Chemischen Industrie,* grants from *Fa. E. Merck, Darmstadt,* und the *University of Milano.*- I am particularly grateful to my capable collaborators whose names are individually mentioned in the references.

Literature Cited

(1)   Schauer, R. *Adv. Carbohydr. Chem. Biochem.* **1982**, *40*, 131-234.
(2)   Okamoto, K.; Goto T. *Tetrahedron* **1990**, *46*, 5835-5957.
(3)   DeNinno, M.P. *Synthesis* **1991**, 583-593.
(4)   Eschenfelder, V.; Brossmer, R. *Carbohydr. Res.* **1980**, *78,* 190-194.
(5)   Van der Vlengel, D.J.M.; Zwikker, J.W.; Vliegenthart, J.F.G.; *Carbohydr. Res.* **1982**, *105*, 19-31.
(6)   Kiso, M.; Nakamura, A.; Hasegawa, A. *J. Carbohydr. Chem.* **1987**, *6*, 411-422.
(7)   Paulsen, H.; von Deesen, U. *Carbohydr. Res.* **1988**, *175*, 283-293.
(8)   Shimizu, C.; Achiwa, K. *Chem. Pharm. Bull.* **1989**, *37*, 2258-2260.
(9)   Sugimoto, M.; Ogawa, T. *Glycoconjugate J.* **1985**, *2*, 5-9.
(10) Numata, N.; Sugimoto, M., Shibayama, S.; Ogawa, T. *Carbohydr. Res.* **1988**, *174*, 73-85.
(11) Kondo, T.; Abe, H.; Goto, T. *Chemistry Lett.* **1988**, 1657-1660; and references therein.
(12) Ito, Y.; Ogawa, T. *Tetrahedron Lett.* **1987**, *28*, 6221-6224; *ibid.* **1988**, *29*, 3987-3990; *Tetrahedron* **1990**, *46*, 89-102.
(13) Prabhanjan, H.; Aoyama, K.; Kiso, M.; Hasegawa, A. *Carbohydr. Res.* **1992**, *233*, 87-99; and refereces therein.
(14) Marra, A.; Sinaÿ, P *Carbohydr. Res.* **1990**, *195,* 303-308; and references therein.
(15) Toepfer, A. *Dissertation*, Universität Konstanz, 1992.
(16) Lönn, H.; Stenvall, K. *Tetrahedron Lett.* **1992**, *33*, 115-116; and references therein.
(17) Wulff, G.; Röhle, G. *Angew. Chem.* **1974**, *86*, 173-187; *Angew. Chem. Int. Ed. Engl.* **1974**, *13*, 157-181.
(18) Paulsen, H.; *Angew. Chem.* **1982**, *94*, 184-201; *Angew. Chem. Int. Ed. Engl.* **1982**, *21*, 155-197.
(19) Schmidt, R.R. *Angew. Chem.* **1986**, *98*, 213-235; *Angew. Chem. Int. Ed. Engl.* **1986**, *25*, 212-235.

(20)   Lemieux, R.U. *Pure Appl. Chem.* **1971**, *25*, 527-548
(21)   Schmidt, R.R.; Rücker, E. *Tetrahedron Lett.* **1980**, *21*, 1421-1424.
(22)   Schmidt, R.R.; Michel, J. *J. Carbohydr. Chem.* **1985**, *4*, 141-169.
(23)   Hashimoto, S.; Hayashi, M.; Noyori; R. *Tetrahedron Lett.* **1984**, *25*, 1379-1382.
(24)   Ito, Y.; Ogawa, T. *Tetrahedron Lett.* **1987**, *28*, 4701-4704.
(25)   Schmidt, R.R.; Gaden, H.; Jatzke, H. *Tetrahedron Lett.* **1990**, *31*, 327-330.
(26)   Hashimoto, S.; Honda, T.; Ikegami; S. *J. Schem. Soc., Chem. Commun.* **1989**, 685-687.
(27)   Schmidt, R.R.; Behrendt, M.; Toepfer, A. *Synlett* **1990**, 694-696.
(28)   Marra, A.; Mallet, J.-M.; Amatore C.; Sinaÿ, P. *Synlett* **1990**, 572-574; Bracini, I.; Deronet, C.; Esnault, J.; Hervé du Penhoat, C.; Mallet, J.-M.; Michou, V.; Sinaÿ, P. *Carbohydr. Res.* **1993**, *246*, 23-41.
(29)   Vankar, Y.D.; Vankar, P.S.; Behrendt, M.; Schmidt, R.R. *Tetrahedron* **1991**, *47*, 9985-9988.
(30)   Schmidt, R.R. in *Carbohydrates - Synthetic Methods and Applications in Medicinal Chemistry* (Eds. Ogura, H.; Hasegawa, A.; Suami, T.) Kodanasha Ltd.; Tokyo **1992**, p. 66-88.
(31)   Bommer, R.; Kinzy, W.; Schmidt, R.R. *Liebigs Ann. Chem.* **1991**, 425-433; and references therein.
(32)   Toepfer, A.; Schmidt, R.R. *Tetrahedron Lett.* **1992**, *33*, 5161-5164; Toepfer, A.; Kinzy, W.; Schmidt, R.R. *Liebigs Ann. Chem.*, submitted.
(33)   Hakomori, S. *Chem. Phys. Lipids* **1986**, *42*, 209-233.
(34)   Hannun, Y.A.; Bell, R.M. *Science,* **1989**, *243*, 500-507.
(35)   Nicolaou, K.C.; Hummel, C.W.; Iwabuchi, Y. *J. Am. Chem. Soc.* **1992**, *114*, 3126-31-28; and references therein.
(36)   Nillson, M.; Norberg, T. *Carbohydr. Res.* **1988**, *183*, 71-82; *J. Carbohydr. Chem.* **1989**, *8*, 613-627.
(37)   Sato, S.; Yukishige, I.; Ogawa, T. *Tetrahedron Lett.* **1988**, *29*, 5267-.5270; and references therein.
(38)   Kinzy, W.; Schmidt, R.R. *Liebigs Ann. Chem.* **1985**, 1537-1545.
(39)   Wegmann, B.; Schmidt, R.R. *Carbohydr. Res.* **1988**, *184*, 254-261.
(40)   Schmidt, R.R.; Toepfer, A. *Tetrahedron Lett.* **1991**, *32*, 3353-3356.
(41)   Sato, S.; Nunomura, S.; Nakano, T.; Ito, Y.; Ogawa, T. *Tetrahedron Lett.* **1988**, *29*, 4097-4100.
(42)   Schmidt, R.R.; Zimmermann, P. *Tetrahedron Lett.* **1986**, *27*, 481-484; *Angew. Chem.* **1986**, *98*, 722-723; *Angew. Chem. Int. Ed.Eng.* **1986**, *25*, 725-726-
(43)   Martin, T.J.; Schmidt, R.R. *Tetrahedron Lett.* **1992**, *33*, 6123-6126.
(44)   Martin, T.J.; Brescello, R.; Toepfer, A.; Schmidt, R.R. *Glycoconj. J.* **1993**, *10*, 16-25.
(45)   Hasegawa, A.; Nagahama, T.; Ohki, H. Kotta, K.; Ishida, H.; Kiso, M. *J. Carbohydr. Chem.* **1991**, *10*, 493-498.
(46)   Sinn, M.M.; Kondo, H.; Wong, C.-H. *J. Am. Chem. Soc.* **1993**, *115*, 2260-2267; and references therein.
(47)   Vlahov, I.R.; Vlahova, P.I.; Schmidt, R.R. *Tetrahedron Lett.* **1991**, *32*, 7025-7028; *ibid.* **1992**, *33*, 7503-7506.
(48)   Greilich, U.; Zimmermann, P.; Jung, K.-H.; Schmidt, R.R. *Liebigs Ann. Chem.* **1993**, 859-864.
(49)   Martin, T.J.; Schmidt, R.R. *Tetrahedron Lett.* **1993**, *34*, 1765-1768.

RECEIVED April 7, 1994

Author Index

Affiliation Index

Subject Index

Carbenes, generation, 202
Carbohydrate(s)
  factors affecting role in biology, 2
  recognition in biological systems, 6–9
  role in biological recognition
    phenomena, 198–199
  roles in life processes, 36
Carbohydrate-based drugs, 19–33
  aminoglycoside antibiotics, 20–21
  anticancer agents, 22–23
  antiviral agents, 22,29–30
  aurothioglucose, 25
  bacterial toxins, 29
  biological screening of derivatives,
    27–29
  cardiac glycosides, 20
  clindamycin, 22
  fertilization, 32
  history, 19
  immunology, 32
  improvement of existing drugs, 26–27
  inflammation, 30–31
  macrolide antibiotics, 21
  origins, 19–20
  polysaccharides, 23–24
  recombinant proteins, 32
  thrombosis, 31
  viral infection, 29–30
Carbohydrate-specific antibodies, binding
  mechanism elucidation, 11–12
Cardiac glycosides, description, 20
Castanospermin, 30
Catalytic activation, accessible sialyl
  donors, 287–291
Cell adhesion molecules, 7
Ceramide derivatives of sialo-
  oligosaccharides, use in study of
  biological functions of gangliosides,
  184–195
Ceredase, carbohydrate structure, 8–9
Chemical modification and probing of
  receptor binding sites, 198–231
Chemical synthesis
  complex oligosaccharides and conjugates,
    applications, 249

Chemical synthesis—*Continued*
  sialylated glycoconjugates
    nitrile effect, 277–287
    sialyl donor accessible to catalytic
      activation, 287–291
  sialylation
    general, 276–277
    phosphates, 289,293–295
Chemically modified sialic acid
  containing sialyl Lewis X ganglioside
  analogues, synthesis, 192–193
Chloralose, 25
N-Chloroacetylated dendrimers,
  solid-phase synthesis, 106–107
Clindamycin, 22
Clobenoside, 25
Compatibility of protective groups,
  oligosaccharide synthesis, 13
Complexity, naturally occurring
  glycoconjugates, 3–6
Conformation, ganglioside lactams,
  235,236*f*,246–247
1,2-*O*-Cyanoethylidene derivatives,
  synthesis, 38–40
Cytarabine, 22

D

Daunorubicin, 23
Dendrimers, sialic acid based,
  synthesis, 104–118
Dendritic α-thiosialosides
  antigenicity, 112–113,114*f*
  solid-phase synthesis, 106–111
Deoxyfucose-containing sialyl Lewis X
  gangliosides, synthesis, 190–192
Deoxynojirimycin, 30
1-Deoxynojirimycin-containing sialyl
  Lewis A and X analogues, 194–195
Dextran-related ligands containing
  deoxyfluoro groups, synthesis,
  169–176
Dialkylstannylene acetals
  intermediates in carbohydrate derivative
  synthesis, 51

Production: Beth Harder & Charlotte McNaughton
Indexing: Deborah H. Steiner
Acquisition: Anne Wilson
Cover design: Pavol Kováč, Alan Kahan

Printed and bound by Maple Press, York, PA